Energy Technology 2015

Carbon Dioxide Management and Other Technologies

TMS2015

144th Annual Meeting & Exhibition

March 15-19, 2015 • Walt Disney World • Orlando, Florida, USA

Energy Technology 2015
Carbon Dioxide Management and Other Technologies

Proceedings of the **Energy Technologies and Carbon Dioxide Management Symposium**, sponsored by the Energy Committee of the Extraction & Processing Division (EPD) and the Light Metals Division (LMD) of The Minerals, Metals & Materials Society (TMS).

Additional papers have been contributed by the following symposia:

Recycling and Sustainability Update

Sponsored by the Recycling and Environmental Technologies Committee of EPD and LMD

Magnetic Materials for Energy Applications V

Sponsored by the Magnetic Materials Committee of the TMS Functional Materials Division (FMD, formerly the Electronic, Magnetic & Photonic Materials Division)

Sustainable Energy and Layered Double Hydroxides

Sponsored by the Chemistry and Physics of Materials Committee of FMD and the TMS Structural Materials Division (SMD)

High-Temperature Systems for Energy Conversion and Storage

Sponsored by the Energy Conversion and Storage Committee of FMD

Structural Materials, Heat Transport Fluids, and Novel System Designs for High Power and Process Heat Generation

Sponsored by the Corrosion and Environmental Effects Committee and Nuclear Materials Committee of SMD, and the Energy Conversion and Storage Committee of FMD

All symposia were held during

TMS2015
144th Annual Meeting & Exhibition

March 15-19, 2015
Walt Disney World • Orlando, Florida, USA

Edited by:
Animesh Jha
Cong Wang • Neale R. Neelameggham • Donna P. Guillen
Li Li • Cynthia K. Belt • Randolph Kirchain
Jeffrey S. Spangenberger • Frank Johnson
Andrew J. Gomes • Amit Pandey • Peter Hosemann

Editors
Animesh Jha
Cong Wang
Neale R. Neelameggham
Donna P. Guillen
Li Li
Cynthia K. Belt
Randolph Kirchain
Jeffrey S. Spangenberger
Frank Johnson
Andrew J. Gomes
Amit Pandey
Peter Hosemann

ISBN 978-3-319-48602-4 ISBN 978-3-319-48220-0 (eBook)
DOI 10.1007/978-3-319-48220-0

Chemistry and Materials Science: Professional

Published by Springer International Publishers, Switzerland, 2016
Reprint of the original edition published by John Wiley & Sons, Inc., 2015, 978-1-119-08240-8

Printed on acid-free paper

This Springer imprint is published by Springer Nature
The registered company is Springer International Publishing AG
The registered company address is: Gewerbestrasse 11, 6330 Cham, Switzerland

TABLE OF CONTENTS
Energy Technology 2015

Energy Technologies and Carbon Dioxide Management Symposium 2015

Carbon Management

Iron & Steel

Metal Processing / Molten Salt / Electrochemistry

Solar Energy

Recycling and Sustainability Update

Waste

Recycling

Poster Session

Magnetic Materials for Energy Applications

Soft Magnetic Materials I

Soft Magnetic Materials II

Poster Session

Sustainable Energy and Layered Double Hydroxides

Sustainable Energy and Layered Double Hydroxides

High-Temperature Systems for Energy Conversion and Storage

Solid Oxide Fuel Cell: Recent Developments II

Innovation in Energy Conversion and Storage I

Structural Materials, Heat Transport Fluids, and Novel System Designs for High Power and Process Heat Generation

Heat Transport Fluids I

EDITORS

Energy Technologies and Carbon Dioxide Management Symposium 2015

Animesh Jha

Animesh Jha has been a professor of materials science at the University of Leeds in the United Kingdom since August 2000. He obtained his Bachelor of Engineering (BE) degree in metallurgy in June 1979 from the University of Roorkee (India); his Master of Engineering (ME) degree in metallurgical engineering in July 1981 from the Indian Institute of Science, Bangalore in India; and Ph.D. and DIC from Imperial College (London) in October 1984 in the area of chemical and process metallurgy. After earning his Ph.D., he did a short postdoctoral research engagement at Purdue University (US) in 1985-86, before returning to Sheffield, UK in December 1986 where he started research on phase transformations in novel salt-based and metallic glass systems. In May 1989, he took his academic position at Brunel University, Uxbridge, UK where he developed his academic career in the areas of IR glasses and mineral processing. He has since 1989 continued research in these two areas and produced over 24 Ph.D. theses from 1992 to 2014.

He has published over 400 research papers in peer-reviewed journals and reviewed conference proceedings. He has been a member of TMS since 1992, a fellow of the Institute of Physics in London, and member of the Optical Society of America and IEEE. He serves as an external reviewer for overseas research agencies including US, Canada, and EU countries. He is also actively involved in translational research, which allows the lab work to reach industry.

Cong Wang

Cong Wang is currently a professor of the School of Materials and Metallurgy, Northeastern University. Previously, he held positions at Northwestern University, Saint-Gobain High Performance Materials Research and Development Center, and Alcoa Technical Center. He is a seasoned metallurgist/materials scientist with rich experience in aluminum alloys, copper alloys, iron and steels, magnesium alloys, and super-abrasives. He obtained his Ph.D. from Carnegie Mellon University; M.S. from the Institute of Metal Research, Chinese Academy of Sciences; and B.S. from Northeastern University with distinctions, respectively.

Dr. Wang is the recipient of the 2014 Outstanding Young Manufacturing Engineer Award and the 2011 TMS Young Leader Award. He is currently a Key Reader for *Metallurgical and Materials Transactions*. He serves on the editorial committees of *International Journal of Refractory Metals and Hard Materials*, *Journal of Materials Science and Technology*, *Journal of Materials Engineering and Performance*, and *Advanced Materials and Processes*. He is a frequent organizer of international conferences, including REWAS 2013 (Global Conference on Sustainability), Energy Technology and Carbon Dioxide Management Symposium (2013, 2014, and 2015), and Materials Processing Fundamentals Symposium (2012, 2013, and 2014).

Neale R. Neelameggham

Neale R. Neelameggham is 'The Guru' at IND LLC, involved in technology marketing and international consulting in the field of light metals and associated chemicals (boron, magnesium, titanium, lithium, and alkali metals), rare earth elements, and battery and energy technologies. He was recently a visiting expert at Beihang University of Aeronautics and Astronautics, Beijing, China. He has over 38 years of expertise in magnesium production and was involved in process development of its startup company NL Magnesium through to the present US Magnesium LLC, UT from where he retired in 2011. He is developing thiometallurgical processes – a new concept of using sulfur as the reductant and or fuel. He has published a heat transfer model for global anthropogenic warming based on thermal emissions independent of energy conversion source.

Dr. Neelameggham holds 16 patents and patent applications, and has published several technical papers. He has served in the Magnesium Committee of the TMS Light Metals Division (LMD) since its inception in 2000, chaired it in 2005, and in 2007 he was made a permanent co-organizer for the Magnesium Technology Symposium. He has been a member of the Reactive Metals Committee, Recycling Committee, and Titanium Committee; and a Programming Committee Representative of LMD. In 2008, Dr. Neelameggham was the inaugural chair of the Energy Committee, a joint committee between the TMS Light Metals and Extraction and Processing Divisions, and has been a co-editor of the Energy Technology symposium through the present. He received the LMD Distinguished Service Award in 2010. He is the Chairman of Hydrometallurgy and Electrometallurgy Committee of the Extraction and Processing Division for the 2013-2015 term, during which time he organized the Rare Metal Technology 2014 and 2015 symposia. He is the co-editor for the *Essential Readings in Magnesium Technology* compendium of TMS papers published in 2014. He is a co-editor of 2015 symposium on *Drying, Roasting and Calcining*.

Donna P. Guillen

Donna Post Guillen is a Distinguished Staff Engineer and Group Lead in the Advanced Process & Decision Systems Department at the Idaho National Laboratory (INL). Dr. Guillen earned a B.S. in Mechanical Engineering from Rutgers University, an M.S. in Aeronautics from Caltech, and a Ph.D. in Engineering and Applied Science from Idaho State University. She is a registered Professional Engineer in Mechanical Engineering in the State of Idaho. She has served as Principal Investigator for several multidisciplinary research projects on the topics of waste heat recovery, synthetic fuels production, nuclear reactor fuels and materials experiments, and waste glass processing. The focus of her research is on multiphase computational fluid dynamics (CFD) and thermal hydraulics for sustainable energy technologies. She applies numerical modeling techniques to provide understanding of a wide variety of complex systems, from greenhouse gas generation/sequestration for dairies to waste vitrification for the Hanford Waste Treatment Plant. Dr. Guillen is especially interested in materials as they relate to heat transfer and fluid flow phenomena. She is experienced with irradiation testing and thermal hydraulic analysis for ATR experiments and serves as Principal Investigator / Technical Lead for the ATR National Scientific User Facility (NSUF) Program. She is the lead inventor on two patents related to the development of a new composite material to produce a fast reactor environment within a pressurized water reactor, such as ATR. She actively mentors students; routinely chairs and organizes technical meetings for professional societies; serves in leadership capacity for the American Nuclear Society (Thermal Hydraulics Executive and Program Committees), The Minerals, Metals & Materials Society (Energy Committee Chair), and the American Society of Mechanical Engineers (Thermal Hydraulics and CFD Studies Track Chair); provides subject matter reviews for proposals and technical manuscripts; has published over 100 conference papers, reports, and journal articles; and has written/edited three books.

Li Li

Li Li is a Senior Research Associate in Department of Materials Science and Engineering of Cornell University. He received his Ph.D. degree in Materials Science and Engineering from Carnegie Mellon University (CMU) in 2012; two master degrees from CMU and Harbin Institute of Technology (HIT); and bachelor degree from HIT, respectively. Dr. Li is a talented materials scientist with expertise in energy materials processing, solar energy conversion, and energy storage. He has authored more than

30 technical journal papers and is the reviewer of more than 30 prestigious journals, including *Energy & Environmental Science* and *Metallurgical and Materials Transactions*. He has frequently chaired symposia at TMS and MRS Annual Meetings and is the member of TMS Energy Committee. Because of his contribution to the materials science, he has been awarded to represent TMS at the Emerging Leaders Alliance Conference in 2014.

Cynthia K. Belt

Cindy Belt has managed energy programs at Superior Industries, Aleris International, and Kaiser Aluminum. She has published multiple papers in the area of energy management in the metals industry and has co-edited several proceedings in energy and recycling.

Cindy is Vice-Chair of the TMS Extraction and Processing Division, Past-Chair of the Energy Committee, and a member of Materials and Society committee within TMS and is involved in numerous energy groups within TMS, AFS, and ASME. Cindy earned her Bachelor of Science in Mechanical Engineering degree from Ohio Northern University.

Recycling and Sustainability Update

Randolph Kirchain

Randolph Kirchain's research and teaching aim to improve materials-technology decisions by characterizing the economic and environmental impact of those decisions. That impact may derive from changes in the performance of the products into which those materials are transformed and/or in the systems in which they are produced, used, and eventually discarded. Dr. Kirchain has authored over 100 publications in refereed journals and conferences. He has been awarded the American Iron and Steel Institute's Top Technical Achievement Award, the General Motors Technical Achievement Award, and the TMS Recycling Technology Award. Currently, Dr. Kirchain serves as the co-director of the MIT Concrete Sustainability Hub.

Jeffrey S. Spangenberger

Jeffrey Spangenberger is a senior engineering specialist at Argonne National Laboratory with a B.S. in chemical engineering from Iowa State University. Working in the Energy Systems Division's Process Technology Group, Jeff has lead research activities ranging in scale from bench top testing through full scale plant installations and process efficiency studies. While he works to find ways of reducing energy demands in many energy intensive processes, much of his work relates to the separation and recovery of materials in waste streams destined for landfill. This research has resulted in the conceptualization, design, and construction of equipment, pilot scale processes and industrial scale plants resulting in numerous patents and awards.

Magnetic Materials for Energy Applications V

Frank Johnson

Frank Johnson joined GE Global Research in 2005 as a materials scientist in ceramic and metallurgy technologies. His current research focuses on the development of magnetic materials for power generation, distribution, and conversion technologies. Frank's area of expertise is the structure-processing-property relationships of crystalline, nanocrystalline, and amorphous soft magnetic alloys, rare-earth permanent magnets, and magnetocaloric materials. Prior to joining GE, Frank received was a post-doctoral fellow in the Magnetic Materials Group, Metallurgy Division of the National Institute of Standards and Technology in Gaithersburg, Maryland. While at NIST, he worked on magnetocaloric materials, magnetic thin-films, and assisted in the development of magnetic standard reference materials.

Dr. Johnson received his Ph.D. in materials science and engineering from Carnegie Mellon University in 2003, and his thesis topic was in the area of Fe-Co based nanocrystalline soft magnetic alloys (HITPERM). He received an M.S. in materials science and engineering from the Massachusetts Institute of Technology in 1999 and a B.S. in materials science and engineering from Carnegie Mellon University in 1996.

Sustainable Energy and Layered Double Hydroxides

Andrew Jewel Gomes

Andrew Jewel Gomes received his doctoral degree in Physical Chemistry from Bergische Universitaet Wuppertal, Germany in 2002. He is currently employed as Research Associate Professor in the College of Engineering at Lamar University in Beaumont, Texas. He has been involved in research for more than two decades in a variety of areas, such as waste water treatment, electrochemistry, air pollution remediation, materials synthesis and characterization, and Gaussian simulation. His number of publication is 70 including 24 peer-reviewed articles. He is co-author of Yaws Handbook, and has contributed as chapter author in the book titled *Hazardous Materials: Types, Risks and Control*. His research group is currently involved in developing NOx and SOx removal device using layered double hydroxides in combustion system.

High-Temperature Systems for Energy Conversion and Storage

Amit Pandey

Amit Pandey's career has been multidisciplinary and he has worked in industry, academics, and national laboratory of higher research and education. Currently he is working as a development lead in reliability at Rolls Royce LG Fuel Cell Systems, Inc., in Ohio. He started working in the field of thermal barrier coatings at Johns Hopkins University as a postdoctoral fellow in the department of mechanical engineering. Later he was awarded the research fellowship at the Oak Ridge National Laboratory in the Materials Science and Technology Division. During his time at ORNL he developed new characterization techniques to study the thermomechanical response of porous and microcracked ceramics systems (Cordierite for filtration, active layers in the solid oxide fuel cells, coatings for turbine blades, etc.).

In the short span of his professional career Dr. Pandey has published aggressively and has received various awards showing his academic excellence and exemplary leadership ability. This includes the recent 2014 TMS Young Professional Award from TMS Functional Materials Division (formerly the Electronic, Magnetic, and Photonic Materials Division) and 2014 American Ceramic Society's Future Leader. He has also been involved as an organizing chair and co-chair of symposia for high temperature materials in particular in the area of functional ceramics for energy conversion and storage.

Dr. Pandey also serves as a director on the executive committee of ASME and ASM International's Akron, Ohio Chapter. The primary focus of these activities is to increase the interest of students toward the STEM field. Dr. Pandey earned his B.S. in 2003 in mining engineering (rock mechanics) from Indian Institute of Technology (IIT), India. He later received his Master's degree in 2005 in civil engineering (geo-mechanics) from University of Arizona and his Ph.D. in 2009 from mechanical engineering (mechanics and materials) from University of Maryland.

Structural Materials, Heat Transport Fluids, and Novel System Designs for High Power and Process Heat Generation

Peter Hosemann

Peter Hosemann is Associate Professor at the University of California Berkeley in the department of Nuclear Engineering. He joined UCB in 2010 and is responsible for the nuclear materials group. The research focus is on materials in nuclear and other extreme environments with a specific focus on structural materials degradation. Prior to this appointment he was post doctorate associate at Los Alamos National Laboratory from 2008-2010. He graduated from the Montanuniveristaet Leoben in Austria in 2008 from the department of chemistry after receiving his M.S. from the same institution in the department of physics.

Dr. Hosemann is also advisor to the editorial board for the *Journal of Nuclear Materials* while being elected the "best reviewer" for this journal. In addition he serves as key reader and guest editor on other journals. He is a frequent organizer of a number of workshops and symposia including but not limited to several TMS and MRS symposia.

SESSION CHAIRS

Energy Technologies and Carbon Dioxide Management Symposium 2015

Carbon Management
Animesh Jha, University of Leeds
Cynthia K. Belt, Consultant

Iron & Steel
Cong Wang, Northeastern University
Zuotai Zhang, Peking University

Metal Processing / Molten Salt / Electrochemistry
Animesh Jha, University of Leeds
Donna P. Guillen, Idaho National Laboratory
Li Li, Cornell University
Shulan Wang, Northeastern University

Solar Energy
Li Li, Cornell University
Shulan Wang, Northeastern University

Recycling and Sustainability Update

Waste
Jeffrey S. Spangenberger, Argonne National Laboratory

Recycling
Randolph Kirchain, Massachusetts Institute of Technology

Magnetic Materials for Energy Applications V

Magnetocaloric Materials I
Frank Johnson, GE Global Research
Victorino Franco, University of Seville

Permanent Magnets I
Raju Ramanujan, Nanyang Technological University
Ivan Skorvanek, Institute of Experimental Physics

Magnetic Materials for Energy Applications Poster Session
Huseyin Ucar, Oak Ridge National Laboratory
Xiujuan Jiang, Pacific Northwest National Laboratory

Permanent Magnets II
Hariharan Srikanth, University of South Florida
Lin Zhou, Ames Laboratory

Magnetocaloric Materials II
Karl Sandeman, Imperial College of London
Alex Leary, Carnegie Mellon University

Soft Magnetic Materials I
Paul Ohodnicki, National Energy Technology Laboratory
Katie Jo Sunday, Drexel University

Soft Magnetic Materials II
Thomas G. Woodcock, IFW Dresden
Orlando Rios, Oak Ridge National Laboratory

Sustainable Energy and Layered Double Hydroxides

Sustainable Energy and Layered Double Hydroxides
Andrew Gomes, Lamar University
Christian Ruby, Université de Lorraine

High-Temperature Systems for Energy Conversion and Storage

Solid Oxide Fuel Cell: Recent Developments I
Paul Ohodnicki, National Energy Technology Laboratory
Kyle Brinkman, Clemson University

High Temperature Ceramic Materials: Response, Modelling and Performance
Amit Shyam, Oak Ridge National Laboratory
Kathy Lu, Virginia Tech

Solid Oxide Fuel Cell: Recent Developments II
Paul Ohodnicki, National Energy Technology Laboratory
Amit Pandey, Rolls Royce LG Fuel Cell Systems Inc.

Innovation in Energy Conversion and Storage I
Soumendra Basu, Boston University
Jung Choi, Pacific Northwest National Laboratory

Innovation in Energy Conversion and Storage II
Amit Shyam, Oak Ridge National Laboratory
Ryan Cooper, Oak Ridge National Laboratory

Structural Materials, Heat Transport Fluids, and Novel System Designs for High Power and Process Heat Generation

Heat Transport Fluids I
Peter Hosemann, University of California, Berkeley
Bruce Pint, Oak Ridge National Laboratory

Heat Transport Fluids II
Kumar Sridharan, University of Wisconsin
Peiwen Li, The University of Arizona

Energy Technology 2015
Carbon Dioxide Management and Other Technologies

Energy Technologies and Carbon Dioxide Management Symposium 2015

Energy Technology 2015
Carbon Dioxide Management and Other Technologies

ENERGY TECHNOLOGIES AND CARBON DIOXIDE MANAGEMENT SYMPOSIUM 2015

Carbon Management

Session Chairs:

Animesh Jha
Cynthia K. Belt

Energy Technology 2015: Carbon Dioxide Management and Other Technologies
Edited by: Animesh Jha, Cong Wang, Neale R. Neelameggham, Donna P. Guillen, Li Li, Cynthia K. Belt, Randolph Kirchain, Jeffrey S. Spangenberger, Frank Johnson, Andrew J. Gomes, Amit Pandey, and Peter Hosemann
TMS (The Minerals, Metals & Materials Society), 2015

A THERMODYNAMIC STUDY OF MIXED CARBON FEEDSTOCK GASIFICATION SLAGS

Jinichiro Nakano[1,2], Marc Duchesne[3], James Bennett[1], Kyei-Sing Kwong[1], and Anna Nakano[1]

[1] US Department of Energy National Energy Technology Laboratory, 1450 Queen Ave., Albany, OR 97321 USA
[2] URS Corp, P.O. Box 1959, Albany, OR 97321 USA
[3] Natural Resources Canada CanmetENERGY, 1 Haanel Drive, Ottawa, ON K1A1M1 Canada

Keywords: Phase diagram, Equilibrium, Gasification, Coal, Petroleum coke, Additive agents

Abstract

Integrated Gasification Combined Cycle used in power and chemical production is considered a clean technology, with the ability to capture almost all CO_2, NO_x, and SO_x emissions. In entrained bed slagging gasifiers, molten slags formed from feedstock's non-volatile impurities contribute to gasifier liner degradation and can cause gasifier clogging, affecting system efficiency and operation. Increased petcoke use as a key feedstock in addition to or as a replacement for coal has drastically modified slag chemistry, leading to unknown chemical/physical slag properties and behavior in the gasifier. In this work, thermodynamic phase equilibria in synthetic slags (Al_2O_3-CaO-FeO-SiO_2-V_2O_3) were evaluated under simulated gasifier conditions to establish an understanding of the phase equilibrium in these slag systems. The effects of V_2O_3 content, slag chemistry, and additives on amorphous and crystalline phases were studied. In this study, increasing calcium oxide and iron oxide additive agents was found to lower the slag melting temperature and caused the karelianite (V_2O_3) crystal size to increase. Equilibrium phase diagrams showing the additive effect on the mixed coal-petcoke slag systems studied were constructed.

Introduction

Integrated Gasification Combined Cycle (IGCC) [1, 2] is a clean technology for chemical and power production, with almost all the harmful emissions effectively captured. During IGCC operation, synthesis gas (syngas) is produced, which consists primarily of hydrogen (H_2) and carbon monoxide (CO). H_2 and CO are used as a fuel in turbines generating power or as a feedstock in chemical production. In entrained flow slagging gasifiers, non-volatile impurities from feedstock ashes form molten slag, which promotes refractory liner degradation and potential gasifier clogging problems. Sufficient slag fluidity, controlled by gasifier temperature, facilitates a downward slag flow toward the exit along the refractory lining of the gasification chamber. As a result, molten slag continuously interacts with the refractory materials. Typical operation temperatures for the entrained flow slagging gasifiers are in the range of 1325 - 1575 °C, with a 10^{-9} to 10^{-7} atm oxygen partial pressure [3]. Currently, petroleum coke (petcoke) feedstock is used as an economical additive or alternative to coal feedstock. The chemistry of the resulting slag is significantly different from coal gasifier slags due to the high vanadium content of petcoke ashes, which leads to uncertainty in gasifier performance. To optimize gasifier design and operation, understanding of the thermodynamic nature of the vanadium-rich slags is important. However, there is no commercially available thermodynamic database for vanadium-containing slags. Chemical composition ranges of selected ash constituents found in the U.S. and Canada are summarized in Table 1.

Table 1. Chemistries of industrial coal and petcoke ashes found in the U.S. and Canada

(wt.%)	Coal ash						Petcoke ash					
	U.S.			Canada			U.S.			Canada		
Reference	[4]			[7, 8]			[5, 6]			[9, 10]		
Ash component	average	min	max	average	min	max	average	min	max	average	min	max
Al_2O_3	25.1	4.1	38.6	18.6	16.3	20.3	5.6	0.5	9.4	19.6	14.4	24.9
CaO	5.8	0.5	45.1	6.9	2.3	13.6	5.4	2.2	11.9	5.1	1.6	15.8
Fe_2O_3	17	2.1	69.7	9.4	2.6	32	9.3	2.5	31.6	9.9	7.4	12.1
K_2O	0.4	0	3.5	1.4	0.3	2.3	0.6	0.3	1.2	1.6	0.9	1.9
MgO	1.2	0	8	1.4	0.9	2.7	1.2	0.2	5.1	1.9	1.3	3.9
Na_2O	0.3	0	6.5	1.9	0.4	6	0.7	0.1	1.8	1.2	0.7	1.7
NiO	-	-	-	0	0	0	8.8	2.9	12	1.9	1.1	2.6
P_2O_5	0.1	0	8	0.2	0	0.6	-	-	-	0.1	0	0.4
SiO_2	43.6	7.1	68.5	52	35.5	67.3	13.6	1.6	23.6	41.4	37	50.1
SO_3	4.1	0	32.3	4.5	1.1	14.3	3.8	0.8	13.8	2.6	1.9	7.1
TiO_2	0.6	0	3.7	0.7	0.5	0.8	0.2	0	0.4	2.3	0.6	4.8
V_2O_5	-	-	-	0	0	0	49.4	19.7	74.5	10.1	3.2	20.3
Total	98.3			97.1			98.6			97.8		
SiO_2/Al_2O_3	1.73			2.79			2.45			2.11		

U.S. coal ash chemical data represents 300 coal samples across the U.S.[4]. Chemistries of U.S. petcoke ashes are reported from delayed coke, shot coke, fluid coke, and flexicoke [5] and [6]. Canadian coal ash compositions are representative of Canadian thermal coals: sub-bituminous coals from Alberta, lignite coal from Saskatchewan and bituminous coals in use in Nova Scotia [7, 8]. Canadian petcoke ash compositions [9, 10] are typical of delayed and fluid cokes from Alberta oil sands operations and petcoke in use at an eastern Canadian power utility.

Phase equilibria for steelmaking slag systems containing vanadium have been reported in the literature [11, 12] and for the Al_2O_3-CaO-V_2O_3 system [13]. In previous work [3] an equilibrium study of the Al_2O_3-CaO-FeO-SiO_2-V_2O_3 system was first conducted at 1500 °C in 10^{-8} atm of oxygen partial pressure with up to 9 wt.% V_2O_3 content. Fluxes such as calcium-based and iron-based oxides have been traditionally used in industry to optimize gasifier operations [14, 15], increasing calcium and/or iron contents in the resulting slags. Basifying gasifier slags would broaden refractory material choices as more economical alternatives are available [16]. Considering the broad ash chemistries listed above, the effective utilization of the full range of petcoke in the U.S. and Canada require development of phase equilibria models containing higher vanadium content at various calcium and iron oxide levels.

The goals of this work were to clarify thermodynamic phase equilibria in vanadium-containing slags and to develop phase diagrams envisaging the thermodynamic nature of the coal-petcoke mixed gasifier slags, which would be used as potential tools for the gasification community. Thermodynamic information would serve as a guide to study slag-refractory interactions.

Materials and methods

Compositions of synthetic slag (Al_2O_3, CaO, FeO, SiO_2, and V_2O_3) studied in this work were formulated to simulate the industrial coal-petcoke mixed ash chemistries found in the U.S. and Canada. In order to construct 5-component phase diagrams on 2-dimensional Cartesian axes, the weight ratio of SiO_2/Al_2O_3 was kept constant at 2.23, while the CaO content was fixed to be 6, 13, or 20 wt.%. V_2O_3 and FeO were varied from 0 to 27 wt.% and 1 to 14 wt.%, respectively. The Al_2O_3 (99.5%, < 37 μm), CaO (99.95%, < 10 μm), Fe_2O_3 (99.945%, < 5 μm), SiO_2 (99.5%, < 44 μm), and V_2O_3 (99.6%, < 100 μm) powders supplied by Alfa Aesar were weighed and mixed by a tumble-mixer for 3 hours. Sample compositions were analyzed after equilibration tests by inductively coupled plasma optical emission spectrometry (ICP-OES) using a Perkin Elmer 7300 DV with a teflon nebulizer and teflon Tracey spray chamber. Uncertainty associated with the ICP-OES equipment was within ± 0.2 %. The experimental setup for equilibration tests is schematically shown in Figure 1.

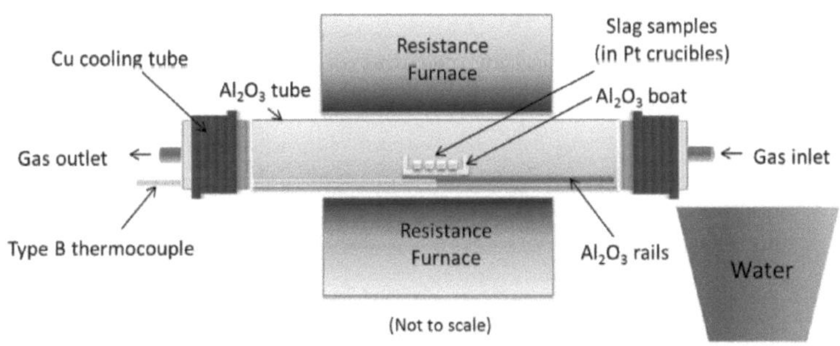

Figure 1. Schematic illustration of an instrument used for the equilibration experiments.

During the equilibration experiments, samples (about 250 mg each) were placed in cylindrical platinum crucibles (99.99%, 10 mm D × 7.5 mm H × 0.05 mm T) on a high-density alumina boat (99.8 wt.%), which was located in a uniform hot zone of a horizontal tube resistance furnace

(CM Rapid Temp 1800) with sealed ends cooled by running water. The alumina boat holding the samples was placed on alumina rails for maneuverability. An oxygen partial pressure of approximately 10^{-8} atm at 1500 °C was predicted for a mixture of 64 vol.% CO-36 vol.% CO_2 (from standard Gibbs free energies [17]). With 30 ml/min constant flow of this gas mixture, the samples were heated from room temperature to 800 °C at 100 °C/hour, and from 800 °C to 1534 °C at 44 °C/hour. Slow heating above 800 °C was implemented to minimize rapid agitation and foaming of the slags. The samples were equilibrated at a sample temperature of 1500 °C ± 0.8 °C for 72 hours. After the furnace was purged with Ar (99.999 %) at 400 ml/min for 3 minutes at the end of the equilibration, the samples were quenched in water. The time taken from opening of the end cap to sample-water contact was less than 30 seconds. Quenched slags were carefully removed from the platinum crucibles and analyzed by X-ray diffractometry (Rigaku Ultima IV XRD spectrometer (Cu Kα, 40 kV, 40 mA, 7 to 80°)), ICP-OES chemical analysis, and by scanning electron microscopy with wavelength dispersive spectroscopy (SEM-WDX) using a JEOL-7000F microscope and an Oxford INCA WAVE spectrometer.

Results

The samples were grouped by CaO contents; the first group contained 6 wt.% CaO ± 0.73, the second 13 wt.% CaO ± 1.54, and the third 20 wt.% CaO ± 2.52. FeO contents were varied from 1 to 14 wt.%. From XRD and SEM-WDX analyses, four distinct phase fields involving amorphous slag, mullite (orthorhombic), and/or karelianite (rhombohedral) were identified in the samples. Representative XRD patterns of the equilibrated samples are shown Figure 2.

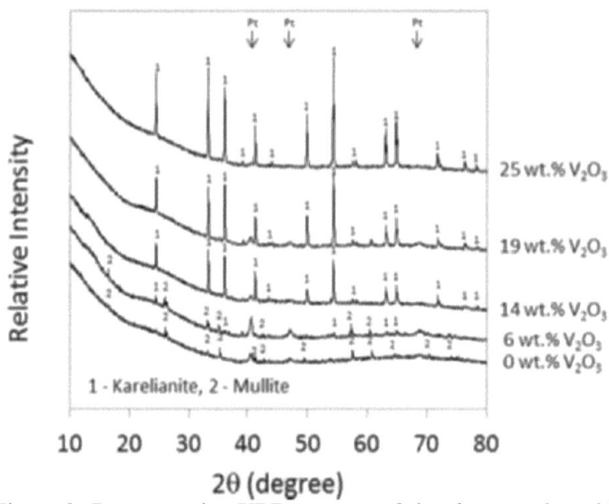

Figure 2. Representative XRD patterns of the slag samples with V_2O_3 (0 to 25 wt.%) at CaO ≈ 6 wt.% and FeO ≈ 7 wt.%.

No crystals were noted at low V_2O_3 and high FeO contents. Mullite crystals were present below 9 wt.% FeO in slags with the lowest CaO content. At higher CaO contents (13 and 20 wt.%), mullite was not present at any FeO content. The karelianite crystal was found at all CaO and FeO contents if the sample contained more than 2 wt.% V_2O_3. At low CaO and FeO contents (6 wt.% and < 8 wt.%, respectively), karelianite tended to coexist with mullite in slags if V_2O_3 content exceeded 2 wt.%. Note platinum residues from the platinum crucibles were detected in some samples (indicated with arrows in Figure 2).

Typical crystal morphology representing different equilibria are presented in Figure 3. It was assumed that amorphous slag found in SEM micrographs had been liquid slag at equilibration temperature (1500 °C) prior to quenching. Four regions were identified in slags with 6 wt.% CaO; including (a) a region where karelianite and mullite coexist in liquid slag, (b) a region where only karelianite exists in liquid slag, (c) a region where only mullite exists in liquid slag, and (d) liquid slag exists without crystals. No clear evidence was noted to show any heterogeneous relationship between karelianite and mullite from SEM analysis. Their nucleation seems to be independent of each other.

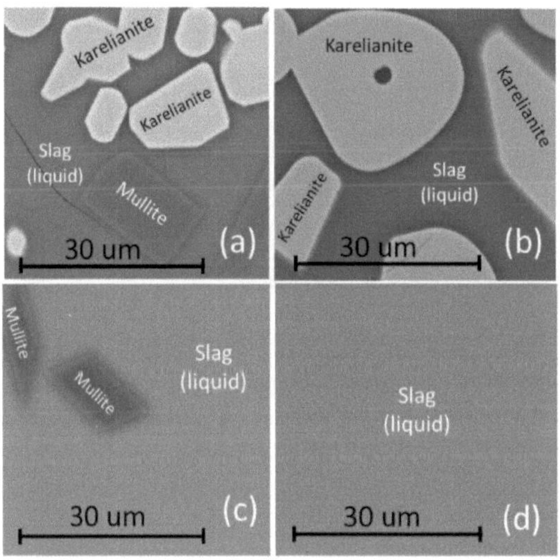

Figure 3. Backscatter electron micrographs showing phase fields of (a) liquid slag + karelianite + mullite, (b) liquid slag + karelianite, (c) liquid slag + mullite, and (d) liquid slag.

Discussion

Phase equilibria

Based on XRD and SEM-WDX analyses, the isothermal phase diagrams of the Al_2O_3-CaO-FeO-SiO_2-V_2O_3 slag system are proposed (Figure 4). The sensitivity of the border locations was evaluated with respect to relative characteristics from XRD and SEM analyses, including peak intensities and crystal fractions. Three isothermal sections at 1500 °C in $Po_2 = 10^{-8}$ atm corresponding to fixed CaO bulk contents (6, 13, and 20 wt.%) were then constructed. Figure 4 shows the isothermal phase diagrams on axes normalized to Al_2O_3, FeO, SiO_2, and V_2O_3. The secondary axes represent approximate bulk compositions [uncertainty = < 0.1 wt.% for (a), ± 0.2 wt.% for (b), and ± 0.4 wt.% for (c)].

From Figure 4, crystal-free slags tend to exist at low V_2O_3 content. The crystal-free liquid slag region expands with higher CaO and/or FeO content. In order to minimize precipitates that would increase viscosity in gasifier slags, more than 8 wt.% FeO is needed at 6 wt.% CaO, or any FeO contents at or above 13 wt.% CaO. Viscosity of liquid slag may decrease with increasing V_2O_3 content, but excessive amounts would cause karelianite crystals to precipitate and may increase the viscosity of the slag mixture [7, 10, 16]. Note tie-lines are not on the plane and V_2O_3 concentrations in 2 or 3 phase regions cannot be determined by liquidus shown on the diagrams.

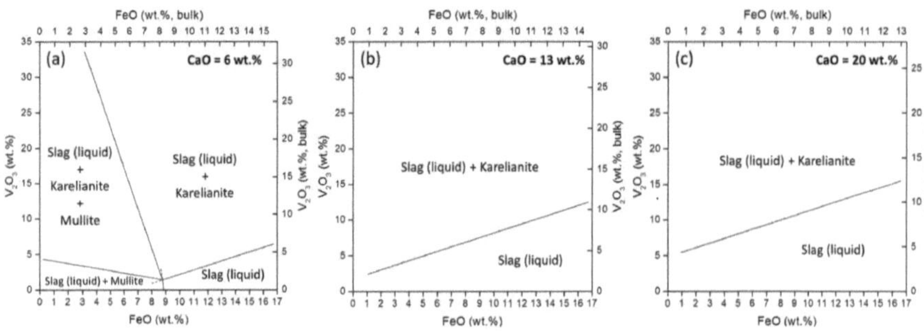

Figure 4. Proposed isothermal phase diagrams of the Al_2O_3-CaO-FeO-SiO_2-V_2O_3 slag system at 1500 °C in $Po_2 \approx 10^{-8}$ atm for: (a) 6 wt.% CaO, (b) 13 wt.% CaO, and (c) 20 wt.% CaO. Secondary axes represent approximate respective concentrations in bulk slag.

V_2O_3 solubility in liquid slag

Vanadium is the key component to precipitate karelianite crystals causing potential flow and clogging issues in gasifiers. Therefore, it is critical to control V_2O_3 solubility in liquid slag by fluxes such as calcium oxide-based and iron oxide-based additives [14, 15]. Based on the present experimental data, a contour map showing the effects of CaO and FeO on V_2O_3 solubility was generated using data analysis computer software, ORIGIN [18] with a total points increase factor = 10,000 and a smoothing parameter = 0.01 (Figure 5). Note, in these calculations, the total points increase factor expands the number of plotted points by multiplying the original number of points by this factor while the smoothing parameter determines the degree of smoothness. At 1500 °C in $Po_2 = 10^{-8}$ atm, the effect of FeO on dissolving vanadium is maximized around 17 wt.% CaO, which only requires 5 wt.% FeO to achieve > 30 wt.% V_2O_3 dissolution in the liquid slags. Note, while an increase in CaO would generally improve V_2O_3 dissolution, excessive CaO supply may decrease V_2O_3 solubility. In the above example, V_2O_3 solubility drops to below 10 wt.% if more than 21 wt.% CaO was added while equivalent solubility may be achieved only by

4 wt.% CaO. An appropriate balance has to be practiced to optimize V_2O_3 solubility and crystal formation. Minimal impact on V_2O_3 solubility was noted by FeO if CaO content was low (< 5 wt.%) and by CaO if FeO content was low (< 3 wt.%). It is note-worthy that V_2O_3 solubility was independent of V_2O_3 bulk concentrations if CaO and FeO bulk contents were fixed. In other words, concentrations of CaO and FeO additives can effectively determine/control how much V_2O_3 would dissolve in the liquid slag regardless of V_2O_3 bulk concentrations.

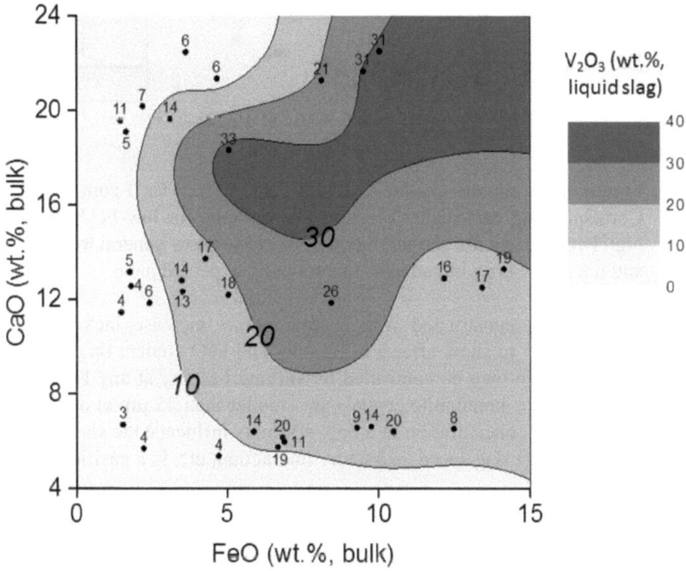

Figure 5. A contour map showing V_2O_3 solubility in liquid slag with respect to bulk CaO and FeO content. Experimental points are shown as solid circles along with the V_2O_3 solubility values (wt.%).

Karelianite crystal size and morphology

More than 10 karelianite particle diameters (longest and shortest dimensions) were measured and averaged per sample. Larger crystal sizes were found in slags with chemistry far from liquidus on the isothermal sections, i.e., higher V_2O_3 bulk concentration. Karelianite crystal size also increased with FeO and CaO content (Figure 6).

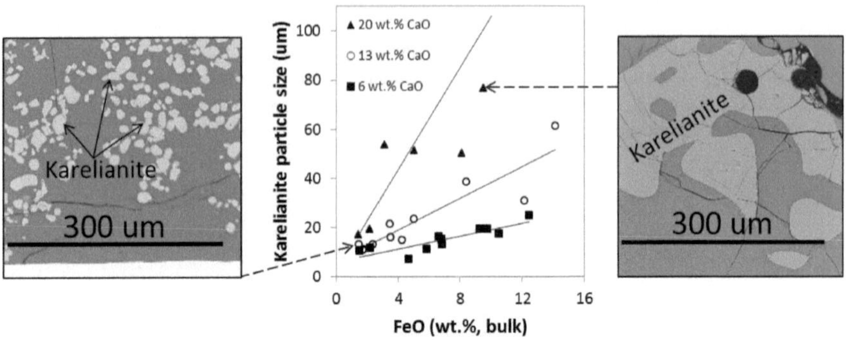

Figure 6. Karelianite crystal size with FeO at fixed CaO contents. Corresponding backscatter electron micrographs for low FeO and high FeO content are shown. Fitted lines are to show general trends and not necessarily intended as theoretically expected paths.

Another contour map was constructed with a total points increase factor = 20,000 and a smoothing parameter = 0.15 to show effects of basicity and FeO content on crystal growth. The map indicates crystal growth may be controlled by varying basicity at any FeO content (Figure 7). At low basicity, however, karelianite crystals are smaller than 35 μm at all the FeO contents studied. Crystal size and its concentration in a bulk slag may influence the slag viscosity. Change in slag viscosity would affect slag speed, refractory interaction, etc. in a gasifier.

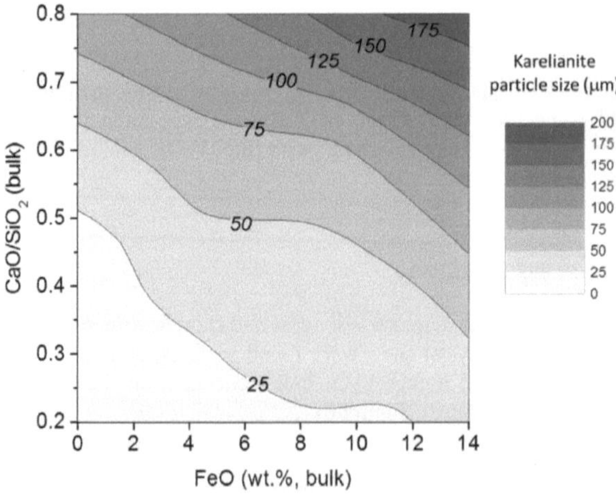

Figure 7. A contour map showing basicity (CaO/SiO_2) and FeO effects on karelianite crystal sizes.

Conclusions

Synthetic slags (Al_2O_3-CaO-FeO-SiO_2-V_2O_3) with 0- 27 wt.% vanadium oxide, corresponding to industrial coal/petroleum coke feedstock blends, were equilibrated in simulated gasifier environments at 1500 °C with an oxygen partial pressure of approximately 10^{-8} atm. Based on XRD and SEM-WDX analyses of these slags after quenching, isothermal phase diagrams were proposed. With increasing CaO and/or FeO content, the homogeneous liquid slag region expanded and composition ranges exhibiting crystals were reduced. Mullite was not present in the slags containing more than 8.1 wt.% FeO, while karelianite was always present if a sufficient amount of vanadium (> 2 wt.%) was present in the slag. Karelianite crystal sizes increased with higher CaO and FeO contents, which may influence the slag viscosity and thus the slag flow nature.

Acknowledgement

This technical effort was performed in support of the National Energy Technology Laboratory's ongoing research under the RES contract DE-FE0004000. Part of this research has been funded by the Natural Sciences and Engineering Research Council of Canada and by Natural Resources Canada through the Program of Energy Research and Development. Funding was also provided by the Clean Energy Dialogue to promote U.S.-Canada collaboration in energy technology development. Mr. Paul Danielson (NETL) must be acknowledged for the quality metallographic work. Authors are grateful to Mr. Keith Collins (NETL) for his SEM-WDX analysis and Ms. Valerie Omatsu-Baas (CanmetENERGY) for her XRD investigation. Anna Nakano acknowledges the support of a postdoctoral fellowship at the U.S. Department of Energy, administered by Oak Ridge Institute for Science and Education.

References

1. *US Department of Energy National Energy Technology Laboratory*. COAL GASIFICATION SYSTEMS; Available from: http://www.netl.doe.gov/research/coal/energy-systems/gasification/.

2. *US Department of Energy National Energy Technology Laboratory.* Gasifipedia; Available from: http://www.netl.doe.gov/File%20Library/Research/Coal/energy%20systems/gasification/gasifipedia/index.html.
3. J. Nakano, et al., "Phase Equilibria in Synthetic Coal-Petcoke Slags (SiO_2-Al_2O_3-FeO-CaO-V_2O_3) under Simulated Gasification Conditions," *Energy & Fuels*, 25(7) (2011), 3298-3306.
4. W.A. Selvig and F.H. Gibson, *Analyses of Ash from United States Coals.* Bureau of Mines Bulletin. Vol. 567. 1956, Washington, USA: United States Government Printing Office.
5. R.W. Bryers, "Utilization of Petroleum Coke and Petroleum Coke/Coal Blends as a Means of Steam Raising," *Fuel Processing Technology*, 44(1-3) (1995), 121-141.
6. M.L. Swanson and D.R. Hajicek. *Advanced High-Temperature, High-Pressure Transport Reactor Gasification.* in *International Conference on Fluidized Bed Combustions.* May 31, 2001. Reno, NV.
7. F. Goodarzi, "Mineralogy, elemental composition and modes of occurrence of elements in Canadian feed-coals," *Fuel*, 81(9) (2002), 1199-1213.
8. M.A. Duchesne, et al., "Fate of inorganic matter in entrained-flow slagging gasifiers: Fuel characterization," *Fuel Processing Technology*, 118 (2014), 208-217.
9. M.A. Duchesne, et al., "Flow behaviour of slags from coal and petroleum coke blends," *Fuel*, 97 (2012), 321-328.
10. E. Furimsky, "Gasification of oil sand coke: Review," *Fuel Processing Technology*, 56(3) (1998), 263-290.
11. N.N. Lopatko, G.G. Mikhailov and S.A. Archugov, "Phase equilibria in the Fe-Si-V-O system," *Russian Metallurgy*, (2) (1996), 1-6.
12. Y. Yang, L. Teng and S. Seetharaman, "Experimental Study of Phase Equilibria in CaO–SiO2–MgO–Al2O3–V2O3 Slags," *Steel Research International*, 85(9999) (2014), 1-9.
13. A. Muan and M.S. Najjar, *Compositions involving V2O3-Al2O3-CaO*, in 1991, Texaco Inc., White Plains, N.Y.: U.S.
14. M.S. Najjar, *Partial oxidation process.* 1987, Texaco Inc., White Plains, NY.: U.S.
15. M.S. Najjar and R.J. Corbeels, *Partial oxidation process.* 1990, Texaco Inc., White Plains, NY.: U.S.
16. K.-S. Kwong, J. Bennett and J. Nakano, *Basic refractory and slag management for petcoke carbon feedstock in gasifiers.* 2014, US 8,703,021 B1, US Department of Energy, Washington, DC (US).
17. O. Kubaschewski, E.L. Evans and C.B. Alcock, *Metallurgical thermochemistry.* 4th ed. 1967: Pergamon Press.
18. OriginLab, *Origin 9.1 Graphing & Analysis.* http://www.originlab.com/.

Energy Technology 2015: Carbon Dioxide Management and Other Technologies
Edited by: Animesh Jha, Cong Wang, Neale R. Neelameggham, Donna P. Guillen, Li Li, Cynthia K. Belt, Randolph Kirchain, Jeffrey S. Spangenberger, Frank Johnson, Andrew J. Gomes, Amit Pandey, and Peter Hosemann
TMS (The Minerals, Metals & Materials Society), 2015

EVALUATION OF HEAT TREATMENT PERFORMANCE OF POTENTIAL PIPE STEELS IN CCS-ENVIRONMENT

Anja Pfennig[1], Helmut Wolthusen[1], Philip Zastrow[1], Axel Kranzmann[2]

[1] HTW University of Applied Sciences Berlin,
Wilhelminenhofstraße 75 A, Gebäude C, 12459 Berlin
anja.pfennig@htw-berlin.de

[2] BAM Federal Institute of Materials Research and Testing,
Unter den Eichen 87, 12205 Berlin

Keywords: corrosion, CCS, carbon storage, aquifer, high temperature, high pressure,

Abstract

To resist the corrosive geothermal environment during carbon capture and storage CCS -such as: heat, pressure, salinity of the aquifer, CO_2-partial pressure, properties of pipe steels-require certain specification. For evaluation samples of differently heat treated high alloyed stainless injection-pipe steels AISI 420 X46Cr13, AISI 420J X20Cr13 as well as X5CrNiCuNb16-4 AISI 630 were kept at T=60 °C and ambient pressure as well as p=100 bar for 700 h up to 8000 h in a CO_2-saturated synthetic aquifer environment similar to a possible geological situation in the northern German Basin. Corrosion rates and scale growth are lowest after long term exposure for steels hardened and tempered at 600 °C to 670 °C and pits -indicating local corrosion- decrease in diameter but increase in number as a function of carbon content of the steel. Martensitic microstructure is preferred with respect to these particular conditions.

Introduction

During the compression and injecting of emission gasses from combustion processes into deep geological saline aquifer reservoirs (CCS Carbon Capture and Storage) [1] CO_2 is dissolved to build a corrosive environment. This may easily cause the failure of pipe steels [2,3]. The following anodic iron dissolution of the pipe steel enhances the growth of a siderite corrosion layer ($FeCO_3$) on the alloy surface [4-6]. Internal corrosion will depend upon the source of the injected gas, its composition and the presence of water and dissolved salts [5,7].

Heat treatment routines such as temperature and time of austenitizing, cooling rate as well as temperature and time of annealing, influence the resistance against corrosive environments. Retained austenite as a microstructural component resulting from the heat treatments applied has a beneficial effect on the pitting corrosion resistance of 13%-chromium steels (13CrNiMo) [8]. A higher Ni and Cr content in the heat treated steels improve the corrosion resistance [8-11]. In general, raising the annealing temperature lowers the pitting potential of lean duplex stainless steels [10,12-13]. The increased corrosion resistance of martensitic stainless steels with 13% Cr at higher austenitizing temperature (980-1050 °C) is related to the dissolution of carbides [14-16]. The precipitation of Cr-rich $M_{23}C_6$ and M_7C_3 carbides reduces the resistance of passive film and pitting corrosion [13] and has high impact on mechanical properties due to secondary hardening [14]. C-Mn (carbon) steels -offering excellent mechanical strength- show highest corrosion rates in a H_2S-containing NaCl solution because martensitic grain boundaries are more reactive [17].

Many authors demonstrate the dependence also on environmental factors, e.g. the composition of surrounding media and alloy, temperature, CO_2 partial pressure, flow conditions, contaminations and formation of protective scales [4,18,19]; in this study the influence of heat treatment of the steels prior to exposure, hydrostatic pressure within the storage site and alloying elements of the steels is analyzed for a critical temperature region well known for severe corrosion processes [3,5,7,20-25] .

Experimental

Exposure tests were carried out using samples made of thermally treated specimen of AISI 4140 (1%Cr) and AISI 420C (X46Cr13, 0.46%C, 13%Cr)), AISI 420J (X20Cr13, 0.20%C, 13%Cr) and AISI 630 (X5CrNiCuNb16-4) with 8 mm thickness and 20 mm width and 50 mm length. A hole of 3.9 mm diameter was used for sample positioning. Heat treatment prior to exposure was done following routine protocols according to the 3 steel qualities (Table 1).

Table 1: Heat treatment of steels prior to exposure to CO_2-saturated aquifer at 100 bar.

material	heat treatment	temperature °C / °C	dwell time min	cooling medium
X20Cr13 1.4021 AISI 420 J	normalizing	785	30	air
	hardening	1000	30	oil
	hardening + tempering 1, 600 °C	1000 / 600	30	oil
	hardening + tempering 2, 670 °C	1000 / 670	30	oil
	hardening + tempering 3, 755 °C	1000 / 755	30	oil
X46Cr13 1.4034 AISI 420 C	normalizing	785	30	air
	hardening	1000	30	oil
	hardening + tempering 1, 600 °C	1000 / 600	30	oil
	hardening + tempering 2, 670 °C	1000 / 670	30	oil
	hardening + tempering 3, 700 °C	1000 / 700	30	oil
	normalizing	785	30	air
X5CrNiCuNb16-4	normalizing	850	30	oil
X5CrNiCuNb16-4	hardening	1040	30	oil
1.4542	hardening + tempering 1, 600 °C	1040 / 550	30	oil
AISI 630	hardening + tempering 2, 670 °C	1040 / 650	30	oil
	hardening + tempering 3, 700 °C	1040 / 755	30	oil

The surfaces were activated by grinding with SiC-Paper down to 120 μm under water. Samples of each base metal were positioned within the vapour phase (1 bar), the supercritical phase (100 bar) and within the liquid phase. The brine (as known to be similar to the Stuttgart Aquifer [6]: Ca^{2+}: 1760 mg/L, K^{2+}: 430 mg/L, Mg^{2+}: 1270 mg/L, Na^{2+}: 90,100 mg/L, Cl^-: 143,300 mg/L, SO_4^{2-}: 3600 mg/L, HCO_3^-: 40 mg/L) was synthesized in a strictly orderly way to avoid precipitation of salts and carbonates. The exposure of the samples between 700 h to 8000 h was disposed in a chamber kiln at 60 °C at 100 bar in an autoclave system and for reference at ambient pressure as well (figure 1). Flow control (3 NL/h) at ambient pressure was done by a capillary meter GDX600_man by QCAL Messtechnik GmbH, München.

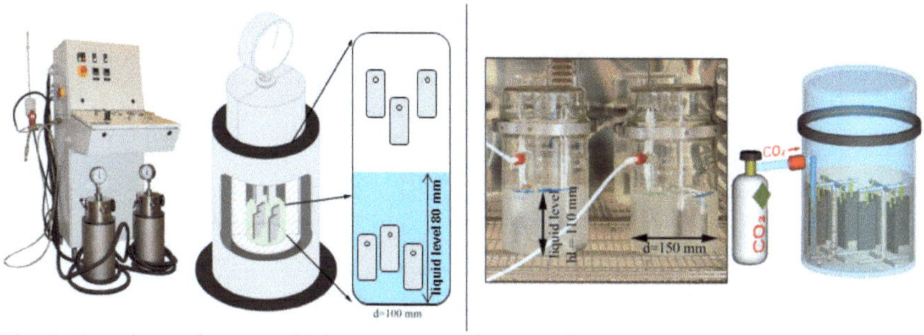

Fig. 1. Experimental set-up of laboratory corrosion experiment: left: autoclaves at 100 bar/60 °C, right: ambient pressure/60 °C [26].

X-ray diffraction was carried out in a URD-6 (Seifert-FPM) with CoKα-radiation with an automatic slit adjustment, step 0.03 and count 5 sec and AUTOQUAN ® by Seifert FPM was used for phase analysis. For gravimetric measurement descaling of the samples (60°C/700 h, 2000 h, 4000 h, 8000 h) was performed by exposure to 37% HCl for 24 hours and mass gain was analyzed according to DIN 50 905 part 1-4. To characterise the pitting corrosion, 3-D-images were realized by the double optical system Microprof TTV by FRT. Scale covered parts of the samples were embedded in a cold resin (Epoxicure, Buehler), cut and polished first with SiC-Paper from 180 μm to 1200 μm under water and then finished with diamond paste 6 μm, 3 μm and 1 μm. The measurement of the layer thicknesses and residual pipe wall thicknesses as well as microstructure analysis were performed via light and electron microscopy techniques using the semi-automatic analyzing program Analysis Docu ax-4 by Aquinto. A set of 100 linescans was measured taking 10 to 20 micrographs per parameter.

Results and Discussion

During normal storage procedures the CO_2 is supposedly injected in its supercritical phase. In the case of intermissions the water level may rise into the injection pipe. This newly two phase regime of water and gas may lead to the precipitation of corrosion products and formation of pits. Specimens exposed to the brine form a carbonate layer as result of the low siderite $FeCO_3$-solubility in CO_2-containing water with rather low pH due to the formation of carbonic acid [24]. These non-uniform corrosion scales are differing in thickness but are present all over the surface. The scale is mainly composed of: siderite $FeCO_3$, goethite α-FeOOH at 100 bar and additionally mackinawite FeS and akaganeite $Fe_8O_8(OH)_8Cl_{1.34}$ as well as spinel-phases of various compositions at ambient pressure. Also carbides, Fe_3C, were identified within the corrosion layer. Pits are covered with the same precipitates of the corrosion products formed on the surface elsewhere [23,24].

surface corrosion

Figure 2 depicts the corrosion rates of the 3 steel qualities as a function of exposure time, heat treatment and pressure. In general the corrosion rate does not exceed 0.02 mm/year and therefore is in good agreement with DIN 6601 allowing for 0.1 mm/year for pressure vessels. The increasing rather high corrosion rate for samples hardened and tempered at 700°C/755 °C may be due to local chromium depletion after carbide precipitation that prohibits the passivation of the surface. This as a consequence leads to the degradation of the base material after long exposure to the CO_2-saturated saline aquifer environment.

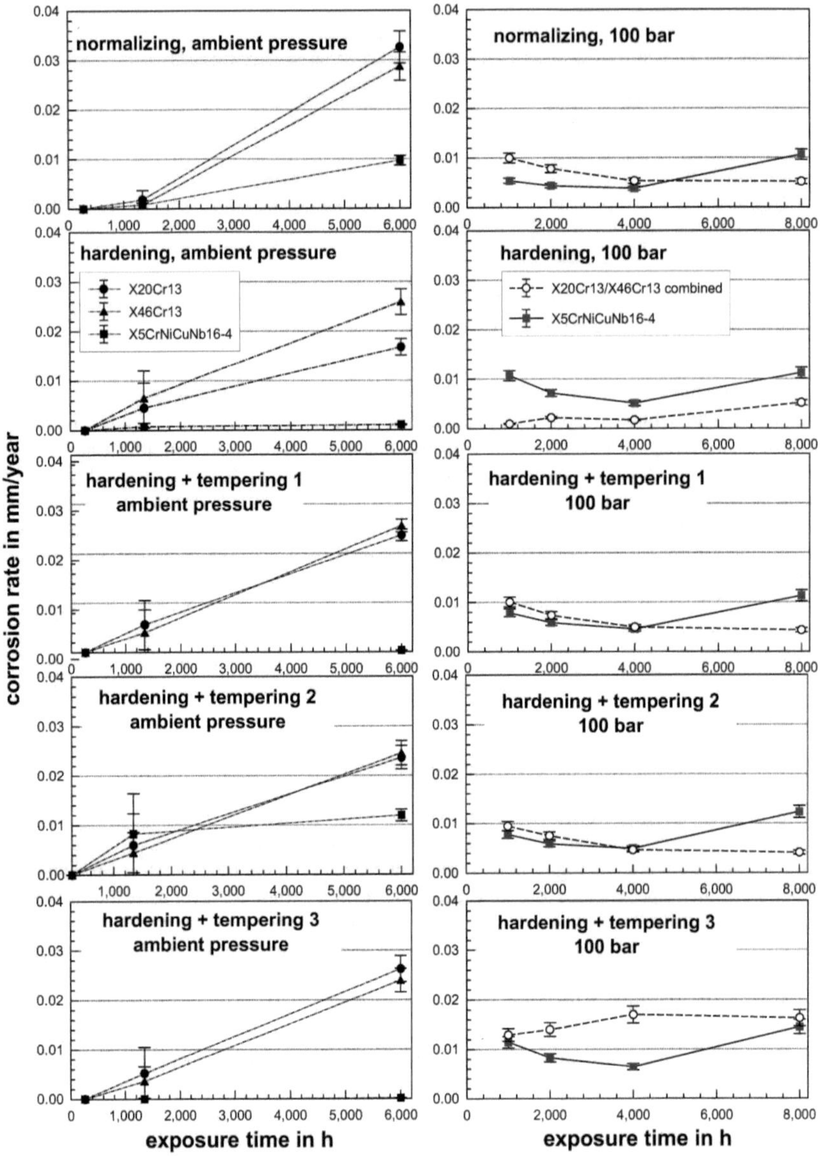

Fig. 2. Corrosion rate after 6000 and 8000 hours of exposure to CO_2-saturated aquifer water at 60 °C and ambient pressure, left and 100 bar, right of X20Cr13, X46Cr13 (combined at 100 bar) and X5CrNiCuNb16-4 heat treated prior to exposure [27] .

The martensitic microstructure of hardened and tempered steels at 600-670 °C offers best corrosion resistance regarding surface corrosion. Between 4000 h and 8000 h of exposure the corrosion rates do not change. This indicates a sufficient thickness of the carbonate layer acting as a diffuiosn barrier and reducing mutual diffusion of ionic species into the base materials (CO_3^{2-}- and O^{2-}-species) and towards the outer surface (Fe-ions).

With corrosion rates obtained via mass gain method about 0.002 mm/year at ambient pressure X5CrNiCuNb16-4 shows the least loss of base material for samples that were hardened only or hardened and tempered (figure 2 left). Normalized samples corrode around 0.01 mm/year determined after 6000 h of exposure. At 100 bar corrosion rates are similar to those obtained at ambient pressure after up to 2000 h, but after 8000 h of exposure the corrosion rates obtained under pressure are significantly lower. Here the pressure influences the corrosion rate. Possibly due to closing of capillary systems after long exposure fast diffusion processes and degradation processes are prevented. Hardening and tempering 2 at 670 °C give best results with lowest corrosion rates in water saturated supercritical CO_2. As a function of time there is no significant influence of exposure time after the initial corrosion reactions have taken place after ca. 1000 h. The slightly lower rates after 4000 h of exposure are most likely due to an increasing corrosion layer and therefore reduced diffusion rates of ionic species out of the steel to condense at the surface. This increase/decrease is much more present when steels are exposed to the CO_2-saturated brine with corrosion rates around 0.004 to 0.014 mm/year (figure 2, right). Additionally the corrosion rate increases further with exposure time, indicating a break-down of the passivating nature of the corrosion layer, possibly due to local lateral detachment of large areas of the corrosion layer. Normalized samples perform best under water. Still hardening+tempering X5CrNiCuNb16-4 would provide suitable corrosion resistance of a CCS-site borehole in saline aquifer environment.

pitting

In general the smallest number of pits is found on steels with martensitic microstructure regarding exposure time and pressure. X46Cr13 reveals the best overall pitting resistance. At ambient pressure as well as 100 bar the heat treatment has little influence on the number of pits per unit area. Still, hardening and tempering (1 and 2) give slightly better results at high pressure (figure 3). Because the number of pots shows no significant difference the number of pits are combined in figure 3 [26], where steels with the same chromium content of 13% : X20Cr13 and X46Cr13 are compared. For both steels hardening and tempering between 600°C and 670°C (1 and 2) show the lowest number of pits per m^2 after 6000 h. The low number of pits for hardened 13%Cr-steels after 8000 h of exposure may be related to the carbide distribution within the distinct microstructure but needs to be discussed further.

At 100 bar 1.4542 developed pitting in CO_2-saturated brine as well as water-saturated supercritical CO_2 with no regard to heat treatment. The number of pits after 8000 h is 50000 - 260000 per m^2 when exposed to the liquid phase (figure 3, lower right) and from 250000 to more than 1000000 per m^2 when exposed to the supercritical phase (not shown in figure 3). The lower amount of pits after 6000 h of exposure is related to the consolidation of pits –then analyzed as surface corrosion phenomena. It is noteworthy to point out that the number of pits counted after exposure at 100 bar exceeds that counted on samples kept at ambient pressure by a factor of 10! This may be due to two reasons: first: the corrosion scale thickness at ambient pressure is much higher after the same exposure time. Therefore pits are a lot harder to detect and process during microscopic study because they lie underneath a continuous scale. Second: Kinetics at 100 bar are faster, pressing CO_2 and water onto the metal`s surface resulting in a lower pH and faster degradation of the steel. This may still offer a suitable explanation although surface corrosion rates at 100 bar are generally lower than at ambient pressure. It shows definitely that local corrosion is not predictable under CCS conditions [27].

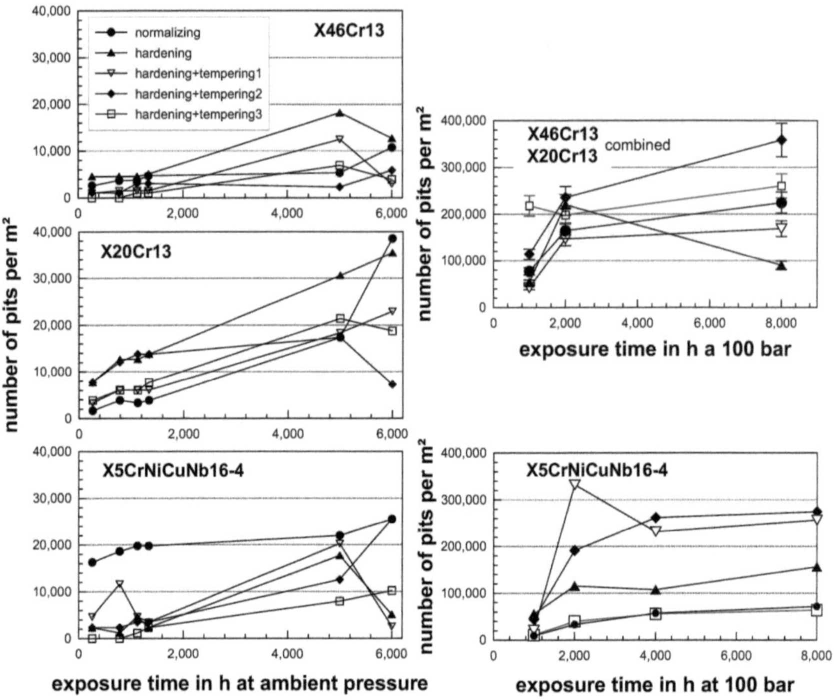

Fig. 3. Number of pits after 6000 hours and 8000 hours of exposure to CO_2-saturated aquifer water at 60 °C and ambient pressure, left and 100 bar, right of X20Cr13, X46Cr13 (combined at 100 bar) and X5CrNiCuNb16-4 heat treated prior to exposure [27].

Conclusions

Differently heat treated steels used as injection pipe with 13% Chromium and 0.46% Carbon (X46Cr13, 1.4034) or 0.2% Carbon (X20Cr13, 1.4021) and 0.05% Carbon (X5CrNiCuNb16-4, 1.4542) were exposed up to 8000 h (approximately 1 year) to supercritical CO_2 and saline aquifer water at ambient pressure as well as 100 bar at 60 °C in laboratory experiments. At ambient pressure the least corrosive attack is achieved by a continuous martensitic microstructure. Regarding steels with similar Cr-content the higher C-content in 1.4034 results in fewer pits and lower maximum intrusion depth compared to 1.4021 at ambient pressure. At 100 bar there is no significant difference. At 100 bar, 60 °C hardening and tempering at low temperatures (600 to 670 °C) for X20Cr13 and X46Cr13 result in lowest corrosion rates. For X5CrNiCuNb16-4 hardening and tempering at 670 °C seems to give best corrosion resistance. Surprisingly, despite its lower surface corrosion rates the local corrosion resistance is not significantly better in CCS environment compared to the much less costly steels X20Cr13 and X46Cr13.

Acknowledgement

This work was supported by the FNK (Fachkonferenz für wissenschaftlichliche Nachwuchskräfte) of the Applied University of Berlin, HTW and by IMPACT (EU-Project EFRE 20072013 2/21).

References

(1) Thomas, D.C., Carbon Dioxide Capture for Storage in Deep Geologic Formations – Results from CO_2 Capture Project, Volume 1: Capture and Separation of Carbon Dioxide form Combustion Sources, CO_2 Capture Project, Elsevier Ltd UK 2005, ISBN 0080445748.

(2) Nešić, S., Key issues related to modelling of internal corrosion of oil and gas pipelines – A review, Corrosion Science 49 (2007) 4308–4338.

(3) Carvalho, D.S., Joia, C.J.B., Mattos, O.R., Corrosion rate of iron and iron-chromium alloys in CO_2-medium, Corrosion Science 47 (2005) 2974-2986.

(4) Cui, Z.D., Wu, S.L., Zhu, S.L., Yang, X.J., Study on corrosion properties of pipelines in simulated produced water saturated with supercritical CO_2, Applied Surface Science 252 (2006) 2368-2374.

(5) Pfennig, A., Kranzmann, A., Reliability of pipe steels with different amounts of C and Cr during onshore carbon dioxid injection, International Journal of Greenhouse Gas Control 5 (2011) 757–769.

(6) Förster, A., Norden, B., Zinck-Jørgensen, K., Frykman, P., Kulenkampff, J., Spangenberg, E., Erzinger, J., Zimmer, M., Kopp, J., Borm, G., Juhlin, C., Cosma, C., Hurter, S., 2006, Baseline characterization of the CO2SINK geological storage site at Ketzin, Germany: Environmental Geosciences, V. 13, No. 3 (September 2006), pp. 145-161.

(7) Wu, S.L., Cui, Z.D., Zhao, G.X., Yan, M.L., Zhu, S.L., Yang, X.J., EIS study of the surface film on the surface of carbon steel form supercritical carbon dioxide corrosion", Applied Surface Science 228 (2004) 17-25.

(8) Bilmes, P.D., Llorente, C.L., Méndez, C.M., Gervasi, C.A., Microstructure, heat treatment and pitting corrosion of 13CrNiMo plate and weld metals, Corrosion Science 51 (2009) 876-882.

(9) Bülbül, Ş., Sun, Y., Corrosion behaviours of high Cr-Ni cast steels in the HCl solution, Journal of Alloys and Compounds 598 (2010) 143-147.

(10) Cvijović Z., and Radenković, G., Microstructure and pitting corrosion resistance of annealed duplex stainless steel, Corrosion Science 48 (2006) 3887-3906.

(11) Hou, B. , Li, Y., Li, Y., Zhang, J., Effect of alloy elements on the anti-corrosion properties of low alloy steel, Bull. Mater. Sci. 23 (2000) 189-192.

(12) Zhang, L., Zhang, W., Jiang, Y., Deng, B., Sun, D., Li, J., Influence of annealing treatment on the corrosion resistance of lean duplex stainless steel 2101, Electrochimica Acta 54 (2009) 5387–5392.

(13) Choi, Y.-S., Kim, J.-G., Park, Y.-S., Park, J.-Y., Austenitzing treatment influence on the electrochemical corrosion behaviour of 0.3C-14Cr-3Mo martnesitic stainless steel, Materials Letters 61 (2007) 244-247

(14) Isfahany, A. N., Saghafian, H., Borhani, G., The effect of heat treatment on mechanical properties and corrosion behaviour of AISI420 martensitic stainless steel, Journal of Alloys and Compounds 509 (2011) 3931-3936

(15) Park, J.-Y., Park, Y.-S., The effects of heat-treatment parameters on corrosion resistance and phase transformation of 14Cr-3Mo martensitic stainless steel, Materials Science and Engineering A 449-451 (2007) 1131-1134

(16) Dyja, D., Stradomski, Z., Pirek, A., Microstructural and fracture analysis of aged cast duplex steel, Strength of Materials, Vol. 40, No. 1 (2008) 122-125

(17) Lucio-Garcia, M.A., Gonzalez-Rodrigueza, J.G., Casalesc, M., Martinezc, L., Chacon-Navaa, J.G., Neri-Floresa, M.A. and Martinez-Villafañea A., Effect of heat treatment on H_2S corrosion of a micro-alloyed C–Mn steel, Corrosion Science 51 (2009) 2380-2386

(18) Banaś, J., Lelek-Borkowska, U., Mazurkiewicz, B., Solarski, W., Effect of CO_2 and H_2S on the composition and stability of passive film on iron alloy in geothermal water, Electrochimica Acta 52 (2007) 5704-5714.

(19) Moreira, R.M., Franco, C.V., Joia, C.J.B.M., Giordana, S., Mattos, O.R., The effects of temperature and hydrodynamics on the CO_2 corrosion of 13Cr and 13Cr5Ni2Mo stainless steels in the presence of free acetic acid, Corrosion Science 46 (2004) 2987-3003.

(20) Seiersten, M., Material selection for separation, transportation and disposal of CO_2, Corrosion paper no. 01042 (2001).

(21) Choi, Y.-Y. and Nešić, S. Determining the corrosive potential of CO_2 transport pipline in high pCO2-water environments, Journal of Green House Gas Control 5 (2011) 788-797.

(22) Han, J., Zhang, J., Carey, J.W., Effect of bicarbonate on corrosion of carbon steel in CO_2-saturated brines, Journal of Green House Gas Control 5 (2011) 1680-1683.

(23) Pfennig, A., Kranzmann, A., 2012, Effect of CO_2 and pressure on the stability of steels with different amounts of Chromium in saline water, Corrosion Science 65 (2012) 441–452

(24) Pfennig, A., Linke, B., Kranzmann, A. Corrosion behavior of pipe steels ex-posed for 2 years to CO_2-saturated saline aquifer environment similar to the CCS-site Ketzin, Germany, Energy Procedia, Vol. 4 (2011) 5122-5129.

(25) Mu, L.J. Zhao, W.Z., Investigation on carbon dioxide corrosion behavior of HP13Cr110 stainless steel in simulated stratum water, Corrosion Science 52 (2010) 82-89.

(26) A. Pfennig, P. Zastrow, A. Kranzmann, Influence of heat treatment on the corrosion behaviour of stainless steels during CO_2-sequestration into saline aquifer, International Journal of Green House Gas Control 15 (2013) 213–224

(27) Anja Pfennig, Helmut Wolthusen, Marcus Wolf, Axel Kranzmann, Effect of heat treatment of injection pipe steels on the reliability of a saline aquifer water CCS-site in the Northern German Basin, Green House Gas Emission Reduction Technologies GHGT11 Conference, Austin, Texas, USA, 5th-9th October 2014

Energy Technology 2015: Carbon Dioxide Management and Other Technologies
Edited by: Animesh Jha, Cong Wang, Neale R. Neelameggham, Donna P. Guillen, Li Li, Cynthia K. Belt, Randolph Kirchain, Jeffrey S. Spangenberger, Frank Johnson, Andrew J. Gomes, Amit Pandey, and Peter Hosemann
TMS (The Minerals, Metals & Materials Society), 2015

ECONOMIC ASSESSMENT OF METHANOL SYNTHESIS BY CO_2 FROM COAL-FIRED POWER PLANTS

Yudong Wang [1*], Yongnian Dai [1]

[1] National Engineering Laboratory of Vacuum Metallurgy,
Kunming University of Science and Technology, Kunming, China

Keywords: Reasonable cost for methanol synthesis, CO_2 from coal-fired power plants, Hydrogen electrolysis

Abstract

The coal-fired power plants are major consumers of fossil fuels and responsible for above 10% of the total Greenhouse gas (GHG) emission to the atmosphere. These GHG can be substantially mitigated by combing multi-step shift with methanol synthesis ($CO_2 \rightarrow CO \rightarrow CH_3OH$). This paper conducted life cycle assessment on this noble methanol synthesis, predicted that process has more sensitivity to cost of electric energy than capital and feedstock costs, electric cost below 0.03 US$ per kWh is acceptable for synthesis methanol, and the spare electric energy from hydraulic and nuclear power appears to be a more attractive option. The O_2/CO_2 recycle combustion of coal-fired power plants would provide sufficient CO_2 resource for producing methanol, and the GHG emission from coal-fired power plants avoidance of about 95%.

1. Introduction

Carbon dioxide from current coal-fired power plants was considered to be a waste with a cost of disposal[1,2,3]. Now, this large amounts of low-cost and relatively pure CO_2 is available to produce methanol[4].

Table 1. Comparison of CO_2 emission from different countries in 2009 [5]

Country	Emission /GT	% Ratio of world
China	7.518	24.2
USA	2.941	19.1
India	1.539	4.9
Russia	1.535	4.5
Japan	1.222	3.9
Germany	0.795	2.6

The consumption of coal was 3.61 billion ton (GT) in China in 2013, released 7.518 GT CO_2, about one third of the world's emission (see Tab.1). Most electric energy is currently produced industrially via coal-fired power plants (see Tab.2) by firing fossil fuel followed by release of huge amount CO_2. The production each kWh electricity requires the consumption of 0.39-0.34 kg of coal. The main drawback of the burning of fossil fuel is the high specific of $CO_{2\text{-equiv}}$

* Corresponding author. Tel: +8613987196558
E-mail: 584772219@qq.com

emissions (3.67kg $CO_{2\text{-equiv}}$ /kg C). This greenhouse gas emission by coal-fired power plants contributed 49.3% to the anthropogenic $CO_{2\text{-equiv}}$ emissions in China. Much effort has been spent to recycle carbon dioxide to fuels, as a complementary technology to carbon sequestration and storage (CSS).

Table 2. Comparison of CO_2 emission in China from different industry in 2010 [2]

Industry	Emission /GT	% Ratio of Chinese CO_2 emission
Power plants	3.559	49.3
Manufacturing	2.014	27.9
Transport	0.486	6.73
Civil service	0.550	7.61
Others	0.585	8.09
Total	7.224	100

The emission of CO_2 by coal-fired power plants is the highest in China. These large amounts of relatively pure CO_2 and nearly zero (or even negative) cost are available for conversion to liquid fuels and chemicals. Using CO_2 and carbon as feedstock to manufacture methanol or dimethyl, the overall reactions can be represented by

$$CO_2+C=2CO \qquad \triangle H^o_{298k}=41.66\text{kcal/mol} \qquad (1)$$

$$CO+2H_2=CH_3OH \qquad \triangle H^o_{298k}=-21.7\text{kcal/mol} \qquad (2)$$

Reaction (1) is thermodynamically favorable above 1280 K, and which is a volume increasing and endothermic reaction, lower pressure and higher temperature is favored. At Kunming University of Science and Technology (KUST), a stack-type vacuum reactor was developed in which a charge of carbon stack was inserted in a high-temperature zone, and CO_2 was send through carbon stack for extraction of relatively pure CO.

Table_3. Energy and mass ratio of reactants and product in reaction (1).

	Molar ratio	Mass Ratio
CO_2	44	1
C	12	0.273
2CO	56	1.273
Require energy 41.66 kcal/mol	Require energy 0.947 kcal/gCO_2	Require energy 0.0011 kWh/gCO_2

The total demand to dispose each gram CO_2 is 0.0011kWh electric energy and 0.273g carbon consumption. If not considering byproducts such as SO_2, NO and H_2, the reaction (2) volume is decreasing and exothermic, which is favor at higher pressure and lower temperature. Normally, the reaction cannot occur without catalysts. The present work at KUST laboratory focuses on examining the thermodynamic constraints for reduction of CO_2 to CO in order to find thermochemical conditions suitable for the application of concentrated solar energy to the production of CO.

2. Producing hydrogen

Many products, from C1-type molecules (CH_3OH ,CH_4) to alkenes, higher molecular weight alkanes and alcohols through C-C bond formation, could be produced by hydrogenation of CO_2 [6]. This class of reaction comprises large-volume chemicals such as methanol and dimethyl ether (DME), and other products such as carboxylic acids (formic and acetic acids). It may be distinguished among processes, where the first step is reduction to monoxide (reaction1) followed by consecutive steps in which CO/H_2 mixtures are converted to methanol or hydrocarbons (reaction2). This class of reaction involves the largest use of hydrogen and maybe limits interest. The hydrogen (or hydrogen sources) is the most important feedstock for producing methanol in reaction (2), the CO_2 emissions from coal fire power plant need huge mounts hydrogen to synthesis methanol. We would chose production of hydrogen by electrolysis of water, which is renewable resource.

Table 4. The costs of electric energy via different resource[4、5]:

Resource	US $/kWh	US $/1kg H_2 (50 kWh)
Hydraulic power	0.013	0.65
Nuclear (USA in 2004)	0.020	0.98
Power plants (USA in 2004)	0.020	0.98
Power plants (China in 2014)	0.036	1.79
Civil price	0.062	3.10
Petroleum, syngas（USA）	0.068-0.117	3.42-5.87
Price of 1kg H_2（2014）	-	3.26

The 718th Research Institute of China Shipbuilding Industry Corporation (718 Institute) is mainly engaged in developing produce line of hydrogen generation [7], which is quite mature technologically and relatively economical.

The price of electricity is the most preliminary factor, because 50MWh electricity is needed to produce per ton hydrogen. In the table 4, the highest cost of the hydrogen is nine times of the lowest one, as well as the different sources of electric energy. The requirement of electric energy is 11 MWh by reduction of 1 ton CO_2 to CO, which equals to the standard-state enthalpy change of reaction (1).

Table 5. Energy and mass ratio of reaction (2)

	Molar mass	Mass ratio	Mass Ratio for 1 g CO_2
CO	28	1	1.273
$2H_2$	4	0.143	0.182
CH_3OH	32	1.143	1.455

It produces 1.273g CO and needs 0.182g H_2 to synthesize methanol by 1g CO_2（see Tab.5）, these H_2 consume 9.1 kWh electric energy. 91 MWh electric energy is needed to produce H_2 for utilization of each ton CO_2. Therefore, The total demand of electric energy, for both standard-state enthalpy change of reaction (1) and electrolysis of water for producing H_2, to synthesize methanol by disposing each ton CO_2 was 102 MWh(11+91=102 MWh), which is significantly huge and expensive.

3. The economic sensitivity analysis

The annual emission of CO_2 from coal-fired power plants was 3.559 giga ton in China, by this huge amount, 5.171 giga ton methanol would be produced and be worth 2249 giga US$ per year, which proved the worth of CO_2 emission from power plants. Assumed annual conversion of 1 million ton CO_2 to methanol, economic evaluation for methanol production from an initial mix of coke (represented by carbon) and CO_2 from coal-fired power plants, initially at 300K and 1 bar, calculated to be transform at 1300 K and 1 bar according to the reaction (1), the resulting CO was mixed with H_2 to be converted to methanol.

Table 6. The economic evaluation for 1 million ton CO_2

Design parameters	
Design CO_2 feed (ton/yr)	$1x10^6$
Annual coke feed (ton/yr)	$0.273x10^6$
Annual H_2 feed (ton/yr)	$0.182\ x10^6$
Annual methanol production (ton/yr)	$1.45x10^6$
Annual costs (million US$)	
Coke cost (US$110/ton [a])	31.40
Electric energy for H_2(million US$)	102-612
Annual sales (million US$)	
Methanol (US$435/ton[b])	630.75

a. August 2014; see www.eia.doe.gov/cneaf

b. September 2014; see www.methanex.com

This process has more sensitivity to cost of electric energy than capital and feedstock costs. Electricity cost below 0.05 US$ per kWh would provide profits for this scheme if only considering the influence of electric cost.

Table 7. Profit analysis for variation of per kWh electric cost

Each kWh electric cost (US$/kWh)	Electric cost (million US$)	Profit of producing Methanol (million US$)[a]
0.01	102	497.36
0.02	204	395.36
0.03	306	293.36
0.04	408	191.36
0.05	510	89.36
0.06	612	-12.65

a. The profit of producing Methanol is that methanol sales subtract coke cost and electric cost.

Assumed the 40% annual sales were for capital, operation and maintenance cost, electric cost below 0.03 US$ per kWh is acceptable for synthesis methanol process. Fortunately, breakthroughs in utilization of superabundant hydraulic power will have the greatest impact on utilization of capture CO_2 to producing methanol.

4. Retrofit of power plants

The CO_2 concentration of existing emission from modern pulverized coal-fired power plants is too low for reduction to CO, the O_2/CO_2 recycle combustion, involves burning the coal with pure oxygen and recycled gas, can be retrofitted to existing units, however it increases operation and maintenance cost for subsequent production of methanol by about 20%.

5. Conclusions

From the works completed in this paper, a number of important conclusions can be drawn with utilization of CO_2 emission from coal-fired power plants to produce methanol.

1. This process have more sensitivity to cost of electric energy than capital and feedstock costs, electric cost below 0.03 US$ per kWh is acceptable for synthesis methanol by CO_2. Breakthroughs in utilization of superabundance hydraulic power will have the greatest impact on reducing costs of CO_2 utilization.
2. This process is significantly expensive; however, the electric energy from hydraulic and nuclear power appears to be a more attractive option.
3. The O_2/CO_2 recycle combustion provided sufficient CO_2 resource for producing methanol, replacing the conventional process with reasonable costs.

Acknowledgements

The authors are grateful to the financial support from the Kunming University of Science and Technology and Yunnan Provincial Science and Technology Department, without which, this work could not have been undertaken.

---

[1] G. Centi, G. Cum, J.L.G. Fierro, and J.M. Lopez Nieto, "in Direct Conversion of Methane, Ethane and Carbon Dioxide to Fuels and Chemicals", *CAP Report, The Catalyst Group Resources,* 2008.

[2] S.K. Ritter, "What can we do with carbon dioxide", *Chem. Eng. News*, 85 (18) (2007) 11.

[3] D. Graham-Rowe, "Turning CO_2 back into hydrocarbons", *New Scientist 2645* (March) (2008) 32.

[4] George A. Olah, A.Goeppert, and G.K.S.Prakash, *Beyond Oil And Gas: The Methanol Economy*, Wiley,2009-09-23.

[5] Li Zhong, ed. Xie Kechang. , "Coal-based alcohol fuel", Beijing, *Chemical Industry Press*, 2011

[6] J.L.G. Fierro, in: J.L.G. Fierro (Ed.), "Metal Oxides Chemistry and Applications", *CRC Taylor & Francis Pub.*, Boca Raton, FL, USA, 2006 , p. 569 (Chapter 18).

[7] Wang Qingbin, Xue Kaku, Zhao Xue, Yang Qi, "The 718th Research Institute of China Shipbuilding Industry Corporation (718 Institute)", *Hebei Industrial of hydrogen production by water electrolysis technology*, Handan 056027, Gas Separation 2011.02 / 47-50.

Energy Technology 2015: Carbon Dioxide Management and Other Technologies
Edited by: Animesh Jha, Cong Wang, Neale R. Neelameggham, Donna P. Guillen, Li Li, Cynthia K. Belt, Randolph Kirchain, Jeffrey S. Spangenberger, Frank Johnson, Andrew J. Gomes, Amit Pandey, and Peter Hosemann
TMS (The Minerals, Metals & Materials Society), 2015

THERMAL WASTES FROM ENERGY CONVERSIONS AND GLOBAL ANTHROPOGENIC WARMING

Neale R Neelameggham [1 a*] Brian R Davis [2 b]
Author Affiliation: [1] IND LLC [2] Brian Davis Associates Consulting
Corresponding Author email: Neelameggham@outlook.com [a*]
Co-author email: bb_dd@sbcglobal.com

Abstract

In today's world of rapidly increasing alternative energy, one should realize what happens to that energy generated and dissipated. These are manmade or anthropogenic. This energy conversion is still mainly for electricity – for chemical and metallurgical processes, for heating and cooling or lighting which also creates thermal wastes or for work involving movement – thus friction and heat. This paper discusses the effect of these thermal wastes into atmosphere – with examples from unavoidable energy inefficiencies and what it can lead to in the meteorological sense of climate change. Results of heat transfer modeling are compared with present day combustion sourced thermal wastes documented by British Petroleum. Evapo-transpiration moderates the atmospheric air temperature increase. Anthropogenic warming with the population increase will continue and is a function of total energy conversions only. This warming can be distributed into ocean and earth by proper engineering. The role of energy efficiency in conversion and usage in moderating the warming re-emphasized.

Introduction

People created activities are called 'Anthropogenic'. It should not come as a surprise that statistically significant correlations can be made of human population growth with time, and consequent energy consumptions. These energy consumptions require energy conversions of stored and un-stored energy from natural power sources. Most of the energy conversions from stored energies, for centuries have been from hydrocarbons, wood and other fossil fuels, in addition to potential energy of elevated water bodies. Combustion based conversion translates to good correlations with the simultaneous effects of water vapor and carbon dioxide. Most of the energy conversions have evolved into secondary conversions into electrical energy and other motive energy. These 'secondary energy sources' in their applications, which improve the comfort of humanity, are converted further to 'tertiary energy form' which is 'heat'. The law of conservation of energy requires that this heat is stored unless dissipated into other forms of storage.

As this heat follows the path of least resistance and warms the atmospheric air mass, it becomes the global anthropogenic air warming or GAW. It is possible to modify the GAW from the air mass into water bodies or other solids by suitable engineering – this modification has been minimal thus far – as can be seen from air conditioning which usually emits thermal emissions into air. GAW is a form of energy storage which is proportional to the total energy conversions made irrespective of the source of that conversion, fuel or a renewable source. Water vapor plays an important role in moderating the atmospheric air temperature in the form increasing specific humidity (Ws) and excess absorption [or precipitation] (Wa) –sum of which is noted as evapotranspiration and represented as Ev . Part of what is called excess precipitation includes the transpiration into plants and increased agricultural biomass –which help sustain the growth. This excess precipitation is over and above what is due to solar insolence radiation on the ocean and other water bodies in the form of latent and specific heat content changes.

As more alternate energy conversion electricity displaces fossil fuel electricity, the atmosphere will still heat up to the extent of total thermal emissions, irrespective of the total carbon dioxide seen in air. These are derived from fundamental thermodynamic heat balance of total energy conversions with atmospheric air mass heating by convection; with the air temperature changes observed with time.

Discussion

Preliminary work confirmed that it is the water evaporation which moderated the atmospheric temperature rise during the past 48 years to an increase of only about 1.2°F instead of the 3.4°F [units used by American Industries] of heating of the atmospheric air by the 'swamp or evaporative cooling' effect from an added evaporation of surface waters [1, 2]. 'Heat Transfer Modeling of Global Anthropogenic Warming' is based on data of worldwide energy conversion data compiled by British Petroleum [3]. While the carbon dioxide content increased by about 75 ppm into the atmospheric mass, the total evapotranspiration equivalent of water vapor increase was a minimum of 740 ppm. and could have been as high as 1200 ppm depending on what proportion of total anthropogenic energy conversion was the total thermal emissions. Our evaluation is a calculating method for the observed specific humidity increases due to anthropogenic events such as noted by Willett, et al [4, 5].

Our preliminary heat transfer modeling of GAW [1,2], was carried out using the cumulative energy consumptions of 48 years of data – applying stoichiometric balances on combustion of oil, natural gas and coal. This provided that the atmospheric increase of carbon dioxide should have been about 156 ppm, while water vapor should have increased by 152 ppm. The fact that only 75 ppm increase of CO_2 was seen in the atmosphere [based on data collected at the Mauna Loa observatory – from about 320 to about 395 ppm, shows the role played by ocean absorption. The absorption of parts of the carbon dioxide by plant matter is discussed in some recent studies on new vegetation in deserts [6]. This has the corresponding associated absorption of water vapor which has to be accounted.

Our heat transfer model [2] emphasized the need for a mathematical expression for the *global mean* temperature changes with the simultaneous effects from both water vapor and carbon dioxide - similar to expressions shown by Hottel [7] starting from his initial work in 1927 and later by Farag [8]. These flue gas radiation emissivity studies, carried out during 1927 to 1977, had ratios of water vapor to carbon dioxide between 2:1 and 1:2, unlike the higher ratios of 50:1 for the GAW heat transfers. F, .A. Brooks measured atmospheric air emissivities using Hottel's experimental approach – indicating difficulties [9]. Recently, Nasif Nahle expanded Hottel's equation to atmospheric air water vapor and CO_2 proportions recently [10].

The radiative – convective modeling of climate correlations started evolving in the mid-1960's such as by Manabe and Wetherald [11]. They used assumptions such as 'heat capacity of earth is zero', in the interest of simplifications and were developed to study the overall solar insolence heating and radiation [cooling] from earth to space without having to worry about mass transfer or stoichiometric balances of how the carbon dioxide comes into the atmosphere. In these studies, the effects of water vapor the major greenhouse gas [80 to 95% of the GHG effect] were difficult to make from the elusive in spite of the observed phenomena on specific humidity and cloud effects.

Easier correlations are made with carbon dioxide which is less elusive to measure as it is a non-condensable gas at atmospheric temperatures. These correlations based on 25 to 30 years time series started to hit a roadblock when the last decade started to show a plateau of global mean temperatures – which is called a 'hiatus' in climate literature during the past 8 years. Figure 1, provided by courtesy of INDLLC ©, shows the global mean temperature plateau while the carbon dioxide in the atmosphere kept increasing since 1997.

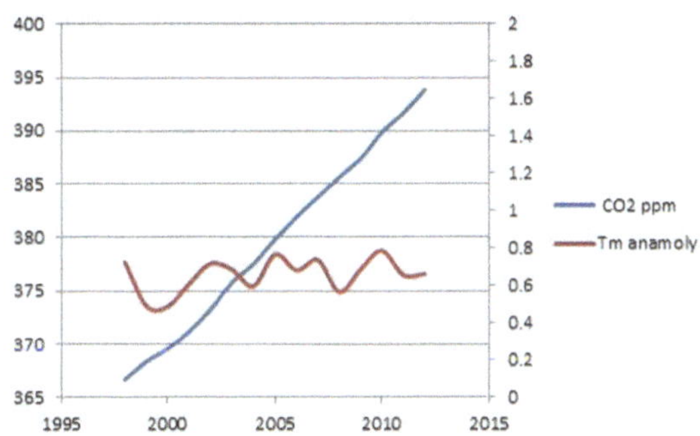

Figure 1. Global Mean Temperature Hiatus Curve Courtesy IND LLC ©

The radiative-convective model continued its evolution by modified assumptions to include measurable cumulated temperature increase in the oceans – this was initiated in the mid-1990s and called the 'OHC [ocean heat content] coupled models' [12]. It should be noted that GAW can be reduced by modifying thermal emissions into air by suitable engineering as thermal emissions into slab ocean or earth with non-zero heat capacity.

Compared to the daily solar insolence onto land and oceans, the power equivalent of energy conversion by humans is a small fraction – similar to the conducted heat from earth's core through the thick insulating mantle. But so is the change in the global mean temperature from one day to the next. It is the cumulative effect which is called the anthropogenic effect – such as global mean temperature increasing by 1.26°F in 48 years - which corresponds to only 75 micro degrees Fahrenheit per day. This should be dubbed an insignificant change in the terminology of climate models which omit human energy conversions as insignificant for convenience.

GAW Model Development by IND LLC

Two quantities reported in the Global Warming representation [GW_R] reports can be used to compare with the GAW energy consumption on an annual basis in BTU's *[units used by American Industries]* or Joules – energy quantity instead of the Watts/m^2 – power units, which is Joules/sec/m^2. The two quantities summed to represent GW are [a] heat required to melting glaciers on land causing ocean level increase averaging 0.071 inches/year since 1965 [b] the heat required for the top 700 m of ocean heating up by 0.2°F in 48 years [11.2 micro degrees Fahrenheit increase per day]. This cumulative total GW_R from 1965 through 2012 is 1.22 x 10^{20} BTU.

The global population statistics are available from United Nations demographic reports on a yearly basis [13] with projections to year 2100. The heat transfer model based on Neale's GAW theory was applied to each of the annual data since 1965 to arrive at additional correlations to arrive at the GAW convective model explaining anthropogenic warming with observed atmospheric temperatures. This is called the Convective model based on mass based heat balances on the atmospheric air mass. This model still includes the minor effects of spectral emissivity of CO_2 which is minor compared to the major water vapor effect.

The heat balances carried out each year by convective heat increase in atmospheric air mass from these thermal emissions provides an initial value, Te for the atmospheric mean temperature estimate. The difference between Te and a constant atmospheric temperature base is noted as DT1 . Cumulatively this showed an anticipated 5°F [maximum] increase since 1965 compared to the observed increase of 1.2°F. The value of observed global mean was computed using published 'anomaly' data from a base [here the base temperature used is 15°C or 59°F]. Observed value anomalies include associated variations from nature, such as sunspots, volcano and other hiatus effects.

This depicted the anthropogenic effect of evapotranspiration – which comes from swamp cooling shown by initial humidity increase which in turn is absorbed without heating the air while maintaining saturation humidity criterion. The data from the 48 years is then used to develop the correlation of thermal emissions, expressed as Te with logarithmic changes in water vapor and carbon dioxide concentrations. This is called the GAW convective equation. It is necessary to use a global mean surface temperature, Ta', projected by a logical or a trend curve to calculate saturation humidity constraints and then the anticipated Px.

This Px is the excess precipitation over and above the average 0.106 inch/day/yr, due to solar insolence on terrestrial bodies of water causing the latent heat evapo-transpiration. Published global energy budgets vary widely on the latent heat conversion of solar energy from a value of 80 W/m^2 per Trenberth, et al [14], to 88 +/- 10 W/m^2 by G.L.Stephens, et al. [15]. Another climate study used an average between these two values. While these quantities have much larger variations there is an insistence of the warming effects from parts per billion changes of manmade events. The elusiveness of measuring water vapor and cloud effects are also adjusted in these models by changing the residence time before precipitation in the radiation-convection models [16].

The calculated Ev starts to become higher than specific humidity average, Ws, shown in the HadCRUT humidity curves. [HadCRUT data base from the United Kingdom is one of the longer term temperature data collections from the mid nineteenth century]. This difference is termed Wa – absorbed water vapor with a concurrent cooling effect- whether from transpiration into plant growth or into the ocean similar to carbon dioxide changes. Our model thus shows a correlation for simultaneous evapotranspiration moisture and carbon dioxide. This can then be used to project future changes in atmospheric temperatures using total energy conversions. Thayer Watkins noted that global mean temperatures showed a cyclic curve of upward increases followed by a period of decreases [similar to recent 'hiatus' data] [17] when analyzing data over a century.

The energy conversions using varying proportions of the six different types – oil, gas, coal, nuclear, hydro and renewables can be projected in proportion to population changes and its usage rates. These proportions are used to compute the varying CO2 emissions net amounts at the same Et –total energy by varying renewable energy. IND LLC's study analyzed the atmospheric temperature anomalies over the period from 1850 to arrive at a statistically significant correlation.

Table1, provided by courtesy of INDLLC ©, summarizes the GAW model projections to year 2100, showing the effects of thermal emissions from renewable energy use along with their conclusions. *These projections are made when none of the thermal air emissions are converted to warming ocean or earth with non-zero heat capacity*

Table 1 GAW -Model Projections to Year 2100 [Courtesy of IND LLC ©]

Energy Conv. source\Year	Base 2013	set1 2100	Base2 2013	set2 2100	set 3 2100	set 4 2100	set 5 2100	set 6 2100	set 7 2100
Fractions									
Oil	0.331	0.331	0.330	0.240	0.189	0.240	0.170	0.100	0.100
natural gas	0.239	0.239	0.240	0.210	0.139	0.210	0.210	0.200	0.200
Coal	0.299	0.299	0.300	0.210	0.171	0.210	0.170	0.100	0.100
Nuclear	0.045	0.045	0.040	0.040	0.023	0.040	0.040	0.040	0.040
Hydroelectric	0.067	0.067	0.070	0.150	0.045	0.150	0.110	0.110	0.110
Renewable	0.019	0.019	0.020	0.150	0.435	0.150	0.300	0.450	0.450←
Total Energy, Et, BTU's[10^{17}]/yr	5.07	9.94	5.00	9.94	9.94	7.78	7.78	6.93	6.08←
CO_2 ppm	397	803	396	713	651	656	634	592	582 ←
H_2O [Ev ppm]*	17265	28048	17269	30774	31731	27523	27843	27480	26664
Te °C [for ref]	17.16	30.41	17.16	30.41	30.41	27.99	27.99	27.37	26.75
Sinusoidal temperature distribution effects.									
Ta °C	15.63	16.81	15.63	16.81	16.81	16.04	16.04	15.73	15.43
H_2O [Ws ppm]	17120	18569	17120	18569	18569	17623	17263	17239	16878
Px 10^{-3} inch/day/yr	0.10	6.50	0.100	7.37	8.02	6.79	7.00	7.01	6.69
[Xs Precip.incl.absorbed]									
Population like Temperature distribution effects.									
Ta °C	15.4	18.0	15.4	18.0	18.0	17.53	17.53	17.41	17.28
H_2O [Ws ppm]	16838	20030	16838	20030	20030	18747	19448	19287	19147
Px 10^{-3}inch /day/yr	0.29	5.50	0.30	7.37	8.02	5.55	5.75	5.59	5.16
[Xs Precip.incl.absorbed]									

*The GAW convective equation. log Ev = a + b Te – c *log C where Ev is ppm initial water vapor in atmosphere [calculated by adding the increase necessary to bring Te to Ta added to a base specific moisture content of 16466 ppm for the year 1964].

These projections show that the observable global mean temperature projections is a function of curve fitting assumptions –sinusoidal with periodic increases and dips, or a polynomial as population growth curves [18]. Still the fact remains the increase of atmospheric temperature is mainly proportional to the Total energy consumed or converted irrespective of the fuel usage. The same Te is possible at different values of CO_2 moderated by different water vapor effect – as can be seen in the table. It is shown that atmospheric mean temperature increase is a function of total anthropogenic energy conversions irrespective of the energy source including renewable alternative energy. While renewable energy may not increase the CO_2 levels, it can increase the evapotranspiration and the specific humidity – resulting in the same Ta'. Details of the study are expected in other publications.

Conclusion

Human population growth with simultaneous increased energy conversions worldwide, will cause more thermal emissions, which are inevitable. These thermal emissions will increase the evapotranspiration irrespective of the fuel source* - causing more precipitation in the future - possible melting of more ocean ice. [* Thermal emission also increases from renewable electricity generated [solar, wind or hydro] as well as fuel combustion based electricity].

Our efforts should be in improving the energy efficiencies which will play an important role in moderating the effects. The growth of population and corresponding food increase needs will require fuel based conversions to supply the botanical carbon cycle along with evapo-transpirations.

Using short term correlations to fear based accusation on fuels is shortsighted and unwarranted. The recent hiatus or the lack of increase in global mean temperatures in spite of increased fuel use worldwide is one such example.

We should also make strides in improving the building architecture to ar*k*itecture to facilitate living with rising water levels, which may happen. It is prudent to analyze 'energy issues' by laboratory based experiments like the studies by Hottel on flue gas radiation – which were done several decades ago.

It is interesting to note that, as the paper is being prepared to be submitted to the symposium, the author, stumbled upon – an analysis which claimed that the problem is not climate change but is the population growth [19] which is a more generalized statement than a detailed analysis done by IND LLC's modeling.

Acknowledgments

We thank the review efforts by Indra Neelameggham, Dilys Davis, Robert E Brown, and Ian Howard Smith. We appreciate insights provided by communications on the understanding climate data, modeling and analysis by Ole Humlum, Nicola Scafetta, and N. Schafer, Tapio Schneider and countless other climate scientists during the development of the independent model by the authors facilitated by Ind LLC.

References

1. Neale R Neelameggham, 'It is the Water Vapour' and Neale's GAW Model.©, Ind LLC, South Jordan, UT.

2. Neelameggham and Davis, 'Heat transfer modelling of global anthropogenic warming', J of Nanomaterials and Energy, September 2014, DOI: 10.1680/nme.14.00012.

3. http://www.bp.com/statistical_review, BP Statistical Review of World Energy June 2013
4. Katharine M. Willett, Nathan P. Gillett, Philip D. Jones & Peter W. Thorne, Attribution of observed surface humidity changes to human influence, Nature 449, 710-712 (11 October 2007).
5. Katharine M Willett, Philip D Jones, Peter W Thorne and Nathan P Gillett, 'A comparison of large scale changes in surface humidity over land in observations and CMIP3 general circulation models', 2010 Environ. Res. Lett. Volume 5 Number 2.
6. Poulter, B, D Frank, P Ciais, RB Myneni, N Andela, J Bi, G Broquet, JG Canadell, F Chevallier, YY Liu, SW Running, S Sitch and GR van der Werf. 2014. The contribution of semi-arid ecosystems to inter annual global carbon cycle variability. Nature, 509, pp.600–603, 29 May 2014.
7. H. C. Hottel, 'Heat Transmission by Radiation from Non–Luminous Gases', Ind. Eng. Chem., 1927, 19 (8), pp 888–894
8. Farag, Radiative Heat transfer from Non-luminous gases Computational study of the emissivities of water vapor and carbon dioxide Table 5.5-1, p. 202 , Doctor of Science Thesis MIT August1976.
9. F.A. Brooks, 'Observations on atmospheric radiations', Papers in Physical Oceanography and Meteorology, Published by Massachusetts Institute Of Technology And Woods Hole Oceanographic Institution, VOL. VIII, NO. 2, 1941
10. Nasif S. Nahle, 'Determination of the total emissivity of a mixture of gases containing 5% of water vapor and 0.039% of carbon dioxide at overlapping absorption bands', http://www.biocab.org/Overlapping_Absorption_Bands.pdf.
11. S. Manabe and R.T. Wetherald, 'Thermal equilibrium of the atmosphere with a given distribution of relative humidity', Vol 24, No. 3 J of atmospheric sciences, 1967, 241-259.
12. Curt Covey, Krishna M. AchutaRao, Steven J. Lambert, and Karl E. Taylor, Intercomparison of Present and Future Climates Simulated by Coupled Ocean-Atmosphere GCMs, PCMDI Report No. 66 UCRL-ID-140325.
13. http://en.wikipedia.org/wiki/World_population_estimates , and http://esa.un.org/unpd/wpp/index.htm
14. Kevin E. Trenberth, John T. Fasullo, and Jeffrey Kiehl., 'Earth's Global Energy Budget' Bull. Amer. Meteor. Soc., 90, 2009, 311–323.
15. Graeme L. Stephens, Juilin Li, Martin Wild, Carol Anne Clayson, Norman Loeb, Seiji Kato, Tristan L'Ecuyer, Paul W. Stackhouse Jr, Matthew Lebsock & Timothy Andrews: An update on Earth's energy balance in light of the latest global observations. Nature Geoscience 5, 2012. 691-696.
16. N. Schaller, J. Cermak, M. Wild, and R. Knutti., 'The sensitivity of the modeled energy budget and hydrological cycle to CO2 and solar forcing', Earth Syst. Dynam., 4, 253– 266, 2013
17. Thayer Watkins, An Estimation of the Period of the Cycle in Global Temperature and the Long Term Trend in Global Temperature, http://www.applet-magic.com/AGT_cycle1st.htm [accessed Sept 3, 2014].
18. Nicola Scaffetta, 'Empirical analysis of the solar contribution to global mean air surface temperature change', Journal of Atmospheric and Solar-Terrestrial Physics, July 2009
19. Paul B. Farrell , 'Opinion: Climate change isn't the problem. A population bomb is killing us', http://www.marketwatch.com/story/climate-change-isnt-the-problem-a-population-bomb-is-killing-us-2014-09-23

Energy Technology 2015
Carbon Dioxide Management and Other Technologies

ENERGY TECHNOLOGIES AND CARBON DIOXIDE MANAGEMENT SYMPOSIUM 2015

Iron & Steel

Session Chairs:

Cong Wang
Zuotai Zhang

Energy Technology 2015: Carbon Dioxide Management and Other Technologies
Edited by: Animesh Jha, Cong Wang, Neale R. Neelameggham, Donna P. Guillen, Li Li, Cynthia K. Belt, Randolph Kirchain, Jeffrey S. Spangenberger, Frank Johnson, Andrew J. Gomes, Amit Pandey, and Peter Hosemann
TMS (The Minerals, Metals & Materials Society), 2015

INFLUENCE OF THE INITIAL SOLIDIFICATION CONTROLLING ON THE ENERGY SAVING DURING CONTINUOUS CASTING

Lejun Zhou[1], Wanlin Wang[1*]

[1]School of Metallurgy and Environment, Central South University, No.932, South Lushan Road, Changsha, Hunan, 410084, China

Keywords: Energy Saving, Continuous Casting, Initial Solidification, Surface Quality

Abstract

Energy saving and greenhouse gas emission reduction are crucial for the future development of advanced continuous casting technology. Careful controlling the initial solidification of molten steel in the mold tends to reduce amount of energy consumption in continuous casting process, as it can improve the quality of casting products and minimize the extra post-cast treatment which will consume lots of extra labor and energy consumption. In this paper, factors such as mold oscillation, cooling potential conditions, mold flux, and fluid flow in the vicinity of meniscus that correlate with the molten steel initial solidification were discussed with the aim to provide strategy and guidelines for the optimization of molten steel solidification and energy saving in the continuous casting.

Introduction

The iron & steel industry is one of the biggest energy-consuming industries, as a large amount of energy is required for the operation of ironmaking and steelmaking. Also energy is dissipated during continuous casting process through mold water cooling, spray water cooling and air cooling.[1] Continuous casting is considered to be one of the most important innovations in steel industry in 20th century. It has many benefits including [2]: increased yield, improved product quality, energy savings, less pollution, reduced costs, and better working condition. The energy saving in continuous casting process is directly through the elimination of soaking pits, reheating furnace and primary rolling mills. Also it can save energy indirectly through substantial increase in the yield attained by the continuous casting process.

With the development of continuous casting technology, advanced innovation technologies, such as the thin slab casting, [3] strip casting, [4] direct rolling after continuous casting, [5] etc., have been developed. Those newly developed technologies are higher efficient and resulted in more energy saving. However, all of them are facing serious problems, which are eagerly needed to be solved, such as the quality of cast product, especially the surface quality of the shell. The bloom, slab or billet with surface defects, like severe oscillation marks;[6] cracks;[7] segregation; [8] entrapped bubbles, [9] inclusions, [10] slags; [11] improper solidification structure distribution, [12] etc., need to be tackled to obtain qualified products, to avoid the extra labor and energy consumption for the post-product treatment.

In fact, the surface quality of continuous casting products are greatly determined by initial solidification of the shell in continuous casting mold, since the first solidified shell in the vicinity of the meniscus forms the surface of the final casting products. While, the initial solidification of the shell is affected by many factors, such as mold oscillation, casting temperature, cooling conditions, and mold flux. Therefore, in this paper, some recent conducted researches regarding the initial solidification of molten metals during continuous casting were summarized with the aim to provide some strategy and guidelines for the optimization of molten steel solidification and energy saving in the continuous casting process.

Methods of controlling the initial solidification of steel

2.1 Mold Oscillation

In continuous casting, the mold is usually oscillated to prevent the newly formed shell from sticking to the copper mold. The periodical oscillation of the mold make the cooling potential in the vicinity of meniscus varied, which will result in the fluctuation of heat flux as shown in Fig. 1. [13] The fluctuation of heat flux will continuously cause the uneven initial solidification of shell and form the oscillation marks, which would tend to associate with other types of surface defects: transverse cracks, entrapped argon bubbles, inclusions, and elemental segregation to further deteriorate surface quality of casting products.

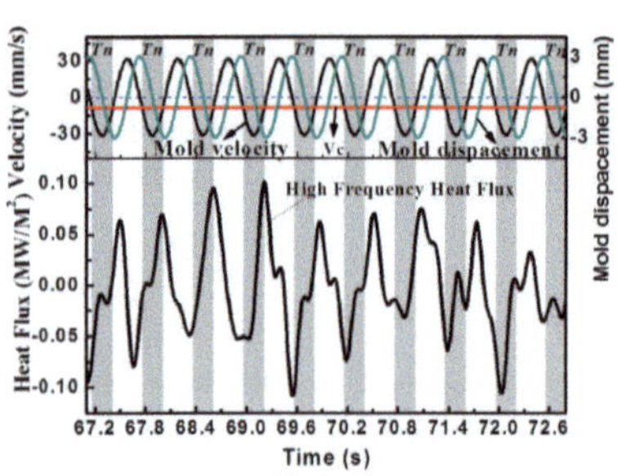

Figure 1

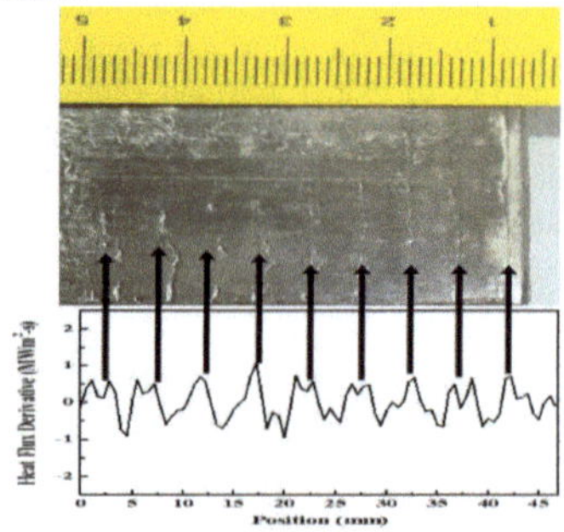

Figure 2

Figure 1. The high frequency heat flux at the meniscus during casting [13]
Figure 2. The relation between oscillation mark and 1st order derivative of heat flux [13]

It can be found from Fig. 1 that the heat flux always increased in the periods of negative strip time (T_n), which may be caused by the following three factors. This first one is that there would be a large part of the cold mold immersed into the hot melt bath and heated during the T_n period as the downward velocity of mold is larger than the casting speed. This could raise the heat transfer from bath to mold. The second one is due to the deformation of meniscus during T_n period, such that the meniscus was getting close to the high cooling potential copper mold, and became being solidified, which would release latent heat, then increased the heat flux during negative strip time. The third factor is the enhancement of mold flux infiltration from the liquid pool into the shell/mold gap, producing a convective cooling effect that leads to the increases of heat flux during negative strip time.

Besides, Fig. 2 shows the profiles of mold oscillation, heat flux in the vicinity of meniscus and oscillation marks appearing on cast alloy surface. The oscillation marks maintain a strong one-to-one correlation with the mold oscillation and heat flux cycles, which gives a more intuitively evidence showing that the formation of oscillation mark and the fluctuation of heat flux are related. More specifically, Fig. 2 also indicates the oscillation marks formed and the heat flux increased in the negative strip time period.

The results of Fig. 1 and 2 suggest decreasing the period of negative strip time (Tn) during the mold oscillation is an effective way to reduce the depth of oscillation marks. Actually, the high frequency, low stroke, and non-sinusoidal oscillation models, [14] which shorten the negative strip time have been applied to inhibit the oscillation marks and improve the surface quality of shell in industry.

2.2 Cooling Potential Conditions

In this section, the superheat of molten steel, flow rate of cooling water, coating of mold wall, etc. are categorized as the cooling potential conditions since they have significant influence on the structure of initial solidification shell through affecting the temperature and cooling rate in molten steel. The oscillation marks are usually categorized into two types: hook and depression oscillation marks. [15] Both of them were observed on the shell in Fig. 3 through metallographic examinations, though most of them are depression type. Actually, the final formation of the mark depends on the strength of the shell relative to the ferrostatic forces interacting with it. If the initial shell is strong enough to withstand the ferrostatic pressure, the liquid steel will overflow the shell tip and form the hook; but if the shell tip is too weak, it will bend back toward the mold and form the depression type oscillation mark. The operation parameters, such as superheat of molten steel, flow rate of cooling water, coating of mold wall, etc. will affect the cooling potential greatly which determines the strength of the shell, then determines the type of the oscillation marks.

There was only one hook formed in this study that occurred in the first oscillation cycle at the beginning of casting (details in reference [13, 16]) when the cooling potential is relatively high as the temperature of mold is still relatively low. A higher cooling potential means a thicker of initial shell, which would be strong enough to withstand the ferrostatic pressure and form the hook. But, the cooling potential of the mold dropped with the dipping into the melt, which warms the mold and resulted in a thinner shell tip that could not withstand the ferrostatic pressure and bend toward to the mold. This consequently forms the depression type oscillation marks.

Besides, the solidification structure of initial shell will also be affected by the cooling potential conditions. The macrostructure of a continuously cast steel semi usually consists of three zones: chilled crystal zone, columnar zone and equiaxed zone from the surface towards the center. Fig. 4 shows the evolution of solidification structure at four typical positions along a line perpendicular to the cast velocity. The chilled zone in Fig. 4 (a) is filled with many very fine equiaxed crystals that can be observed on the shell near the cooled mold due to the high cooling potential as well as the heterogeneous substrate of mold wall, which make nucleation in this zone very fast. Next to the chilled zone, the parallel large columnar grains were appearing and grew toward the melt, as shown in Fig. 4 (b). The formation of a columnar zone results from the steep temperature gradient causing the grains to grow preferentially in the opposite direction to the heat flux. The columnar to equiaxed transition structure was observed in Fig. 4 (c), which shows the equiaxed grains occurred at end of columnar tip. It should be fully equiaxed zone after the C-E zone if the melt solidified totally as the industrial billet, bloom, or slab. However, the fully equiaxed zone did not exist here due to the limitation of the mold simulator. The mixed columnar and equiaxed grains could be observed at the edge close to the melt as shown in Fig. 4 (d).

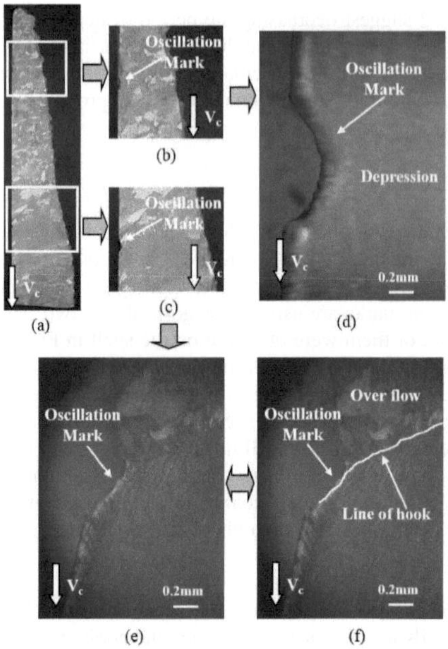

Figure 3. The depression and hook oscillation marks

(a) the shell, (b) the position of the typical depression oscillation mark on shell, (c) the position of hook oscillation mark on shell, (d) the profile of depression oscillation mark, (e) the profile of hook oscillation mark, (f) the line of hook

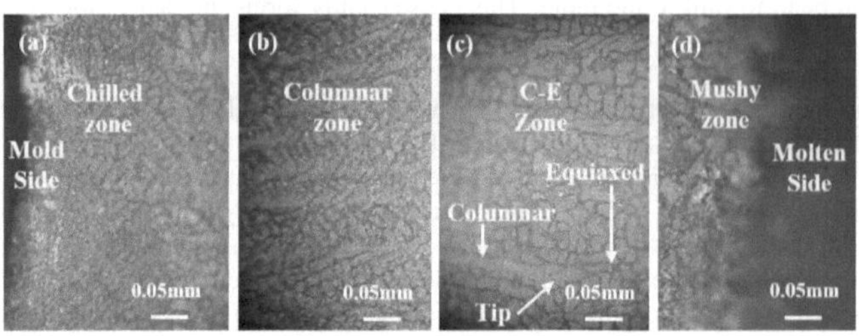

Figure 4. The solidification structure

According to Hunt' C-E Transition theory, [34-36] the fully equiaxed grains grow in the region of high growth velocity and low temperature gradient, while the fully columnar grains occur in the small velocity and large gradient region, and the mixed grains form in between them. Also, the growth velocity and temperature gradient are undercooling dependent, [34, 37-38] which means the solidification structure of initial shell is greatly affected by cooling potential conditions, likes superheat of molten steel, flow rate of cooling water, coating of mold wall , etc.

Therefore, through adjusting those cooling potential conditions, the proper solidification structure and oscillation marks profiles of initial shell can be obtained, consequently improve the surface and internal quality of slab, billet, bloom.

2.3 Mold flux

Mold flux plays a very important role in the continuous casting of steel. Liquid slag on the top of molten steel pool infiltrates into the gap between mold and shell to lubricate the newly formed steel shell and control the heat transfer from molten steel to mold wall. As the initial solidified shell with low strength and ductility in the high temperature zone, [17] can be torn very easily due to the improper lubrication or the thermal stress caused from the uneven heat transfer.

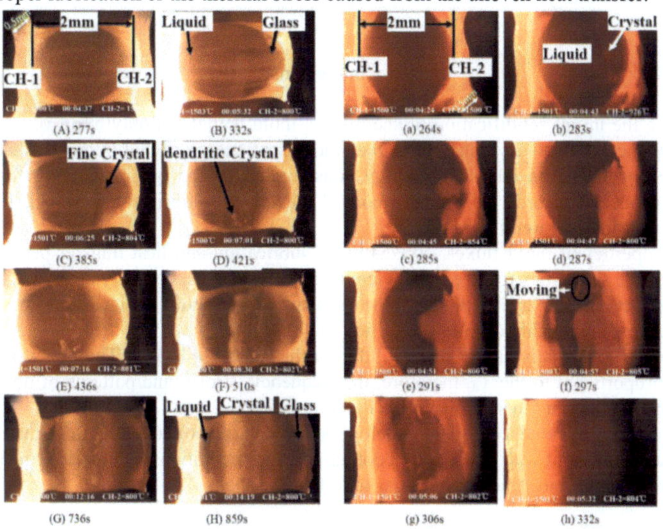

Figure 5. Phases evolution of LC mold flux and MC mold flux under the simulated temperature of meniscus [23]

The lubrication performance of mold fluxes is mainly determined by: 1) Slag film structure, [41] such as liquid, crystalline and glassy layers distribution in the gap of mold/shell; where the liquid layer has been approved to own the best lubrication ability; 2) The viscosity of mold flux, as the frictional force of molten slag is directly related to viscosity. [18-19] While, the heat transfer performance of mold fluxes is also mainly determined by slag film structure, as the thermal resistant of crystalline mold flux is obviously larger than glass or liquid mold flux since a large quantity of the thermal radiation will be blocked due to the scattering of grain boundaries, defects and surfaces in crystalline mold flux. [20-21] Another very important parameter to affect the heat flux transferring from shell to mold is the interface thermal resistance between mold wall and mold flux which results from the shrinkage and deformation of mold flux when it gets crystallized.

Fig. 5 shows the phases evolution of two typical mold fluxes for casting LC (low carbon steels) and MC (middle carbon steels) in continuous casting under the simulated temperature of meniscus by using the double hot thermocouple technique. [22-23] Among them, Fig. 5 (A)-(H) are 8 images of the crystallization process of LC mold flux; while Fig. 5 (a)-(h) are 8 images corresponding to the crystallization process of MC mold flux. It can be found from Fig. 5 that the crystalline layer of LC mold flux crystal layer formed in the middle of mold fluxes first, and then grew towards both sides of thermocouple. Finally, the three layers structure were formed where the glass layer is close to the cold thermocouple (CH-2), the liquid layer is close to the hot thermocouple (CH-1), and the crystalline layer is formed in between them. However, for the MC mold flux, the liquid MC mold flux was crystallized directly from the cold side CH-2, and no glass phase formed during the whole process. Also, the MC mold flux can get crystallization much quicker than the LC mold flux. This is a benefit for the initial solidification of middle carbon steel, as it is of crack-sensitive due to the δ to ϒ phase transformation, which will lead to relatively larger volumetric shrinkage. The quickly formed crystalline layer can achieve mild cooling in the vicinity of meniscus that helps to reduce the stress on the initial solidification shell.

Conclusions

Controlling the initial solidification of shell in the mold would not only improve the quality of casting products, but it will realize the large amounts of energy saving in continuous casting to greatly reduce the extra labor and energy consumption of the post-processing for obtaining qualified products. In order to properly control the initial solidification of shell, four aspects should be considered carefully.

(1) The properties of mold fluxes, especially the lubrication and heat transfer performances of mold flux. The slag film structure in the vicinity of meniscus, the viscosity of liquid slag, and the formation of crystalline layer will determine the heat transfer, mold lubrication that in turn would affect the stress and quality of initial solidified shell.

(2) The period of negative strip time (T_n) during the mold oscillation. The depth of oscillation marks is proportional to the T_n, therefore, the frequency, stroke, and patterns of non-sinusoidal oscillation are important to shallow the oscillation marks.

(3) The cooling potential conditions, such as superheat of molten steel, flow rate of cooling water, coating of mold wall, etc. Those cooling potential conditions mainly determine the types of oscillation marks and structure of initial solidification shell.

References

[1] S. Mazumdar and S. K. Ray, "Solidification control in continuous casting of steel", *Sadhana*, 26 (1) (2001), 179-198.

[2] Hirohiko Okumura, "Recent trends and future prospects of continuous casting technology", *Nippon Steel Technical Report*, 1994, No.(61), 9-14.

[3] Amit Chatteree, Sanjay Chandra, "Thin-slab casting: new possibilities", *Sadhana*, 26 (1-2) (2001), 163-178.

[4] N. Zapuskalov, "Comparison of continuous strip casting with conventional technology", *ISIJ International*, 43 (8) (2003),1115-1127.

[5] G. S. Nikitin et al., "Highly efficient casting-rolling units for making wire rod and reinforcement bar", *Metallurgist*, 52 (1-2) (2008), 86-94.

[6] E. Takeuchi and J. K. Brimacombe, "The formation of oscillation marks in the continuous casting of steel slabs", *Metallurgical Transactions B*, 15 (3) (1984),493-509.

[7] S. Harada et al., "A formation mechanism of transverse cracks on CC slab surface", *ISIJ International*, 30 (4) (1990), 310-316.

[18] G. Krauss, "Solidification, segregation, and banding in carbon and alloy steels", *Metallurgical and Materials Transactions B*, 34 (6) (2003), 781-792.

[9] Z. Wang, K. Mukai and I. J. Lee, "Behavior of fine bubbles in front of the solidifying interface", *ISIJ International*, 39 (6) (1999), 553-562.

[10] B. G. Thomas, et al., "Transport and Entrapment of Particles in Steel Continuous Casting", *Metallurgical and Materials Transactions B*, 45 (1) (2014), 22-35.

[11] D. Gupta, and A. K. Lahiri, "Water-Modeling Study of the Surface Disturbances in Continuous Slab Caster", *Metallurgical and Materials Transactions B*, 25 (2) (1994), 227-233.

[12] J. W. Elmer, S. M. Allen, and T. W. Eagar, "Microstructural development during solidification of stainless steel alloys", *Metallurgical Transactions A*, 20 (10) (1989), 2117-2131.
[13] H. Zhang et al., "A Study for Initial Solidification of Sn-Pb Alloy during Continuous Casting: Part I. The Development of the Technique", *Metallurgical and Materials Transactions B*, 45 (3) (2014), 1038-1047.
[14] M. Suzuki et al., "Development of a new mold oscillation mode for high speed continuous casting of steel slabs", *ISIJ International*, 31 (3) (1991), 254-261.
[15] A. Badri et al., "A mold simulator for continuous casting of steel: Part II. The formation of oscillation marks during the continuous casting of low carbon steel", *Metallurgical and Materials Transactions B*, 36 (3) (2005), 373-383.
[16] D. Zhou, et al., "A Study for Initial Solidification of Sn-Pb Alloy During Continuous Casting: Part II. Effects of Casting Parameters on Initial Solidification and Shell Surface", *Metallurgical and Materials Transactions B*, 45 (3) (2014), 1048-1056.
[17] M. R. Ridolfi et al., "Mathematical Modeling of Hot Tearing in the Solidification of Continuously Cast Round Billets", *Metallurgical and Materials Transactions B*, 41 (6) (2010), 1293-1309.
[18] Y. Meng, and B. G. Thomas, "Interfacial friction related phenomena in continuous casting with mold slags", *2003 ISSTech Steelmaking Conference*, Indianapolis, IN, USA, April 27-30, (2003), 1-18.
[19] M. Yao, X. Wang, and B. Du, "Study on lubrication and friction between strand and mould in continuous slab casting", *Dev. Chem. Eng. Mineral Process*, 14 (3/4) (2006), 459-472.
[20] W. Wang et al., "Radiative heat transfer behavior of mold fluxes for casting low and medium carbon steels", *ISIJ International*, 51 (11) (2011), 1838-1845.
[21] L. Zhou et al., "A Kinetic Study of the Effect of Basicity on the Mold Fluxes Crystallization", *Metallurgical and Materials Transactions B*, 43 (2) (2012), 354-362.
[22] L. Zhou et al., "In Situ Observation and Investigation of mold flux crystallization by using double hot thermocouple technology", *Metallurgical and Materials Transaction B*, 43 (4) (2012), 925-936.
[23] L. Zhou et al., "Computational Modeling of Temperature, Flow, and Crystallization of Mold Slag During Double Hot Thermocouple Technique Experiments", *Metallurgical and Materials Transaction B*, 44 (5): (2013),1264-1279.
[24] B. G. Thomas, "Modeling of continuous casting defects related to mold fluid flow", *AIST Transactions*, 3 (5) (2006), 2-17.
[24] A. Ray et al., "Microstructural features of sliver defects in hot-rolled low-carbon steel sheets", *Journal of Materials Science Letters*, 12 (14) (1993), 1148-1150.
[26] L. Zhang, J. Aoki, and B. G. Thomas, "Inclusion removal by bubble flotation in a continuous casting mold", *Metallurgical and Materials Transactions B*, 37 (3) (2006), 361-379.
[27] P. Kaushik, J. Lehmann, and M. Nadif, "State of the Art in Control of Inclusions, Their Characterization, and Future Requirements", *Metallurgical and Materials Transactions B*, 43 (4) (2012), 710-725.

Energy Technology 2015: Carbon Dioxide Management and Other Technologies
Edited by: Animesh Jha, Cong Wang, Neale R. Neelameggham, Donna P. Guillen, Li Li, Cynthia K. Belt, Randolph Kirchain, Jeffrey S. Spangenberger, Frank Johnson, Andrew J. Gomes, Amit Pandey, and Peter Hosemann
TMS (The Minerals, Metals & Materials Society), 2015

PERFORMANCE OF TWIN OXYGEN-COAL LANCES FOR PCI OPERATION IN BLAST FURNACE IRON MAKING

Fanqing Meng[1], Huiqing Tang[1*], Zhilong Zhao[2]

[1]State Key Laboratory of Advance Metallurgy, University of Science and Technology Beijing; 30 Xueyuan Rd.; Beijing, 100083, China
[2]MCC Capital Engineering and Research Incorporation Limited, 7 Jiananjie Rd; Beijing Economic- Technological Development Area, Beijing 100176, China

Keywords: Pulverized coal injection; Oxygen-coal lance; Blast furnace; Iron making

Abstract

A three-dimensional multi-phase model with Discrete Particle Model (DPM) approach has been developed to examine the performance of the twin oxygen-coal lances for PCI operation. Simulation conditions and properties of the PC were obtained from the real operation conditions of the blast furnace of 580 m^3 with molten iron output of 1775 t/d. The simulation results of novel PCI style were also compared with the results of conventional PCI operation. Results show that the twin coal-oxygen lances changed the gas flow pattern near the tuyere, which results in a different mixing and burning process of the pulverized coal. The coal combustion efficiency near tuyere could be improved by applying coal-oxygen lance. For the PC adopted in the simulation, coal BR at the exit of the simulation region increased from 30% to 60% and the maximum temperature of the gas phase increased from 2200K to 2600K by applying the twin coal-oxygen lances and the high temperature zone is in a safe distance from the nozzle.

Introduction

Pulverized coal injection (PCI) is widely adopted for blast furnaces (BF) iron making and is one of key steps for cost effective BF iron making [1]. In the conventional PCI operation, Pulverized coal (PC) particles are supplied to the hot blast by a simple coal lance in the blowpipe of the tuyere. Its injection rate is on the level of 100-200 kg-coal/tHM. However, with the increase of PCI rate, negative effects of this PCI pattern emerged such as combustion efficiency decrease of PC. In general, pulverized coal combustion in the blowpipe and tuyere is likely to be influenced by important BF operating parameters as lance design, blast parameters , coal type and so on.[1-2] . Industrial practices indicate that and several modelling works have been reported.[3-5] Recently, Authors of current study envisaged novel twin coal-oxygen lances for PCI operation. The present work aims to provide a reference for improving PCI using the twin coal-oxygen lances by numerical simulation.

Description of PCI process

When PC particles are injected by carrier gas (N_2, 1kg N_2/20kg PC) from the lance end into the blowpipe of the nozzle in a continuous operation, they get dispersed due to the turbulent nature of the hot blast and reach the raceway by a projectile path under the influence of drag force. While they are traversing the projectile path, the coal particles get heated up and start releasing volatiles which then undergo homogeneous combustion; the PC particles undergo heterogeneous combustion depending upon the availability of oxygen in the hot blast. Hence they provide both heat and carbon source in BF iron making. The model geometry of the investigated nozzle is based on an engineering design for the BF of 580 m^3 with an output of 1775 tHM/day. Its schematic diagram and major specification are shown in Fig. 1. Specification of the cross section of the designed coal-oxygen lance is shown in Fig. 1 as well.

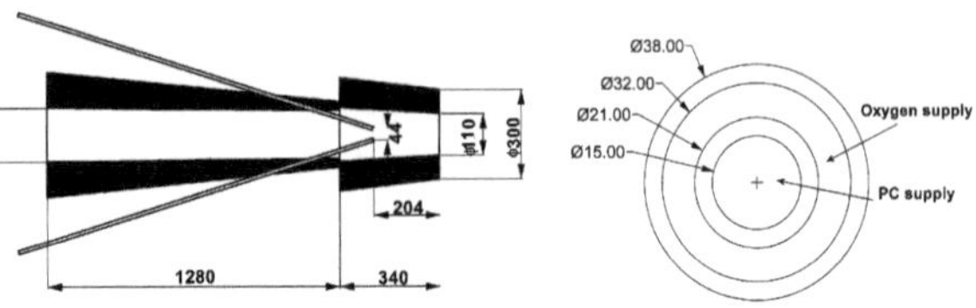

Figure.1 Schematic diagram of the nozzle equipped with novel oxygen-coal lances (dimension: mm).

Mathematical Model

Governing Equations

The simulation of PCI process in tuyere and its nearby surroundings is based on the comprehensive representation of the features of gas phase fluid dynamics, combustion of coal particles, heat transfer including radiation and chemical reactions. It is to investigate the velocity profile, temperature and emissions at various locations in the nozzle and its nearby area. It is based on the numerical solution of three-dimensional differential equations for conservation of mass, momentum and energy for both gas and solid phases. The gas phase is treated as a continuous phase; its general governing equation is Eq. (1) and details of Eq. (1) are listed in Table 1. The solid phase consisted of injected PC particles is solved using particle transport model in which a representative number of particles are tracked through the computational domain in a Lagrangian way; details are listed in Table 2. The two phases are coupled through momentum interaction (drag) and energy interphase interaction, heterogeneous combustion, and heat transfer. Full coupling of mass, momentum and energy of particles with the gaseous phase are carried out. Considering the operation of PCI, when the coal particles are heated, volatile matters will be released to react with oxygen, resulting in diffusion flame combustion. The combustion mainly includes the following reactions: coal devolatilization, volatile combustion, and char reactions. At the same time, gaseous combustion of CO is considered as well. The devolatilization process is modeled by the single step model and the volatile combustion, char reactions and gas combustion process are modeled by the eddy break-up model. All the reaction rates follow Arrhenius formula and are listed in Table 3. The governing equations are solved using the commercial CFD software FLUENT and our own UDFs. A residual value of 1.0E-5 was used as the convergence criterion for mass-momentum equations in all simulations. DPM is used to track the coal particles which interact with the continuous gas phase flow. Wall temperature of the zone in BF is fixed to be 1973 K. The operation pressure is 3.0e5 Pa.

$$div(\rho_g U_g \Phi) = div(\Gamma grad\Phi) + S \tag{1}$$

Table 1. Terms used in Eq. (1) for the gas phase

Equation	Φ	Γ	S
Mass	1	0	$\sum_{n_p}\Delta m_p / V_{cell}$
Momentum	u_g	μ_{eff}	$-\frac{\partial P_g}{\partial x}+S_x+\sum_{n_p}F_x\Delta t/V_{cell}$
	v_g	μ_{eff}	$-\frac{\partial P_g}{\partial y}+S_y+\sum_{n_p}F_y\Delta t/V_{cell}$
	w_g	μ_{eff}	$-\frac{\partial P_g}{\partial z}+S_z+\sum_{n_p}F_z\Delta t/V_{cell}$
Turbulent kinetic energy	k	$\mu_g+\frac{\mu_t}{\sigma_k}$	$G_k-\rho_g\varepsilon$
Turbulent dissipation rate	ε	$\mu_g+\frac{\mu_t}{\sigma_\varepsilon}$	$\frac{\varepsilon}{k}(C_1G_k-C_2\rho_g\varepsilon)$
Energy	H_g	$\frac{\mu_{eff}}{Pr}$	$\sum_{np}Q/V_{Cell}-\Delta H_5 r_5-\Delta H_6 r_6$
	y_{VM}	$\frac{\mu_{eff}}{Sc}$	$\sum_{n_p}\Delta m_{vm}/V_{Cell}-r_5$
	y_{H_2O}	$\frac{\mu_{eff}}{Sc}$	$\frac{1.5M_{H_2O}}{M_{vm}}r_5$
Species	y_{CO}	$\frac{\mu_{eff}}{Sc}$	$\sum_{n_p}\Delta m_{CO}/V_{Cell}-r_6$
	y_{O_2}	$\frac{\mu_{eff}}{Sc}$	$\sum_{n_p}\Delta m_{O_2}/V_{Cell}-\frac{1.7M_{O2}}{M_{VM}}r_5-\frac{M_{O2}}{2M_{CO}}r_6$
	y_{CO_2}	$\frac{\mu_{eff}}{Sc}$	$\sum_{n_p}\Delta m_{CO_2}/V_{Cell}+\frac{M_{CO_2}}{M_{CO}}r_6+\frac{M_{CO2}}{M_{VM}}r_5$

Table 2. Equations for PC particle phase

Equation	Φ	Γ	S
mass	$\frac{dm_p}{dt}=-r_1-(r_2+r_3+r_4)M_C$		
Momentum	$m_p\frac{d\overline{U}_p}{dt}=F_D+m_p\overline{g}$ $F_D=\frac{1}{8}A_p\rho_gC_D\left\lvert\frac{\overline{U}_g}{\varepsilon}-\overline{U}_P\right\rvert(\frac{\overline{U}_g}{\varepsilon}-\overline{U}_P)$, $C_D=\max(24(1+0.15\mathrm{Re}^{0.687})/\mathrm{Re},\ 0.44)$		
Energy	$m_pC_p\frac{dT_p}{dt}=hA_p(T_g-T_s)+\sum_{i=1,4}(-\Delta H_{reac,i})r_i+A_p\varepsilon_p\sigma_p\left(T_g^4-T_p^4\right)$;		

Table 3. Chemical reactions involved in models

No.	Reaction	Reaction Rate Expression
		PC Combustion
1	Coal=VM+Char	$r_1 = -k_1\left[m_p - \left(1-f_{v,0}\right)\left(1-f_{w,o}\right)m_{p,0}\right]$, $k_1 = 3.75\times10^5 \exp(-\frac{7.336\times10^4}{RT_p})$, (kg/s)
2	$2C+O_2=2CO$	$r_i = -\pi d_P^2 \frac{\rho_g R T_g y_{O_2}}{M_{O_2}} \frac{D_0 k_i}{D_0 + k_i}$ $D_0 = \frac{D_{ref}}{r_p}(\frac{T_p+T_g}{2T_{ref}})^{0.75}\frac{P_A}{P_g}$ $P_A = 1.05\times10^5 \text{pa}, T_{ref} = 298K$ $k_2 = 1.813\times10^3 \exp\left(-\frac{1.089\times10^5}{RT}\right)$ (kmol/s)
3	$C+O_2=CO_2$	$k_3 = 1.225\times10^3 \exp\left(-\frac{9.977\times10^4}{RT}\right)$
4	$C+CO_2=2CO$	$k_4 = 7.351\times10^3 \exp\left(-\frac{1.380\times10^5}{RT}\right)$
		Gaseous Combustion
5	$VM+1.7O_2=CO_2+1.5H_2O$	$r_{Arr} = 2.11\times10^{11} \exp\left(-\frac{2.02\times10^5}{RT}\right)(\rho_g y_{VM})^{1.3}(\rho_g y_{O_2})^{0.2}$ $r_{EBU} = 4.0\rho_g \frac{\varepsilon}{k}\min(y_{VM}, \frac{y_{O2}}{1.70})$ $r_5 = \min(r_{Arr}, r_{EBU})$ (kg/m^3·s)
6	$2CO+O_2=2CO_2$	$r_{Arr} = 7.0\times10^4 \exp\left(-\frac{6.651\times10^4}{RT}\right)\rho_g^2 y_{CO} y_{O2}$ $r_{EBU} = 4.0\rho_g \frac{\varepsilon}{k}\min(y_{CO}, y_{O2})$ $r_6 = \min(r_{Arr}, r_{EBU})$ (kg/m^3·s)

Results and Discussion

PC for this simulation has an average size of 0.080 mm and its chemical composition is 78.5%(ad) of fixed carbon, 10%(ad) of ash, 10% (ad) of volatile and 1.5% of moisture. Three coal injection patterns were examined. The operation parameters are listed in Table. 4. Operation data of the base case (Case A) is given in accordance with the typical running conditions of the BF in MCC industrial base. The base case was also used for model validation. The operation condition of PCI using twin coal-oxygen lances (Case B) is on the basis of BF iron making design.

As the measurements of in-furnace thermal and chemical properties are very difficult and only combustion temperature at the outlet of the tuyere is available. The measured gas temperature in the raceway is 2300 K and the temperature predicted using the model is 2200 K. The model validity could then be confirmed.

In the following discussion, the combustion characteristic of PC particles, oxygen concentration profile and gas temperature profile will be examined and compared. The horizontal plane across the tuyere axis is selected for analyzing.

Table 4. Simulation conditions

	Hot blast temperature (K)	O_2 supply in hot blast (Nm^3/s)	N_2 supply in hot blast (Nm^3/s)	O_2 supply in oxygen lance (Nm^3/s)	PC supply in coal lance (kg/s)	Carrier gas in coal lance (kg/s)
Case A (Base)	1423	0.350	0.99	0.0	0.257	0.01
Case B	1473	0.330	0.790	0.030	0..200	0.01

*Case A: single coal lance (conventional PCI 7% oxygen enrichment in hot blast ,1775 tHM/day) , Case B: twin oxygen-coal lance (15% oxygen enrichment in , 1856 tHM/day)

Trajectories and Combustion of PC

Fig.2 demonstrates trajectories of PC particles of the two cases. In the conventional PCI pattern (Case A), trajectories of coal particles are nearly along the axis of the nozzle after leaving the outlet of nozzle though the coal lance is installed at an angle of 12 degree to the axis. This is due to that , in this case, PCI rate is in normal level at some 200 kg-coal/tHM, and therefore PC particles could be dispersed in the hot blast within a very short flying distance, which is some 10 cm from the lance tip as shown in Fig. 2. The trajectories of dispersed PC particles are then adjusted to be approximately along the nozzle axis by the drag force. In Case B, because PC particles are supplied by twin coal-oxygen lances installed symmetrically to the nozzle axis, PC particles fly along the nozzle axis even better than they behave in Case A. To evaulate the performance of the twin oxygen-coal lances, a parameter of burnout ratio (BR) of PC is evaluated, which is defined as the ratio between the PC mass loss at the designated flying distance or time to the original PC mass at the inlet. BR variation of typical PC particle of different cases are shown Fig. 3, it could be seen that, ignition time of the PC particle is delayed as room temperature oxygen was introduced near the lance tip. In case A, coal particles start to devolatize at the flying time of 0.008s, however, in case B it takes place at some 0.010s; PC residence time of the base case is 0.040s, and that of Cases B is 0.050s, which is elongated. This mainly attributed to the reduction of hot blast supply and introducing room temperature oxygen in Cases B. Fig. 3 shows that applying coal-oxygen lances could improve BR of PC. Compared to the base case, increase of final BR is 0.25 in Case B. Regarding the final BR, the PCI pattern in Case B is much better.

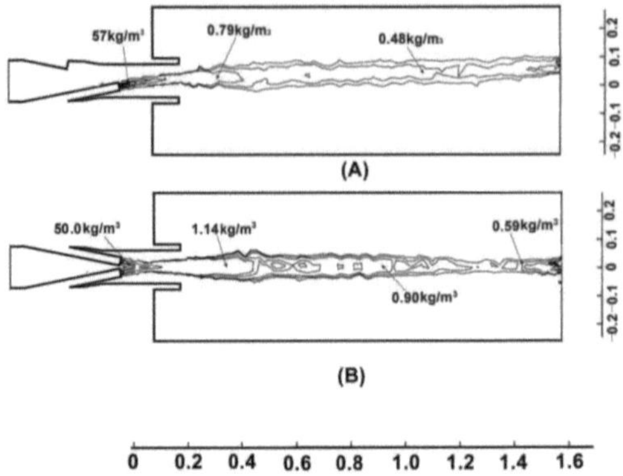

Fig.2 DPM (PC) concentration profiles

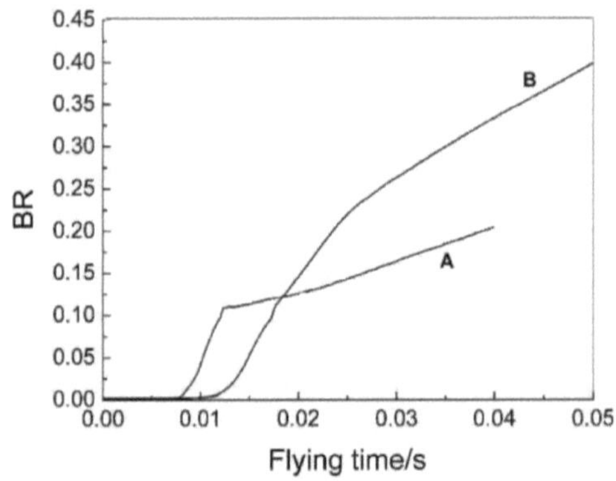

Figure.3 BR variation with PC flying time

Oxygen Concentration Profile

Intuitively, PC combustion in tuyere and raceway depends on surrounding oxygen concentration in the hot blast. Cases B are intended to increase oxygen concentration around the PC flow. Simulation results of oxygen concentration profile for the two cases are given in Fig. 4. In Case A, oxygen concentration profiles symmetrically to the tuyere axis and it drops from 0.20 in the tuyere to 0.10 at the outlet of the raceway. In Case B, oxygen presents a similar concentration profile as in Case A. Oxygen concentration drops from 0.40 in the tuyere to 0.05 at the outlet of the raceway. This indicates that PCI pattern in this case is superior to that in Case A.

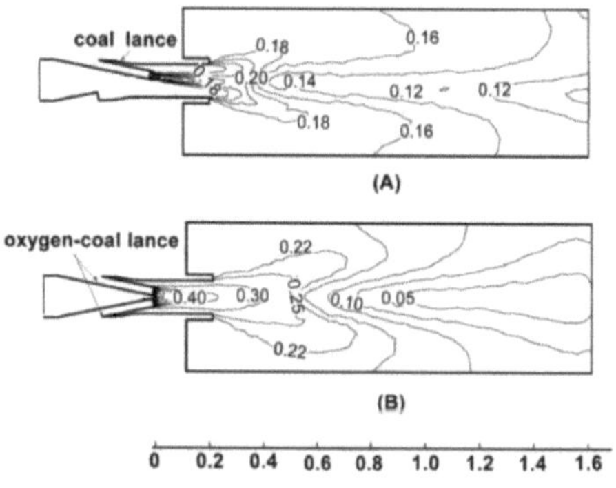

Figure.4 Oxygen volume fraction profile

Gas Temperature Profile

In PCI operation, high temperature zone is expected to have a certain distance from the tuyere and temperature near the tuyere must be controlled or else the refractory of the tuyere can not stand. Fig.5 displays gas temperature profiles for the cases. In the base case, it shows that maximum temperature is some 2200K and the hot temperature zone is some 0.2m from the tuyere outlet. Since its max temperature is not high, the distance is not important. Maximum combustion temperatures of cases B could reach 2800K as shown in Fig 5, which is much higher than that of the base case and indirectly indicates that the PC combustion is intensified in Case B. however, in Case B, The hot temperature zone in on the axis of nozzle and its distance from the nozzle is 1.2 m, which is a safe distance to protect the tuyere. This still indicates the twin oxygen-coal lances design has the effect to push the high combustion temperature zone away from the tuyere.

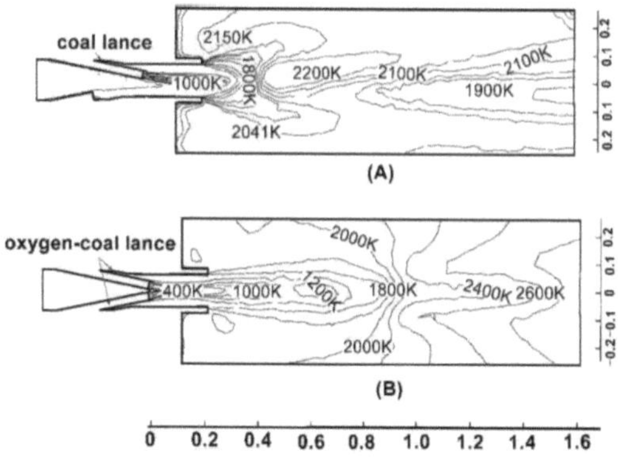

Figure.5 Gas temperature profile

Conclusions

Performance of the conventional PCI mode and the envisaged PCI mode applying twin oxygen-coal lances has been numerically analyzed. Their results on PC combustion behavior, oxygen concentration and combustion temperature in tuyere and raceway were compared. The conclusions are: PCI using twin oxygen-coal lances is superior to the conventional PCI mode. Its advantages include doubled PC BR, satisfying oxygen consumption level by PC in the raceway and high temperature zone in a safe distance from the nozzle.

Acknowledgements

The authors wish to thank MCC Capital Engineering and Research Incorporation and State Key Laboratory of Advanced Metallurgy (USTB) for their financial supports.

Meaning of symbols in this paper could be referred to FLUENT(V6.3.26) USER MANNUAL

References

1. J. Gorres, U. Schnell, K. Hein, "Trajectories of burning coal particles in highly swirling reactive flows", *International Journal of Heat Fluid Flow*, 16 (1995), 440-450.

2. Y. Shen, A. Yu, and P. Zulli, "CFD Modelling and analysis of pulverized coal injection in blast furnace: an overview", *Steel Research International*, 82 (2011): 532-542.

3. S. Du, W. Chen, "Numerical prediction and practical improvement of pulverized coal combustion in blast furnace", *International Communication of Heat Mass Transfer*, 33 (2006), 327-334.

4. M. Gu,, G. Chen, M. Zhang, D. Huang, P. Chaubal, C. Zhou, "Three-dimensional simulation of the pulverized coal combustion inside blast furnace tuyere", *Applied Mathematica Model,* 34 (2010) ,3536-3546.

5. S. Du, W. Chen, J. Lucas, "Performances of pulverized coal injection in blowpipe and tuyere at various operational conditions", *Energy Conversion Management.* 48(2007): 2069-2076.

Energy Technology 2015: Carbon Dioxide Management and Other Technologies
Edited by: Animesh Jha, Cong Wang, Neale R. Neelameggham, Donna P. Guillen, Li Li, Cynthia K. Belt, Randolph Kirchain, Jeffrey S. Spangenberger, Frank Johnson, Andrew J. Gomes, Amit Pandey, and Peter Hosemann
TMS (The Minerals, Metals & Materials Society), 2015

ENERGY SAVING AND CO_2 EMISSION REDUCING ANALYSIS OF CHINESE IRON AND STEEL INDUSTRY

ZHANG Qi, LI Yu, SHEN Yinghua, DONG Hui

SEPA Key Laboratory on Eco-industry; Northeastern University; Shenyang, Liaoning, 110819, China

Keywords: iron and steel industry; energy efficiency; energy saving; energy technologies, CO_2 emission reduction

Abstract

In this paper, the energy saving processes and main affecting factors of energy consumption and CO_2 emission were analyzed, based on the energy consumption and CO_2 emission status of Chinese iron and steel industry (CISI), The process which is mainly based on BF-BOF route and adopt coal as the main energy sources is the main factor to result in high energy consumption and CO_2 emission in CISI. During the 30 years development of energy saving since 1980, the direct and indirect energy savings was analyzed and compared with the energy saving potential analysis. To comparing the energy use in different works and improve energy efficiency, the energy consumption and CO_2 emission index from different analysis system is discussed. The analysis results will contribute to the development of energy saving and help to achieve the sustainable development of the CISI.

Introduction

Iron and steel industry is an important pillar and basic industry of the national economy and it's also an important supporting industry in the new industrialization process and urbanization process of China. Since the 1990s, Chinese iron and steel industry has sustained a rapid development and has achieved remarkable achievements. Especially average annual growth rate of crude steel production has reached 18.5% since the 21st century [1]. In 2012, crude steel production of China has already reached 716.5 million tons which accounts for 46.3% of world's crude steel production, and China has become a truly steel superpower.

The rapid development of Chinese iron and steel industry has played an important role in the development of economic for many years. Iron and steel industry is a resource and energy intensive industry, however Chinese iron and steel industry has been walking a way of high energy consumption, high resource consumption and high pollution for a long period. Resources, energy and environment have become an important factor in its development [2-4].

1. The development of Chinese iron and steel industry

The production of China's crude steel exceeds 100 million tons since 1996, and since that China keeps the highest crude steel production. Meanwhile, the overall development level of Chinese steel industry has also been greatly improved, which embodies in the adjustment and optimization of technological process and product structure and the improvement of international competitiveness [5].

In the 21st century, the level of Chinese iron and steel industry's technology and equipment has been improved greatly. During "the Eleventh Five-Year Plan" period, the proportion of blast furnace with a production capacity over 1000 m^3 has been increased from 48.3% to 56.7%. Most of enterprises have been equipped with hot metal pretreatment facilities and molten steel secondary refining facilities which made the refining ratio reach 70%. System of steel rolling basically achieved wholly continuous rolling, and chronic shortage of hot continuous rolling, cold-continuous rolling steel rolling have been increased from 26 sets and 16 sets to 72 sets and 50 sets respectively. Baosteel, AnSteel, WuSteel, ShouSteel and other large enterprises' technology and equipment have reached the international advanced level [6].

With the emphasis of "the Eleventh Five-Year Plan", energy saving technology penetration has been improved. New technology represent by blast furnace slag and converter slag solid waste utilization technology and energy management systems and the "three dry an electric" which mainly composed of Coke Dry Quenching (CDQ), dry dedusting technique for blast furnace gas, dry dedusting technique for converter flue gas and gas and waste heat integrated power, have been in the promotion and application among the iron and steel enterprises, which also promote the development of steel industry and the energy conservation.

2. Energy saving process of Chinese iron and steel industry

Energy conservation work of Chinese iron and steel industry started relatively late and had developed from single device and process energy consumption to system energy conservation stage. In recent years, with the development of technology, structural optimization, especially the popularization of advanced energy saving technologies, energy saving level enhanced unceasingly and energy consumption greatly reduced. Since 2000, energy consumption per ton steel of the entire industry was no longer counted in China, and medium-sized enterprises were only counted, as shown in figure 1. The comprehensive energy consumption per ton of steel of large and medium enterprises decreased from 0.92tce/t to 0.741tce/t between 2000 and 2005, droop rate reached 19.457%. The electricity standard coal coefficient was revised from 0.404kgce/kWh to 0.1229kgce/kWh by the National Bureau of Statistics since 2006, which resulting in energy consumption per ton steel of iron and steel industry dropped deeply, and which also makes the energy efficiency index no longer comparable. Comprehensive energy consumption per ton of steel of large and medium enterprises descended from 0.645tce/t to 0.6027tce/t during 2006 and 2012, rate of descend was 6.558%.

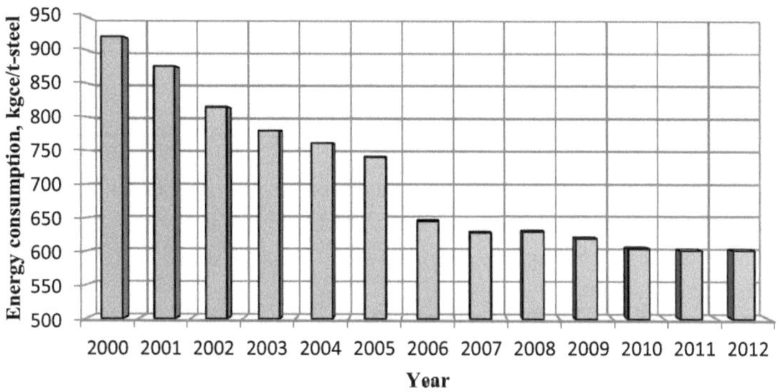

Figure 1. Energy consumption of per ton steel in Chinese iron and steel industry between 2000 and 2012

The reduction of energy consumption of iron and steel industry benefited from the development and widely application of energy conservation and emission reduction technologies. For instance, CDQ device increased from 36 sets to 112 sets during 2005 and 2010, and another 50 sets are under construction. Penetration of CDQ device among key iron and steel enterprises had reached 85%. By the end of 2010, more than 600 blast furnaces had been equipped coupled with 597 sets of top pressure recovery turbine unit (TRT) which increased 357 sets compared to 2005 and produced a huge benefit.

In the key iron and steel enterprises, the penetration rate of TRT among the blast furnace which is over 1000m^3 have reached 98%, tons of iron power generation capacity has reached 32kWh. The use of metallurgical coal gas' gas steam combined cycle power plant (CCPP) technology has been applied, by 2010, 15 sets of units have been built, the total installed capacity has reached 2200MW, and there are another 15 sets are under construction. At the same time, other technology which mainly including sensible heat recovery technology of sintering ore, coal moisture control technology and furnace dehumidified blast technology have been applied in various degree . The status of energy conservation for iron and steel industry has been greatly improved. Changes of specific energy consumption (SEC) per ton steel in Chinese iron and steel industry from 1980 to 2010 of and energy saving effect are shown in Table 1.

Table 1. Changes of energy consumption per ton steel in Chinese iron and steel industry and its energy saving effect during 1980 and 2010

Items		1980~1985	1985~1990	1990~1995	1995~2000	2000~2005	2005~2010	Energy saving
SEC	kgce	-179.0	-89.0	-37.0	-199.0	-67.0	-24.0	-595.0
Change	%	30.08	14.96	6.22	33.45	11.26	4.03	100.0
Direct	kgce	-124.3	-56.5	-20.4	-103.3	-38.3	-16.5	-359.3

energy saving	%	69.44	63.48	55.14	51.91	57.16	68.78	60%
In-direct	kgce	-54.7	-32.5	-16.6	-95.7	-28.7	-7.5	-235.4
energy saving	%	30.56	36.52	44.86	48.09	42.84	31.25	40%

As shown in Table 1, the energy consumption per ton steel of Chinese iron and steel industry has decreased from 1285kgce/t to 690kgce/t in the past 30 years. The total energy saving is 595kgce/t, including direct energy conservation 359.3kgce/t, which accounting for 60%.

3. CO_2 emission situation in Chinese iron and steel industry

The CO_2 emission in iron and steel industry takes over a very large proportion of total CO_2 emission of a country. In 2005, the CO_2 emission of Japanese steel industry accounts for 13% of the total emissions of Japan' while the CO_2 emission of Chinese steel industries takes over 37.2% of the total emission of the world steel industries and which reached 51% on 2012[1-3]. With the yearly increase of steel production in China, the amount of CO_2 emission increases dramatically, just shown in figure 2 [7].

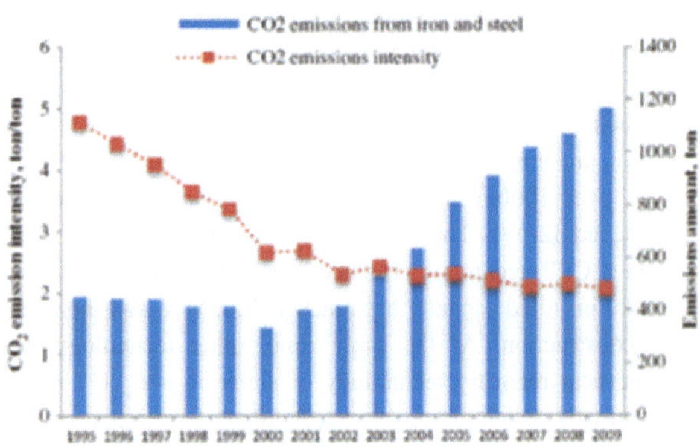

Figure 2. CO_2 emissions amount and intensity of the Chinese iron and steel industry

Considering the steel production process, all of the processes contain CO_2 emission. In the Blast Furnace- Basic Oxygen Furnace (BF-BOF) process, iron ore are reverted to liquid iron in blast furnace almost completely built on the basis of coal as the main energy products. Therefore, steel production process inevitably emits a large amount of CO_2, which occupying 5-7% the total amount of CO_2 from human activities.

The amount of CO_2 emission vary from production processes, the amount of CO_2 emission per ton steel is 1.7t in blast furnace process in traditional western production enterprise; which is only 0.4t if scrap steel fully uses as materials to product in steelmaking enterprise in electric furnace.

The work of the energy conservation becomes the vital basis of CO_2 emissions reduction. Steel production process produces more than 95% of CO_2 which derives by energy consumption, therefore, energy saving and CO_2 emission reduction is inseparable in the iron and steel industry.

4. Energy consumption analysis of iron and steel industry

1) Impact on energy consumption of production processes

Chinese steel production process mainly refer to BF-BOF long process, which accounts for over 80% steel production of the total crude steel production; the steel production of Electric Arc Furnace (EAF) process occupy only 17%, the proportion of which is more than 30% of the world's crude steel production in overseas .Compared with foreign countries, large gap exist among EAF process, as shown in Figure 3. Different from the BF - BOF process which takes iron ore as the source, scrap is the main source of EAF process. In EAF process, every produce 1 ton steel, 1.3t ore can be saved, 350kgce energy consumption can be reduced and 1.4t CO_2 emission (excluding short process CO_2 emissions from electricity) and 600kg waste emission can be avoided. Due to the poor scrap accumulation of our society, and the rapid growth in iron and steel production, the scrap resources are relative shortage, the proportion of EAF is below 20% from long-term, which was only 9.8% in 2010.The utilization of scrap resources helps to reduce the energy consumption and environmental impact of iron and steel industry, especially the CO_2 emissions, but due to the lack of resources and higher prices of scrap the use of scrap encountered numerous obstacles .Governments should make some policy to encourage enterprises to import more scrap and multi-purpose steel scrap, and then increase the use of scrap steel and attaches importance to the development of the EAF process.

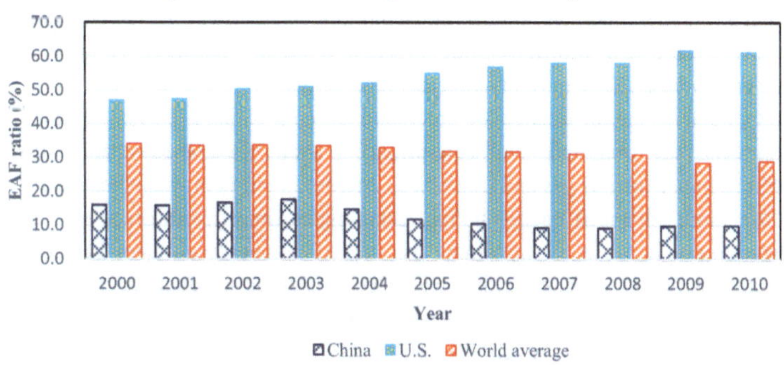

Figure 3 Share of EAF in total steel production in China and the U.S. and world average values

Recycling scrap steel as much as possible, and bring back to the iron and steel industry, as a raw material. In this way more steel can be produced while less energy is used [8-9].

2) The influence of iron and steel ratio on energy consumption

The energy consumption of the blast furnace occupies the largest proportion of integral iron and steel production process. The amount of pig iron consumption for steel making is the main factor affecting the ratio of iron to steel. Iron and steel ratio is defined as the ratio of the consumption of molten iron and molten steel by producing 1 ton steel. It is one of the most important factors for energy consumption per ton steel. The higher the iron and steel ratio is, the more energy consumption there will be, otherwise means steelmaking using less molten iron and more scrap steel.

In 2010 the world average level of iron and steel ratio is 0.7388, and 0.9440 in China. The world's average level of iron and steel ratio is 0.5734, with the deduction of Chinese iron and steel industry [10]. According to the statistics, in 2010, American iron and steel ratio is 0.327, the European Union's 27 countries are 0.5446, Germany is 0.6461, the UK is 0.7451.With the iron and steel ratio raised by 0.1, energy consumption per ton steel will be raised around 50kgce/t. Because of the high iron and steel ratio, the comprehensive energy consumption per ton steel in China is about 110~250kgce/t higher than the international advanced level. Due to the huge demand of iron and steel, and the lack of scrap steel, in the short term to reduce iron steel ratio is not reality. Extensive use of scrap steel can also decrease the quality of the steel, in the high quality steel production which use scrap as resources, the relevant technology need to be developed.

3) The influence of energy structure on energy consumption

Steel manufacturing process is deeply affected by the energy structure; the energy structure of China is rich coal, oil-poor and little gas. In Chinese Iron and steel industry, coal and power take up the highest share in the energy structure. Due to coal's energy conversion efficiency and energy efficiency in the application process is lower than the oil and gas, China's iron and steel industry energy consumption is 15 ~ 20 kgce/t higher than the energy consumption of industrial developed countries. Because of the low utilization rate of coal, pollutant discharge is large, energy consumption and CO_2 emissions increased every year.

Compared the difference of international and China's energy structure, we recommend the iron and steel enterprises to adopt diversified energy structure, preferred to use natural gas and other clean energy. In terms of waste heat and energy, the use more advanced energy-saving technologies, improve the recovery rate and utilization rate, such as the remaining gas implement energy efficient transformation and optimization of integration, the new water consumption reduction, etc.

4) The influence of the level of energy conversion on energy consumption

The level of energy conversion has a great impact on energy consumption; table 2 shows the comparison of different companies for China's power generation and oxygen consumption. As can be seen, if the power consumption per ton steel is calculated according to 600kWh, due to the different levels of power consumption the energy consumption per ton steel may increase 84.6kgce, if 90m^3 oxygen is consumed per ton steel, the different levels of oxygen consumption may bring another 15kgce energy consumption per ton steel.

Table 2 Energy conversion level of different enterprises in our country

Name	A	B	C	D	E	F	G	H	different
Energy use of power generation (kgce/kWh)	0.327	0.456	0.448	-	-	0.461	0.468	-	0.141
Energy use of oxygen produce (kgce/m^3)	0.266	0.432	0.423	0.298	0.347	0.408	0.344	0.292	0.166

The higher the conversion efficiency is, the smaller standard coal coefficient will be. In 2005,China adjust electricity standard coal coefficient from 0.404kgce/kWh to 0.1229kgce/kWh, the power energy consumption dropped by 69.57%which seriously obstructed the assessment of China's steel industry total energy level, so as to the energy consumption statistics and analysis. WEI Bingxian analyzed the advantages and disadvantages of this adjustment [10].

The current electricity standard coal coefficient of China is 0.1229kgce/kW·h, which is the theoretical ratio of 1kg standard coal calorific value and 1kW·h electricity. In fact, when calorific value of coal is converted into electrical energy, the conversion efficiency is limited by the second law of thermodynamics. The average thermal power generation standard coal consumption for China is 0.4kgce/kW·h, according to the current electric power industry in China; the advanced level of the power supply coal consumption calculation is 0.2909kgce/kW·h. The theoretical power conversion coefficient is 0.1229kgce/kW·h, and it has large gap with the most advanced technological level of standard coal consumption which is 0.2909 kgce/kW·h. The electricity standard coal coefficient of Japanese steel industry and Western European steel companies is 0.357kgce/kW· h and 0.320kgce/kW· h is 0.357kgce/kW· h. The adjustment of Chinese electricity standard coal coefficient, affect the statistical base, hence, the energy consumption of China's steel industry and foreign countries become incomparable.

5) The influence of technical equipment of energy consumption

Metallurgical technology and equipment is toward large-scale, efficient, automated, continuous, compact, longevity and environmentally friendly direction. At present, the development of China's metallurgical technology and equipment level is unbalanced, and the industry is in a multi-level and different equipment level, all kinds of technical and economic indexes of coexisting stage. Account for about one-third of their properties to international level. About 20% of companies are relatively backward in the technology and equipment and production index, which should be eliminated.

5. Conclusions

(1) The iron and steel industry is one of highest energy and resource consumption industry which also cause lot pollution. Adopting advanced technologies for energy conservation, improving the level of energy conversion process and optimizing material and energy flow structure are the main measures to reduce the energy consumption and CO_2 emission.

(2) Higher iron and steel ratio brings more energy consumption, otherwise to reduce the ratio need more scrap steel. Limited by the lack of scrap steel, to reduce the ratio in the short term is not reality. In fact using more scrap steel may decrease the quality of the steel, related technologies still need to be refined.

(3)The level of energy conversion which affects the energy consumption a lot closely interrelate with standard coal coefficient. The adjustment of Chinese electricity standard coal coefficient made the energy consumption of China incomparable with other countries.

Acknowledgements

This work was supported by the Fundamental Research Funds for the Central University (N130402008, N110702001) and National Natural Science Foundation (51274065).

References

1 Li Xiu-ping, Zhang Chun-xia Zhou Ji-cheng, "Steel industry development challenges and technology application of energy saving and emission reduction," Power Demand Side Management, 3 (2011), 4-9.

2 Liu Jun, Xie Rui, and Wang La-fang, "International Comparison on the Added Value from Industry Value Chain in China's Steel Industry," The Theory and Practice of Finance and Economics, 170(2011), 115-119.

3 Cai Jiu-ju et al. "Material Flows and Energy Flows in Iron and Steel Factory and Their Influence on CO_2 Emissions," Research of Environmental Sciences, 21(2008), 790-794.

4 Zhang Qi, Cai Jiu-ju, Shen Feng-man. "Application of Process Integration on Systemic Energy Saving and Emission Reduction in Iron and Steel Works," China Metallurgy, 21(2011), 3-6.

5 Ran Rui, Weng Duan. "Current Situation of CO_2 Emission in Iron and Steel Producing and its Controlling Methods," Science and Technology Review, 24(2009), 53-56.

6 Li Shi-qi. "Status of energy saving and emission reduction and countermeasures for iron and steel industry in China," Research on Iron and Steel, 3(2011), 1-8.

7 Zhang Hui, Dong Liang, Li Huiquan, etc. "Analysis of low-carbon industrial symbiosis technology for carbon mitigation in a Chinese iron and steel industrial park," Energy Policy, 61(2013),1400-1411.

8 Larsson M, Dahl J. "Reduction of the specific energy use in an integrated steel plant----the effect of an optimization model," ISIJ Int, 43 (2003), 1664-1673.

9 Ryman C., Larsson M. "Reduction of CO_2 emissions from integrated steelmaking by optimized scrap strategies: Application of process integration models on the BF-BOF system," ISIJ Int, 46 (2006), 1752-1758.

10 Wang Wei-xin, "Energy consumption status and energy saving potential analysis for Chinese iron and steel industry China Steel," 4(2011), 19-22.
11 Wei Bin-xian. "Several issues of the standard energy calculation," Energy of China, 2(1992),24-26.

Energy Technology 2015: Carbon Dioxide Management and Other Technologies
Edited by: Animesh Jha, Cong Wang, Neale R. Neelameggham, Donna P. Guillen, Li Li, Cynthia K. Belt, Randolph Kirchain, Jeffrey S. Spangenberger, Frank Johnson, Andrew J. Gomes, Amit Pandey, and Peter Hosemann
TMS (The Minerals, Metals & Materials Society), 2015

INVESTIGATION ON THE HEAT RECOVERY FROM HIGH TEMPERATURE SLAGS

Yongqi Sun[1], Zuotai Zhang[1]
Department of Energy and Resources Engineering, College of Engineering, Peking University, Beijing 100871, P.R. China

Keywords: Heat recovery; High temperature slags; Al_2O_3; Multi-stage; Glass forming ability

Abstract

High temperature slags carry a substantial amount of high quality thermal energy (up to 1550°C), and represent the largest undeveloped energy sources in steel industry. However, the waste heat recovery rate of slags is less than 2% in China. So there is a great potential of waste heat recovery. Our study mainly investigated the cooling processes of high temperature slags using Single Hot Thermocouple Technique (SHTT) and Computational Fluid Dynamics (CFD) method. A series of slags with the constant CaO/SiO_2 ratio and different Al_2O_3 contents were designed and analyzed. The SHTT results indicated that the variation of Al_2O_3 content greatly changed the slag properties, including crystallization ability and glass forming ability, and this had a great effect on heat recovery of slags. We proposed a multi-stage control method of heat recovery from slags. The CFD results indicated that the temperature differences could be over 100°C between core and surface of slag droplets with the diameter of 3 mm. These large temperature differences could cause crystallization inside the slag droplets, which should be considered in the actual processes of heat recovery.

Introduction

In steel industry, although numerous advanced technologies have been implemented during past decades, the energy intensity is still quite high. It is generally believed that the heat recovery from high temperature slags represents the last potential to remarkably decrease the energy intensity and reduce the CO_2 emission in the steel industry [1-3]. The liquid slags with the temperature up to 1550°C, carry a high quality thermal energy required to be effectively recovered, especially in China. In 2012, the output of the crude steel is more than 700 million tons and accordingly around 200 million tons of blast furnace slags (BFS) is produced, the waste heat of which is equivalent to 16 million tons of standard coal. However, the recovery ratio of the high quality heat is less than 2% because of the unreasonable treatment of these slags [4].

Conventionally, the BFS can be disposed through three methods. First, the slags are poured into the slag yard and naturally cooled by the ambient air. This method is at the lowest cost, however, the high quality heat of the slags is wasted and solid slags obtained this way can only be used for low value aims, such as landfill. Thus, this method is nonscientific in views of the serious situation of energy saving and emission reduction in China. Second, the slags can be rapidly quenched by the cool water to obtain a glassy phase of the slags with high hydraulicity, which can be subsequently used for cement manufacture [5]. Nowadays, water quenching method has been increasingly utilized to treat the slags because of the high commercial value of the glassy slags obtained this way. However, the high quality heat is wasted through water quenching method in addition with a series other problems including water consumption, H_2S and SO_2 release and alkalis leaching [6]. To deal with the foregoing problems, dry granulation method has been proposed and extensively investigated [2-3], through which the liquid slags is first broke into the small droplets and particles and the heat in the slags is then exchanged with the medium (air) and therefore the waste heat is recovered. Dry granulation shows the research trend in this field, in spite of the fundamental challenges, i.e., low thermal conductivity of the slags. The low thermal conductivity make it difficult for the heat to be exchanged out; and, if the slags cannot be cooled rapidly enough, the crystal can precipitate inside the slags, which means that the solid slags cannot be used in the cement industry and the recycling of the material resources of the slags cannot be achieved. To meet these challenges and design a reasonable dry granulation method, the clear clarification of the property variation during the cooling path of the slags is fundamental. Therefore this study is motivated.

Furthermore, the combination of dry granulation method of slags and chemical reaction has been numerously studied recently, such as coal gasification [7-8] and methane reforming [9-10]. Through chemical method, the high temperature slags can act heat carrier and the waste heat untapped in the slags can be transformed into the high value chemical energy in the produced syngas (CO, H_2 and CH_4). The development of these advanced methods provides more possibilities of the waste heat recovery and makes the processes of waste heat recovery more commercially valuable. In addition, the degraded iron ore leads to an increase of Al_2O_3 content in the BFS [11], and therefore a series samples with varying Al_2O_3 content were designed in this study. The aim of this study is to explore the property variation during the cooling path of a slag particle, which may provide the fundamental clues of waste heat recovery from slags.

Experimental

Sample preparation

In this study, three samples with a CaO/SiO_2 ratio of 1.05 and Al_2O_3 contents of 10-25 wt. % were prepared using analytically regent (AR) oxides. These oxides were fully mixed and pre-melted in a molybdenum (Mo) crucible (Φ40 mm × 45 mm × H40 mm) under an argon atmosphere at 1550 °C for 1 h to homogenize the chemical compositions. Then the liquid slags were rapidly poured into the cold water to obtain glassy phases. Subsequently, the glassy slags were dried at 120 °C for 12 h, crushed and ground into 300 meshes for compositional analysis and crystallization experiments. The chemical compositions were measured by the X-Ray fluoroscopy (XRF) technique, as listed in **Table 1**. X-ray diffraction (XRD) tests were conducted to confirm the glassy phases of slags.

Table 1. Chemical compositions of samples (wt. %, XRF)

Samples	Basicity（B）	CaO	SiO_2	MgO	Al_2O_3
A1	1.05	40.7%	38.9%	9.2%	11.3%
A2	1.02	35.8%	35.1%	9.1%	20.0%
A3	1.02	34.3%	33.5%	9.0%	23.1%

Crystallization experiments

The crystallization behaviors of the slags were measured through a series of isothermal and non-isothermal experiments using a Single Hot Thermocouple Technique (SHTT) [12], as shown in **Figure 1**. During the isothermal experiments, around 10 mg sample was heated to 1500 °C, fully melted and held for 120 s to homogenize the chemical composition. The liquid slags were rapidly quenched to a set temperature at 50 °C/s and held at this point for crystalline evolution. It takes some time for the crystal to occur in the melts, defined as incubation time. The relationship between incubation time and holding temperature is generally presented in the Time Temperature Transformation (TTT) curves, which was obtained in this study. During the non-isothermal experiments, the samples were melted in a similar way, but the liquid slags were cooled at various cooling rates. The time and temperature where crystal is observed to form in the melts can be recorded, and the relationship of them is generally shown in the continuous cooling transformation (CCT) curves.

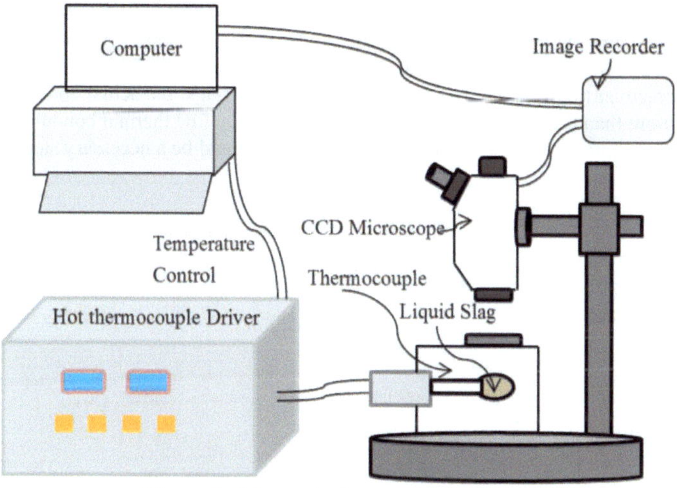

Figure 1. Schematic diagram of the SHTT

Simulation methods

We established a 3-D model to simulate the heat transfer between a slag droplet and the ambient air using a computational fluid dynamics (CFD) package, Fluent [13]. The simulation domain was composed of a slag droplet with r_0 in radius and the ambient air ($10r_0$). The boundary condition of the simulation domain was constant pressure (atmospheric pressure p_0) and the initial temperature of the slag droplet and the ambient air were assumed to be 1500 °C and 25 °C, respectively.

Results and discussions

CCT curves and control of cooling path of slags

For a reasonable method of heat recovery, the cooling path of the slag droplets or particles should be designed in advance. A CCT curve presents the crystallization temperature at different cooling rates, which provides important clues to control the cooling path of slags. The CCT curves for different samples are shown in **Figure 2**, from which several characteristics could be clarified. First, with the increase of cooling rate, the crystallization temperature became lower for each sample. As generally accepted, the viscosity of the liquid slags remarkably increased with the decreasing temperature [14]; as the cooling rate increased, the degree of viscosity increase became larger, which in turn indicated that a bigger undercooling rate was required for crystalline formation; i.e., a lower crystallization temperature occurred. Second, CCT curves moved to left with increasing Al_2O_3 content. This trend indicated that the crystallization ability was enhanced by the increasing Al_2O_3 content; in other words, it was harder to obtain glassy slags for cement manufacture, which could make up a new challenge for heat recovery from slags.

To further effectively control the cooling path of slags, the crystal formation inside the slags should be avoided; i.e., the cooling rate should be larger enough. The smallest cooling rate that crystal could be avoided was defined as the critical cooling rate, which was also the smallest cooling rate required to get glassy slags. In this study, the critical cooling rates for these samples were measured, which were 22, 15 and 26 °C/s for sample A1, A2 and A3, respectively. To effectively recover the heat and avoid crystal formation in slags, the actual cooling rate of slags should be more than the critical cooling rate. Considering the low thermal conductivity, the slags should be broke into small droplets; i.e., dry granulation should be a necessary step.

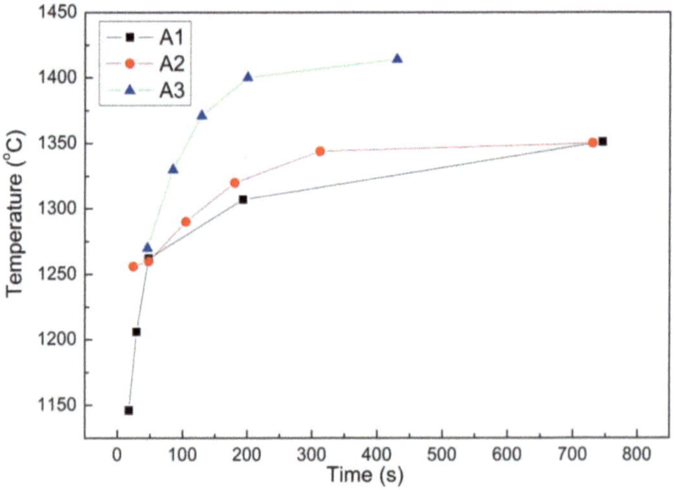

Figure 2. CCT curves of the samples

<u>TTT curves and multi-stage control of heat recovery</u>

TTT curves present the relationship between holding temperature and incubation time during isothermal experiments, as shown in **Figure 3**. There were two characteristics could be deduced from the TTT curves. First, these TTT curves showed a shape of double "C" from high temperature to low temperature, which indicated that two different crystallization events occurred. Actually, at high crystallization temperatures, pure crystal could form in the slags, namely primary crystalline phase; and several crystalline phases could coexist in the slags with the decreasing temperature. Second, an apparent trend could be observed that the TTT curves moved to the left side in high temperature zones with increasing Al_2O_3 content, which was in agreement with the variation trend of TTT curves and also suggested that the crystallization ability of the BFS was improved by the increasing Al_2O_3 content. Thus, perhaps a different process should be performed for the high Al_2O_3 content slags, such as a higher flow rate of exchange medium, a smaller particle size, etc.

From **Figure 3**, several temperature points related the crystallization behavior of the BFS could be identified and based these temperature points, a multi-stage control of heat recovery could be proposed. The highest temperature where crystal could form was defined as the break temperature (T_b); the lowest temperature where crystal could form was defined as the solid temperature (T_s). The temperature region higher than T_b is a region where the slags were in liquid phase and there was enough time for heat exchange. The temperature region between T_b and T_s is a region where crystal could form. In views of material recycling, the crystallization behavior should be avoided in this region and therefore the cooling rate in this region must be larger than the critical cooling rate. The temperature region below T_s is a region the slags were in glassy phase obtained by a rapid cooling and the heat in this region can be fully exchanged to other mediums or materials, such as coal and air. Based on the aforementioned analysis, a multi-stage of the heat recovery from slags was effective and scientific, the schematic of which was shown in **Figure 4**.

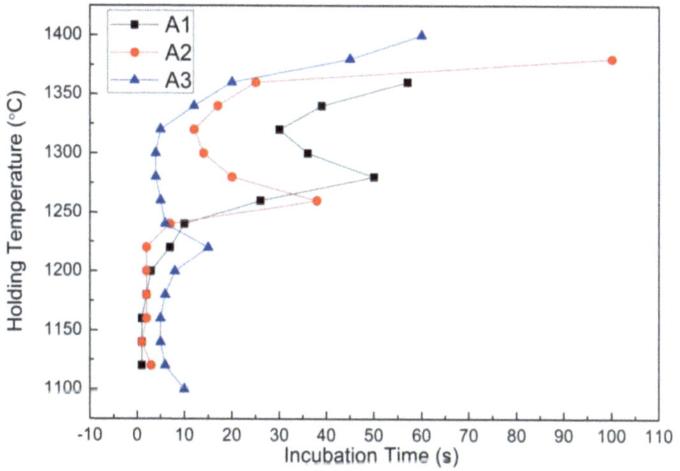

Figure 3. TTT curves of the samples

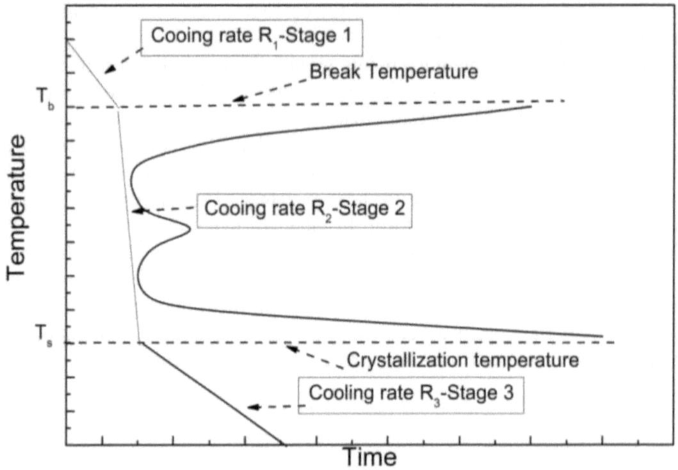

Figure 4. Schematic of multi-stage control method

Simulation analysis

By CFD simulation, the temperature evolution versus time on core and surface of the slag droplet can be obtained. As an example, **Figure 5** depicts the variation of core and surface temperature for sample A2. As can be noted, the temperature became lower with time with a smaller cooling rate. With the decrease of the temperature, the intensity of radiation heat transfer became smaller, thus, the cooling rate reduced. These samples showed a similar variation trend of temperature evolution. Another interesting phenomenon observed was the temperature difference between core and surface. Initially, the surface temperature decreased quite faster than the core temperature, which caused an increasing temperature difference. As the temperature became lower, the cooling rate on the core and surface became more similar; therefore the temperature difference became lower again. This variation trend leaded to a maximum value of temperature difference during the cooling process. The relative low cooling rate on the core of slag droplet could make the crystal form inside the slags, which should be taken into account for an actual heat recovery process.

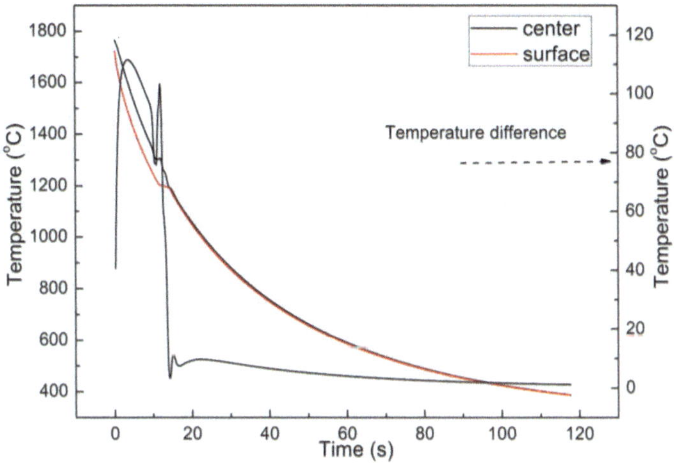

Figure 5. Cooling process for sample A2

Conclusions

Our study provided fundamental clues for heat recovery from high temperature BFS using SHTT experiments and CFD simulations. The crystallization behaviors of the slags were investigated and it was found that an increasing Al_2O_3 content enhanced the crystalline propitiation in the melts. Based on TTT curve, we proposed a multi-stage control of the heat recovery. The simulation results indicated the big temperature difference between the core and the surface of the slags, which could cause the crystal formation in the slags and should be considered in an actual heat recovery process.

References

1. G. Bisio, "Energy recovery from molten slag and exploitation of the recovered energy," *Energy*, 22 (1997), 501–509.
2. M. Barati, S. Esfahani, and T.A. Utigard, "Energy recovery from high temperature slags," *Energy*, 36 (2011), 5440–5449.
3. H. Zhang, et al., "A review of waste heat recovery technologies towards molten slag in steel industry," *Applied Energy*, 112 (2013), 956–966.
4. J.J. Cai, et al., "Recovery of residual heat integrated steelworks," *Iron Steel*, 42 (2007), 1–6.
5. A.A. Usón, et al., "Uses of alternative fuels and raw materials in the cement industry as sustainable waste management options," *Renewable and Sustainable Energy Reviews*, 23 (2013), 242-260.
6. T. Mizuochi, et al., "Feasibility of rotary cup atomizer for slag granulation," *ISIJ International*, 41 (2001), 1423-1428.
7. P. Li, et al., "Kinetics of CO_2/Coal gasification in molten blast furnace slag," *Industrial & Engineering Chemistry Research*, 51 (2012), 15872-15883.
8. P. Li, et al., "Adaptability of coal gasification in molten blast furnace slag on coal samples and granularities," *Energy & Fuels*, 25 (2011), 5678-5682.

9. N. Maruoka, et al., "Feasibility study for recovering waste heat in the steelmaking industry using a chemical recuperator," *ISIJ International*, 44 (2004), 257-262.
10. H. Purwanto, and T. Akiyama, "Hydrogen production from biogas using hot slag," *International Journal of Hydrogen Energy*, 31 (2006), 491-495.
11. K. Yajima, H. Matsuura, and F. Tsukihashi, "Effect of simultaneous addition of Al_2O_3 and MgO on the liquidus of the CaO-SiO_2-FeO_x system with various oxygen partial pressures at 1573 K," *ISIJ International*, 50 (2010), 191-194.
12. Y. Kashiwaya, et al., "Development of double and single hot thermocouple technique for in site observation and measurement of mold slag crystallization," *ISIJ International*, 38 (1998), 357-365.
13. ANSYS FLUENT 13.0 theory guide. Pittsburgh, U.S.A: ANSYS, Inc; 2010.
14. S. Sridhar, et al., "Break temperatures of mould fluxes and their relevance to continuous casting," *Ironmaking & steelmaking*, 27 (2000), 238-242.

Energy Technology 2015: Carbon Dioxide Management and Other Technologies
Edited by: Animesh Jha, Cong Wang, Neale R. Neelameggham, Donna P. Guillen, Li Li, Cynthia K. Belt, Randolph Kirchain, Jeffrey S. Spangenberger, Frank Johnson, Andrew J. Gomes, Amit Pandey, and Peter Hosemann
TMS (The Minerals, Metals & Materials Society), 2015

Optimal Distribution of Byproduct Gases in Iron and Steel Industry Based on Mixed Integer Linear Programming (MILP)

Xiancong Zhao[1,2], Hao Bai[1,2,*], Qi Shi[1,2], Jiehai Han[3] , Hongxu Li[2]

[1]State Key Laboratory of Advanced Metallurgy, University of Science and Technology Beijing, 30# Xueyuan Road, Beijing 100083, China
[2]School of Metallurgical and Ecological Engineering, University of Science and Technology Beijing, 30# Xueyuan Road, Beijing 100083, China
[3]Handan Iron and Steel Company Limited, 232# Fuxin Road, Handan 056003, China
Corresponding author: baihao@metall.ustb.edu.cn

Keywords: iron and steel industry; byproduct gases; MILP; penalty value; optimal distribution

Abstract

In China, in steel works, byproduct gases account for almost 40% of total energy consumption. With the rise of energy price, increasing attention is paid on efficient and optimized utilization of byproduct gases recently. In this paper, a MILP model for optimization of byproduct gas is proposed. The objective function of this model is minimization of operation cost. Compared with previous optimization model, this proposed model chose short time period and considered the influence of Boiler Penalty Value (BPV) and Gasholder Penalty Value (GPV) on optimization results. The results based on real production data indicate GPV and BPV are sensitive to optimization results. Thus it is necessary to evaluate GPV and BPV reasonably. Compared with manual operation results, the optimal calculation results can reduce the fluctuation of gasholder and boiler so as to make the operation of byproduct gas system safe and stable.

Introduction

The iron and steel industry is one of the most energy-intensive industries in China. Byproduct gases are important secondary energy resources which account for 40% of total energy consumption in steel works [1]. With the rise of energy price, increasing attention is paid to efficient and optimized utilization of byproduct gases.

In iron and steel making process, there exist imbalance between the production and consumption of byproduct gases, and gasholders serve as buffer units to solve the temporary imbalance. Since the capacity of a gasholder is limited, temporary excess or shortage of byproduct gas occurs. Such imbalance will cause an increase in operation cost, pollution to the environment and even a threat to proper production [2]. Thus it is of great importance to optimize the scheduling and distribution of byproduct gas system to reduce byproduct gas emission or shortage.

Many efforts have been conducted on the research of scheduling and production planning problems in iron and steel making process [3-5]. However, little was focused on the optimal distribution of byproduct gases. Akimoto et al. [6] first brought MILP model to iron and steel industry, and he proposed a MILP model that implemented gasholder safety level control and optimal distribution on different users. Sinha et al. [7] implemented MILP model and hierarchical optimization for profit maximization in Tata Steel. Significant amount of benefit (US $ 73 million) was reported in the first year of implementation (1986-1987). Zhang X.P. et al. [2] developed a hybrid parameter optimization algorithm to optimize the gasholder prediction model for high prediction accuracy. Results demonstrate that the proposed approach can precisely predict and adjust gasholders. Kim et al. [8] formulated a mathematical model to minimize total operation cost over multi-period by considering different fuel load change and running cost of boiler system. However, all the previous researchers hypothesized that Boiler Penalty Value (BPV) and Gasholder Penalty Value (GPV) in objective function are all decided beforehand, i.e. they ignored the influence of BPV and GPV on optimization results. Thus whether the results of previous researches reasonable or not are unknown.

In this paper, a MILP model is proposed for solving the byproduct gas distribution problem. Compared with the previous approach, this model considered the influence of Boiler Penalty Value (BPV) and Gasholder Penalty Value (GPV) on optimization results. The aim is to reduce operation cost in the byproduct gas system for multi-period operation through optimizing byproduct gas distribution in an iron and steel plant. Furthermore, a case study was performed to verify the suitability of this proposed approach.

Problem analysis

Fig. 1 shows the typical byproduct gases flow in an iron and steel plant. The byproduct gas system consist of byproduct gas production and storage system, iron and steel making system and power & thermal cogeneration system. In the byproduct gas production and storage system, byproduct gases are produced in Blast Furnaces (BF), Coke Ovens (CO) or Linze-Donawitz Converters (LDC), and stored in gasholders if necessary. Since the consumption of byproduct gases in iron and steel making system is according to the steel production schedule, there is little flexibility to change it. In the power & thermal cogeneration system, byproduct gases are transported to boilers in power plant to generate electricity and steam.

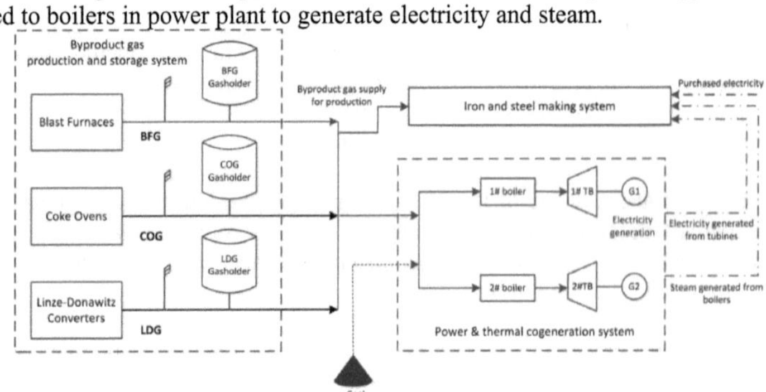

Figure 1 Schematic view of a typical byproduct gas system

Generally, the consumption rate of byproduct gas must equal the production rate for the sake of guaranteeing the safety and the stability of iron and steel making process. Since the capacity of byproduct gasholder is rather limited, byproduct gas imbalance frequently occurs. More byproduct gas supply than demand makes the byproduct gasholder exceed its capacity, and thus lead to gas emissions, which means an economic loss and pollution to the environment. On the other side, the shortage of byproduct gas takes mechanical trouble to byproduct gasholder and affects production process. In this paper, one gasholder is divided into four sections, namely, emission section, high section, normal section and low section (Figure 2). As a result, the level of gasholder for byproduct gases should be maintained around the normal section as long as possible (neither too high nor too low).

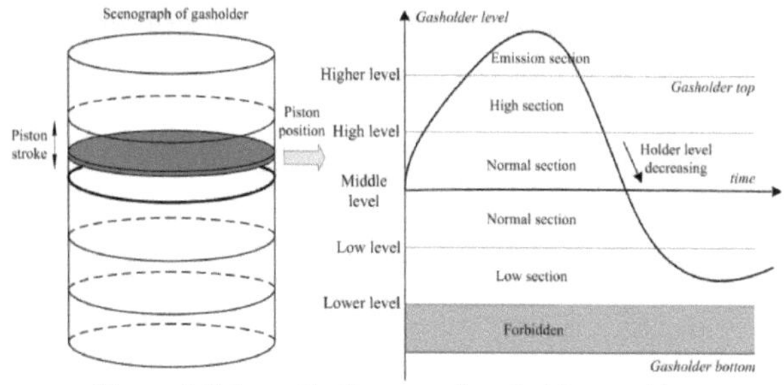

Figure 2 Schematic diagram of gasholder partition

In order to minimize the operation cost, the holder level of each byproduct gas should be kept as constant as possible within the normal section in order to avoid unfavorable byproduct gas emission and shortage. In addition, the following conditions are needed for optimal distribution of byproduct gases:

(1) Minimization of boiler switching in cogeneration system, extending the service life of boiler.
(2) Minimization of coal and water consumption in cogeneration system.
(3) Maximization of electricity production in cogeneration system.

Mathematical model

Objective function

The objective function (Eq. (1)) is to minimize operation costs during the time period P. The first term in the objective function is purchased coal consumption cost for all the boilers. The second term is penalty cost for byproduct emissions. The third, fourth and fifth term is the penalty cost for byproduct gas deviation from middle level. The sixth and seventh term represents the boiler switching penalty respectively. The eighth term is water consumption cost for all boilers. The ninth term is electricity production benefit in power plant.

$$Y = min\{C^{Coal}\sum_{t=1}^{P}\sum_{i=1}^{B} f_{i,t}^{Coal} + \sum_{t=1}^{P}\sum_{G} W_{HH} S_{HH,t}^{j} + \sum_{t=1}^{P}\sum_{G} W_{H} S_{H,t}^{j} + \sum_{t=1}^{P}\sum_{G} W_{L} S_{L,t}^{j} + \sum_{t=1}^{P}\sum_{G} W_{d} S_{d,t}^{j} + \sum_{t=1}^{P}\sum_{i=1}^{B}\sum_{G} W^{1s} \Delta n_{i,t}^{j}$$
$$+ \sum_{t=1}^{P}\sum_{i=1}^{B}\sum_{G} [W^{2s}(ibn_{2,i,t}^{j+} + ibn_{2,i,t}^{j-}) + W^{3s}(ibn_{3,i,t}^{j+} + ibn_{3,i,t}^{j-})] + C^{Water}\sum_{t=1}^{P}\sum_{i=1}^{B} f_{i,t}^{Water} - C^{Elec}\sum_{i=1}^{P}(E_t - E_{Dem}) \quad (1)$$

Material and energy balance for byproduct gas system

The detailed material and energy balance equations for byproduct gas system can be seen from Kong's work [9].

Other equations

Eq. (2)-(4) is equations for burners switching in boilers. $\Delta n^{j}_{i,t}$ stands for the number of burner change at boiler i. $sw^{j+}_{i,t}$ stands for the number of burner turn on at time t, which equals to the sum of one burner, two and three burners turn on during the same period. $sw^{j-}_{i,t}$ stands for the number of burner turn off at time t, which equals to the sum of one burner, two and three burners turn off during the same period. $ibn^{j+}_{1,i,t}$, $ibn^{j-}_{1,i,t}$, $ibn^{j+}_{2,i,t}$, $ibn^{j-}_{2,i,t}$, $ibn^{j+}_{3,i,t}$ and $ibn^{j-}_{3,i,t}$ are binary variables. When they are equal to 1, means burner switching (switch on or off) happen. sw_1 represents the sum of one burn switching on or off during six periods (Eq. (5)). sw_2 and sw_3

represents the sum of two and three burners simultaneously switching on or off during six periods (Eq. (6)-(7)).

$$\Delta n_{i,t}^{j} = n_{i,t}^{j} - n_{i,t-1}^{j} = sw_{i,t}^{j+} + sw_{i,t}^{j-} \tag{2}$$

$$sw_{i,t}^{j+} = ibn_{1,i,t}^{j+} + ibn_{2,i,t}^{j+} + ibn_{3,i,t}^{j+} \tag{3}$$

$$sw_{i,t}^{j-} = ibn_{1,i,t}^{j-} + ibn_{2,i,t}^{j-} + ibn_{3,i,t}^{j-} \tag{4}$$

$$sw_{1} = \sum_{t=1}^{P}\sum_{i=1}^{B}(ibn_{1,i,t}^{j+} + ibn_{1,i,t}^{j-}) \tag{5}$$

$$sw_{2} = \sum_{t=1}^{P}\sum_{i=1}^{B}(ibn_{2,i,t}^{j+} + ibn_{2,i,t}^{j-}) \tag{6}$$

$$sw_{3} = \sum_{t=1}^{P}\sum_{i=1}^{B}(ibn_{3,i,t}^{j+} + ibn_{3,i,t}^{j-}) \tag{7}$$

Case study

To verify the above method, a case study was performed by using this model in a steel plant in China. This plant consists of four blast furnaces (BF), six coke ovens (CO), four Linze-Donawitz converters (LDC), one BFG gasholder, one COG gasholder and one LDG gasholder. There are two boilers and two turbines in cogeneration system, namely boiler 1#, boiler2#, turbine 1# and turbine 2#. The maximization consumption for BFG, COG and LDG in each boiler was set to be 1.8×10^5, 1×10^4 and $3\times10^4 m^3/h$. The efficiency of boiler 1#, boiler 2#, turbine 1# and turbine 2# was set to be 0.85, 0.8, 0.4 and 0.4. Lower Heating Value (LHV) of BFG, COG and LDG was kept as 3140, 16590 and 7000 kJ/Nm^3. Unit cost for coal, electricity and fresh water was set as 100 US dollar/ton, 0.083 US dollar/kWh and 0.5 US dollar/ton. The planning period is composed of six time periods, and each is 6 min in length. The number of time steps is randomly chosen from real production data. The capacity bound of gasholders is shown in Table 1.

Table 1 Capacity bound of gasholders / m^3

holders	Lower level	Low level	Middle level	High level	Higher level
BFG holder	30000	75000	150000	225000	300000
COG holder	12000	30000	60000	90000	120000
LDG holder	8000	20000	40000	60000	80000

In the previous researches [6-9], Boiler Penalty Value (BPV) and Gasholder Penalty Value (GPV) in objective function are all decided beforehand, that is, GPV and BPV were set as fixed parameter. Whether the values of GPV and BPV were reasonable or not was not discussed. To investigate the influence of BPV and GPV, two different modes are studied in this paper: (1) GPV remains fixed while BPV changes, referred as mode 1. (2) BPV remains fixed while GPV changes, referred as mode 2. The aim is to select reasonable value for GPV and BPV in gas system optimization.

The effect of BPV on gasholder level

The combination of penalty values for mode 1 is shown in Table 2. The first four items represent GPV, which are penalty factor for byproduct gas emission W_{HH}, penalty factor for high level storage W_H, penalty factor for normal level deviation W_d and penalty factor for low level storage W_L. The last three items represent BPV, which are penalty factor for one burner switching W^{1s}, two burners simultaneously switching W^{2s} and three burners simultaneously switching W^{3s}.

Table 2 The combination of penalty values for mode 1

Penalty values		A1	A2	A3	A4	A5	A6
GPV	W_{HH}	100	100	100	100	100	100
	W_H	2	2	2	2	2	2
	W_d	0.2	0.2	0.2	0.2	0.2	0.2
	W_L	1	1	1	1	1	1
BPV	W^{1s}	1	10	50	100	500	1000
	W^{2s}	2	20	100	200	1000	2000
	W^{3s}	5	50	250	500	2500	5000

To start up, we must give an initial value to GPV to ensure the calculation program running well. The initial value for GPV is set as W_{HH}=100, W_H=2, W_d=0.2, W_L=1, according to Kong's work [9]. In the problem analysis part we stated that the gasholder level should be kept close to the middle level as long as possible to avoid unfavorable byproduct gas emission and shortage. Thus, the effectiveness of BPV on gasholder level is determined by the closeness of calculation curve to middle level of gasholder. Figure 3(a) shows BFG gasholder level before and after optimization in mode 1. When penalty value A1 is taken, the calculation curve is closest to middle level. With the increase of BPV, calculation curves gradually deviate from middle level of BFG holder. Figure 3(b) shows COG gasholder level before and after optimization in mode 1. The gasholder level by manual operation is extremely low in six periods, nearly close to lower level (1.2×10^4 m^3). However, the middle level of COG gasholder is 6.0×10^4 m^3, much higher than actual holder level. Thus, the calculation curve should be kept away from lower level as far as possible to avoid supply shortage of COG. According to calculation result, the optimal solution emerged when BPV penalty value A2 or A3 is taken. Figure 3(c) shows LDG gasholder level before and after optimization in mode 1. According to manual operation result, LDG holder suffered from great fluctuation in six periods. When penalty value A1~A4 is taken, calculation results are rather better.

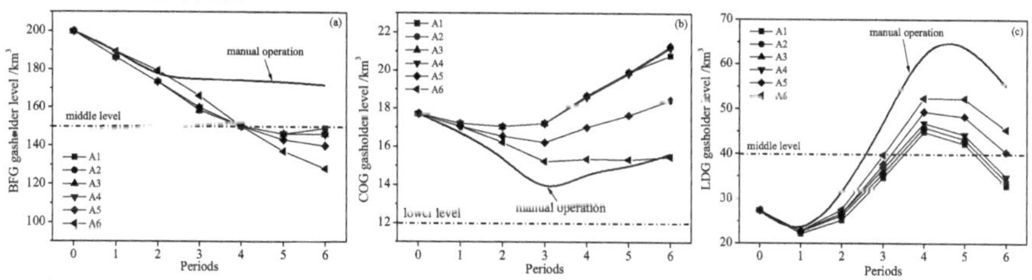

Figure 3 Gasholder level before and after optimization during each time period in mode 1. (a)BFG gasholder; (b) COG gasholder; (c) LDG gasholder

In all, the characteristic of mode 1 is that reducing BPV contributes to gasholder stability. Considering the optimization results of three different gasholders, the optimal penalty value is A2 or A3.

<u>The effect of BPV on boiler burner switching times</u>

In order to study the effect of penalty value on boiler stability, the Total Switching Times (TST), is defined in Eq. (8). TST refers to the sum of boiler burner switching (including one burner, two and three burners switching times) in six periods. The definition of sw_1, sw_2 and sw_3 is presented in Eq. (5)-(7). As a result, the smaller the TST, the less frequent the boiler switches.

$$TST = \sum_{t=1}^{P}\sum_{i=1}^{B}\Delta n_{i,t}^{j} = 1\times sw_1 + 2\times sw_2 + 3\times sw_3 \tag{8}$$

Table 3 Boiler switching times before and after optimization in mode 1

item	A1	A2	A3	A4	A5	A6	Manual
sw_1	8	6	6	6	3	1	3
sw_2	2	6	5	10	13	4	10
sw_3	16	10	10	6	0	0	11
TST	60	48	46	44	29	9	56

Boilers switching times before and after optimization in mode 1 are shown in table 3. With the rise of BPV, TST gradually decrease from 60 to 9, sw_3 decrease from 16 to 0, which means increasing BPV contributes to the less switching times of boiler burner, that is, high GPV is in favor of stability of boiler. As mentioned above, the optimal penalty value is A2 or A3, since A3 suffered less boiler fluctuation than A2, A3 is chosen as the optimal BPV for the following discussion.

The effect of GPV on gasholder level

The combination of penalty values for mode 2 is shown in Table 4, that is, BPV remains fixed while GPV changes. Penalty value B3 is the same as A3.

Table 4 The combination of penalty values for mode 2

Penalty value		B1	B2	B3(A3)	B4	B5	B6
GPV	W_{HH}	37.5	50	100	250	500	1000
	W_H	0.75	1	2	5	10	20
	W_d	0.075	0.1	0.2	0.5	1	2
	W_L	0.375	0.5	1	2.5	5	10
BPV	W^{1s}	50	50	50	50	50	50
	W^{2s}	100	100	100	100	100	100
	W^{3s}	250	250	250	250	250	250

Figure 4(a) shows BFG gasholder level before and after optimization in mode 2. With the increase of GPV, calculation curves are gradually approaching the middle level of BFG holder. When penalty value B3~B6 is taken, the calculation curve becomes better. Figure 4(b) shows COG gasholder level before and after optimization in mode 2. When penalty value B1~B5 is taken, the calculation curve is much far away from lower level. Figure 4(c) shows LDG gasholder level before and after optimization in mode 2. When penalty value B5~B6 is taken, the calculation curve is relatively better.

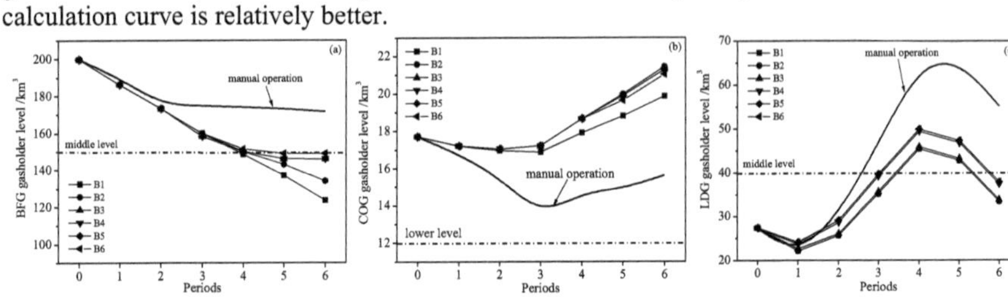

Figure 4 Gasholder level before and after optimization during each time period in mode 2. (a) BFG gasholder; (b) COG gasholder; (c) LDG gasholder

In all, the characteristic of mode 2 is that increasing GPV contributes to gasholder stability. Considering the optimization results of three different gasholders, the optimal penalty value is B5 or B6.

The effect of GPV on boiler burner switching times

Table 5 Boiler switching times before and after optimization in mode 2

item	B1	B2	B3	B4	B5	B6	Manual
sw_1	4	5	6	6	8	9	3
sw_2	12	9	5	5	3	6	10
sw_3	5	8	10	10	12	13	11
TST	43	47	46	46	50	60	56

Boilers switching times before and after optimization in mode 2 are shown in table 5. With the rise of GPV, TST gradually increase from 43 to 60, sw_3 increase from 5 to 13, which means decreasing GPV contributes to the less switching of boiler burner, that is, low GPV is in favor of stability of boiler.

The selection of reasonable GPV and BPV

According to the analysis result above, the values of BPV and GPV affect the stability of gasholder and boiler heavily. This influence can be studied by defining Standard Deviation Volume (SDV) and TST. SDV_j refers to the average deviation of holder level to middle level of j gasholder in six periods (Eq. (9)). $V_{j,t}$ stands for the holder level of j gasholder at period i. $V_{j,center}$ referred to the middle level of j gasholder. In order to evaluate the overall impact of penalty value on three gasholders, SSDV (Sum of Standard Deviation Volume) is defined (Eq. (10)). SSDV stands for the sum of SDV_j for three different gasholders. In a word, the overall stability of three gasholders is depended on the value of SSDV.

$$SDV_j = \sqrt{\frac{1}{n}\sum_{1=1}^{n}(V_{j,t} - V_{j,center})^2} \tag{9}$$

$$SSDV = \sum_{j=1}^{3} SDV_j \tag{10}$$

The SDV, TST and SSDV before and after optimization under penalty A1~A5 are shown in Figure 5. We can learn from Figure 5(a) that, the SDV is relatively low when penalty value A1~A4 is taken. According to Figure 5(b), the SSDV of penalty value A2 and A3 is the smallest, which means the overall stability of three gasholders is the best among them. Considering the TST of A2 is a little higher than A3 (46 to 44), penalty value A3 is chosen as the best BPV, and used for the selection of GPV.

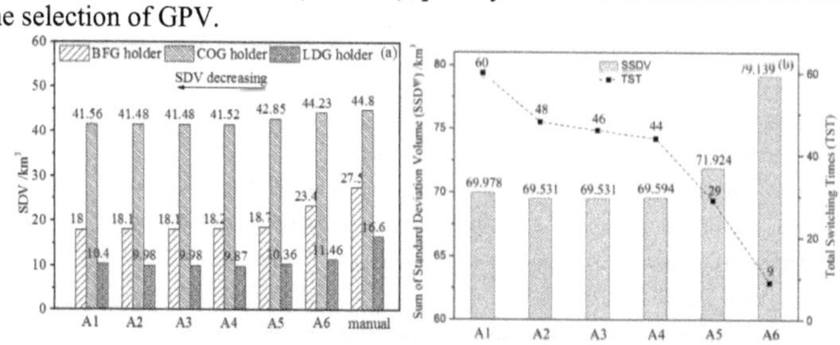

Figure 5 SDV, TST and SSDV before and after optimization under penalty A1~A5: (a) SDV and TST (b) SSDV.

The SDV, TST and SSDV before and after optimization under penalty B1~B5 is shown in Figure 6. We can learn from Figure 6(a) that, the SDV is relatively low when penalty value B5~B6 is taken. According to Figure 6(b), the SSDV of penalty value B5 is the smallest, which means the overall stability of three gasholders is the best among them. Considering the TST of B5 is a 50 (less than manual operation value 56), penalty value B5 can be selected as appropriate GPV. Thus, the reasonable combination of penalty value in this paper is B5. Compared with manual operation results, the SDV of BFG holder, COG holder and LDG holder of optimal calculation result reduced by 34.25%, 7.41% and 43.61%.

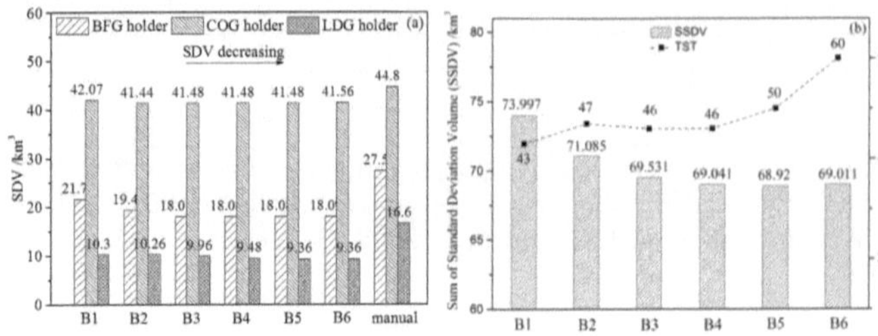

Figure 6 SDV, TST and SSDV before and after optimization under penalty B1~B5: (a) SDV and TST (b) SSDV.

Conclusion

Penalty value will affect optimization results. GPV should not be too small, while BPV should not be too big. When GPV increased or BPV decreased, calculation results reached optimum gradually. Thus, it is necessary to evaluate penalty values reasonably. Results of case study shows, the best combination of penalty value is B5. The Standard Deviation Volume of BFG, COG, LDG gasholder is decreased by 29.77%, 2.12% and 37.28% after optimization. The stability of gasholder is improved.

Acknowledgment

The authors gratefully acknowledge the financial support from the research funds for State Key Laboratory of Advanced Metallurgy of China [41603006].

REFERENCES

[1] Xiaoping Zhang, Jun Zhao, Wei Wang, et al. An Optimal Method for Prediction and Adjustment on Byproduct Gas Holder in Steel Industry [J]. *Expert Systems with Applications*, 2011, 38(4): 4588-4599.

[2] Zhang JD, Rong G. An MILP model for multi-period optimization of fuel gas system scheduling in refinery and its marginal value analysis. *Chem. Eng Res Des* 2008; 86(2):141–151.

[3] Larsson M, Wang C, Dahl J. Development of a method for analyzing energy, environmental and economic efficiency for an integrated steel plant. *Appl. Therm. Eng* 2006; 26(13):1353–1361.

[4] Glankwamdee W, Linderoth J, Shen J, et al. Combining optimization and simulation for strategic and operational industrial gas production and distribution. *Computers & Chemical Engineering*, 2008, 32(11): 2536-2546.

[5] Fukuda, K, Makino, H, Suzuki, et al. Optimal Energy Distribution Control at the Steel Works // *IFAC Simulation of Control Systems*. Vienna, 1986: 337

[6] Akimoto K, Sannomiya N, Nishikawa Y, et al. An optimal gas supply for a power plant using a mixed integer programming model. *Automatica*, 1991, 27(3) : 513

[7] Sinha G P, Chandrasekaran B S, Mitter N. Strategic and operational management with optimization at Tata steel. *Interfaces*, 1995: 25(1): 6

[8] Kim J H, Yi H S, Han C. A novel MILP model for plant-wide multi-period optimization of byproduct gas supply system in the iron and steel making process. *Chem. Eng Res and Des*, 2003, 81(8): 1017

[9] Kong H N, Qi E S, Li H, et al. An MILP model for optimization of byproduct gases in the integrated iron and steel plant. *Appl. Energy*, 2010, 87(7): 2156

Energy Technology 2015
Carbon Dioxide Management and Other Technologies

ENERGY TECHNOLOGIES AND CARBON DIOXIDE MANAGEMENT SYMPOSIUM 2015

Metal Processing / Molten Salt / Electrochemistry

Session Chairs:

Animesh Jha
Donna P. Guillen
Li Li
Shulan Wang

Energy Technology 2015: Carbon Dioxide Management and Other Technologies
Edited by: Animesh Jha, Cong Wang, Neale R. Neelameggham, Donna P. Guillen, Li Li, Cynthia K. Belt, Randolph Kirchain, Jeffrey S. Spangenberger, Frank Johnson, Andrew J. Gomes, Amit Pandey, and Peter Hosemann
TMS (The Minerals, Metals & Materials Society), 2015

ENERGY EFFICIENT SEPARATION OF MAGNETIC ALLOY FROM THE CARBOTHERMIC REDUCTION OF NKANA Cu-Co CONCENTRATES

Yotamu R.S. Hara and Animesh Jha

The Institute for Materials Research, Houldsworth Building
Clarendon Road, Leeds University, Leeds LS2 9JT (UK), England

Keywords: Carbothermic reduction, energy, concentrates, Nkana, mineral sulphides and separation

Abstract

The Nkana Cu-Co mineral sulphide concentrates containing 40 wt.% SiO_2 were carbothermically reduced in the presence of CaO by following the equilibrium: MS + CaO + C = M + CaS + CO(g), where M represents the metallic Cu, Co and Fe, over a temperature range of 1073 K – 1323 K. We investigated the magnetic separation of alloy phases into copper-cobalt rich and iron-cobalt rich fractions. The proposed approach appears more energy efficient than melting a mixture which yields a ternary alloy composed of Cu, Co and Fe, and a non-magnetic fraction (CaS, excess CaO and SiO_2). The extent of magnetic separation was affected by the temperature and mole ratios of MS/CaO and C/CaO, used for the reduction of mineral concentrate. The reduced and magnetically separated samples were characterised by XRD, SEM-EDX and XRF techniques. The alloy phase contained ~92 wt% of metallic constituents, with remaining 8 wt.% as mixture of CaS and SiO_2. The utilisation of this gangue mixture in the process is briefly discussed.

1.0 Introduction

The Nkana Cu-Co mineral sulphide concentrates contains Cu, Co and Fe in the form of chalcopyrite ($CuFeS_2$), bornite (Cu_5FeS_4), pyrite (FeS_2), cobaltian pyrite ($(FeCo)S_2$) and carrolite ($CuCo_2S_4$) mineral phases. The major gangue minerals are silica (SiO_2) and dolomite ($CaMg(CO_2)_2$). The conventional treatments of these concentrates include smelting and roast - leach - electro-winning processes.

In conventional smelting, the concentrates are smelted to produce blister copper and $FeSiO_3$ rich slag phases [1]. The main disadvantage of the conventional smelting process is that cobalt and part of copper are oxidized and hence lost into the slag phase [1]. The slag containing Cu and Co is smelted with carbon (reductive smelting) at about 1823 K to obtain a metallic iron rich Cu-Co-Fe liquid and a slag phase [2, 3]. The temperature of liquid Cu-Fe-Co alloy is raised to 1873 K for atomization by subjecting the liquid stream in a high pressure of water jet which yields fine particles of alloy. It is worth noting that reductive smelting of the slag is an energy intensive process, requiring 0.8 kwh per kg of feed material [4]. Moreover, the iron-rich Cu-Co alloy produced from the reduction smelting requires hydrometallurgical treatment such as high pressure leaching, purification of the leach solution and electro-winning [3].

In the roast - leach - electro-winning process, the concentrates are roasted below 923 K to minimize the formation of copper and cobalt ferrites ($CuFeO_2$ and $CoFe_2O_4$). The roasting is carefully controlled for making Cu/Co sulphates and oxides and, Fe_2O_3 [1]. The roast calcine is leached in sulphuric acid and, Cu and Co elements recovered from the leach solution via electro-winning process [1]. The main disadvantages are:

(i) Production of the dilute SO_2 off-gas due to slow reactions when roasting below 923 K. The dilute SO_2 off-gas cannot be converted into sulphuric acid or elemental sulphur which results into a serious environmental pollution.

(ii) The gangue minerals dissolve into the leach solution needs additional filtration prior to electro-winning process [1].

1.1 Carbothermic reduction of mineral sulphides in the presence of CaO

The Cu-Cu-Fe mineral sulphide phases may be reduced to metallic state in the presence of lime (CaO) and carbon, explained in equation 1 [5-7]. In essence, the exchange reaction occurs between mineral sulphide (MS) and CaO to yield metal oxide (MO) and CaS phases (equation 2) [7]. The metal oxide is then reduced to metallic state via equation 3 [7].

$$MS(s,l) + CaO(s) + C(s) = M(s,l) + CaS(s,l) + CO(g) \quad 1$$

$$MS(s,l) + CaO(s) = MO(s,l) + CaS(s,l) \quad 2$$

$$MO(s,l) + C(s)/CO(g) = M(s,l) + CO(g)/CO_2(g) \quad 3$$

The Cu-Co-Fe-Ca-O-S predominance area diagram at 1073 K is shown in figure 1 and it is evident that the metallisation of Cu, Co and Fe occurs in the presence of CaO at lower partial pressure of O_2 gas (log10(P(O_2)(atm) < -19). The thermodynamic analysis in figure 1 also shows that preferential metallisation of Co and Co/Fe is possible in regions D (-17.8 < log10(P(O_2)(atm) < -16.2) and B (log10(P(O_2)(atm) < -19), respectively, at 1073 K from which it is apparent that either a complete metallisation of Cu, Co and Fe or partial metallisation of Co and, Co and Fe depends on the partial pressure of O_2 gas (figure 1).

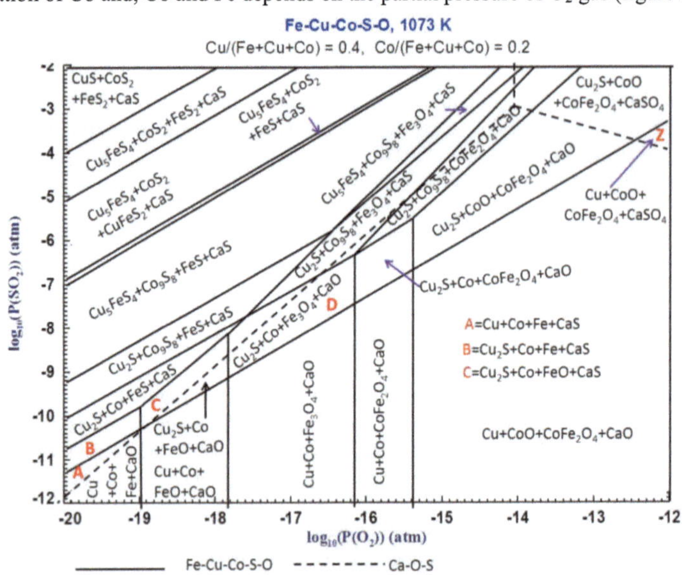

Figure 1 – The computed predominance area diagram for Cu-Co-Fe-Ca-O-S at 1073 K, using FactSage software 6.1 [8]

1.1.1 Magnetic separation As shown in equation 1 and figure 1, the reduction of the mineral sulphide concentrates in the presence of CaO yields a mixture of metallic constituents Cu, Co and Fe. Since the cobalt and iron are magnetic in nature, these two elements may be preferentially separated magnetically after reduction. By comparison, copper is diamagnetic [9] suggesting that if a magnetic separation step is implemented for the reduced sample, the yield will be a magnetic fraction containing Co and Fe and a non-magnetic fraction of Cu, CaS and gangue minerals (SiO_2, CaO, MgO, Al_2O_3). However in a previous investigation, the Cu-rich alloy was magnetically separated from the strongly magnetic alloy Co and Fe [10]. The magnetic separation and subsequent microscopic analysis showed that Co and Fe forms within a Cu-rich phase [10], during cooling of the reduced material. This is due to: (i) increased solubility of Co and Fe in Cu at the reduction temperature (> 1073 K) [10] and (ii) limited solid solubility of Co and Fe in Cu at room temperature [10].

Based on the analysis of process reported in literature [11] it is proposed that the production of Cu-Co-Fe powder via carbothermic reduction (< 1323 K), followed by magnetic separation is a more direct and economic process (figure 2a) than the conventional smelting (≈ 1573 K) – slag reduction smelting (≈ 1823 K) process (figure 2b). In figure 2a, the magnetic fraction containing Cu, Co and Fe may also be processed via leach – electro-winning process to produce copper and cobalt cathodes. The non-magnetic fraction containing SiO_2 and CaS may be utilized for the treatment of the Cu-Co slag as explained in equation 4. The reaction products (MS and $CaSiO_3$) in equation 4 may then be separated from each other due to the difference in the densities, as in the copper matte smelting [10]. The $CaSiO_3$ rich slag may be used as a construction material whereas the matte may be treated in the presence of CaO, via equation 1.

$$MO + CaS + SiO_2 = MS + CaSiO_3 \qquad 4$$

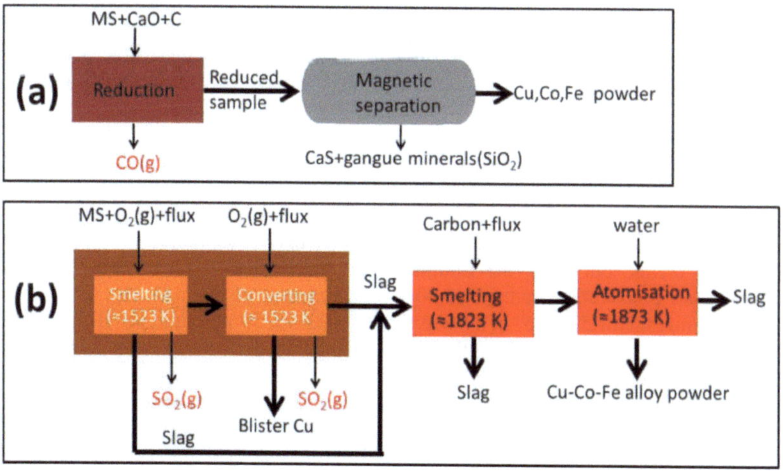

Figure 2 – A simplified process sheet for the pyro-metallurgical treatment of the Nkana Cu-Co mineral sulphide concentrates; (a) proposed new process and (b) conventional process

2.0 Experimental

2.1 Materials and Reduction Experiments

The mineral sulphide concentrates were obtained from Nkana concentrator, on the Copperbelt region in Zambia. The composition of concentrates derived from froth flotation is given in table 1. From detailed analysis reported earlier [6, 12], cobalt is contained in the carrollite ($CuCo_2S_4$) and pyrite (FeS_2) minerals [6]. Lime (CaO) with better than 98% purity was obtained from Alfa Aesar UK. Carbon black used in this investigation was supplied by Evonik Company (Essen, Germany).

For the reduction experiments, the mineral sulphide concentrates were mixed with CaO and carbon by following the stoichiometric conditions in equation 1 [6, 12]. Each powder mixture was ground for about 15 minutes in a mortar and pestle and then transferred in a cylindrical alumina crucible. The reduction experiments were carefully designed so as to minimise oxidation of the reaction products. The cylindrical alumina crucible containing the sample was placed inside the graphite – clay (salamander) crucible. The gap between the cylindrical alumina crucible and salamander crucible was filled up with some graphite blocks. All reduction experiments were carried out isothermally inside an elevating hearth furnace, under the flow of argon maintained at 2 lit min^{-1}.

Table 1 – Mineralogical composition for Nkana froth flotation concentrates in weight %

Cu_2S	$CuFeS_2$	Cu_5FeS_4	$CuCo_2S_4$	FeS_2	SiO_2	Other
0.7	8.6	5.0	2.2	10.1	40.0	33.4

2.2 Magnetic Separation

The reduced samples were ground for 20 minutes in a Mortar Grinder RM 200 in order to liberate any sintered metallic phases. Each sample was separated on a high intensity wet magnetic separator. The magnetic and non-magnetic fractions were dried at 333 K, weighed and characterised.

2.3 Characterisation

The reacted samples were characterised by the X-ray powder diffraction (XRD), X-ray fluorescence (XRF) and scanning electron microscopy (SEM) techniques. X-ray diffraction analysis was carried out using the Philips X'pert machine with a Cu Ka (0.15417 nm) radiation, at an acceleration voltage of 40 kV and current of 40 mA. The samples were scanned for 2θ range of 5 ° - 85° in 2θ step size of 0.0334225 ° and time per step of 200.7 s. The XRD patterns were analysed by the X'pert high score plus software, supported by the JCPDS PDF-4z data base (RDB 2013). For SEM analysis, a representative portion of the sample was placed inside the plastic mould. A mixture of 3 parts epoxy resin and 1 part epoxy hardener were poured into the mould and cured for 8 hours at 298 K. The cured samples were released from the mould and gently ground onto silicon carbide papers. The ground sample was polished down to 3 μm, after which the surfaces for microscopic examination was coated with 8 nm platinum layer to prevent charging during SEM analysis. The SEM equipped with an energy dispersive X-ray (EDX) detector was used for semi-quantitative analysis of a range phases under investigation.

3.0 Results and Discussion

3.1 Carbothermic Reduction

The XRD pattern for the sample after reaction for 2 hours at 1273 K is shown in figure 3a, from which it is evident that the main phases are metallic/alloy (Cu, Fe, Co-Fe), CaS, excess CaO and SiO_2. The presence of the metallic/alloy and CaS confirms that the mineral sulphide ($CuFeS_2$, Cu_5FeS_2, FeS_2 etc.) mixture was reduced in the presence of CaO and C via equation 1. The metallisation was lower at 1073 K than at 1273 K, as shown in figures 3a and 3b. The main phases in figure 3a in the reduced material at 1073 K are Fe_3O_4, Fe_2O_3, $Cu_{2-x}S$, CaS, SiO_2, $CaCO_3$ and excess CaO, as shown in figure 3a. Some Fe_2O_3 formed via external oxidation of the Fe_3O_4 phase, when removing the sample from the furnace.

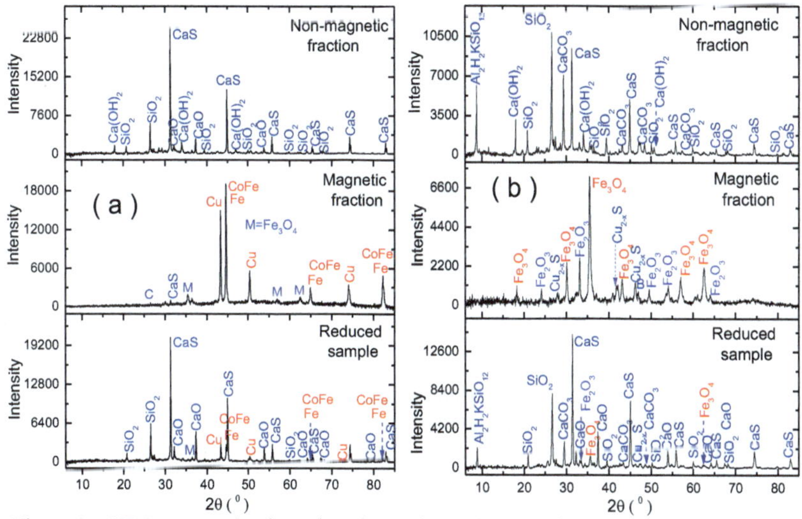

Figure 3 – XRD pattern for the reduced samples and, magnetic and non-magnetic fractions, the samples were reduced at; (a) 1273 K and (b) 1073 K

The SEM images of the sample after reduction at 1273 K are shown in figures 4a and 4b. The composition of phases were analysed by the SEM-EDX technique and these are inserted in the electron micrographs. The metallic iron resulting from the reduction of $CuFeS_2$ and Cu_5FeS_4 remained inside the copper matrix (figure 4a), whereas the cobalt once formed rapidly dissolved in the iron matrix to form a Co-Fe alloy, as shown in figure 4b. The EDX analysis confirmed that the concentrations of cobalt in the alloy phase at 1073 K may be as high as 98 wt.%, which suggests that a low temperature reduction at 1073 K may further optimize for a preferential separation of cobalt from the remaining copper and iron.

3.2 Magnetic Separation of the Samples

The magnetic and non-magnetic fractions of the reduced sample at 1273 K are shown in figure 3a. It is evident from figure 3a that the main phases in the magnetic fraction are Cu, Fe and CoFe. As explained above, although metallic copper is diamagnetic, it was possible to magnetically separate it from more magnetic Fe/Co alloy due to differential magnetization in the presence of Fe and Co inside the Cu matrix (see figure 4a). The non-magnetic fraction mainly consists of CaS, SiO_2 and $Ca(OH)_2$. The formation of $Ca(OH)_2$ occurs due to atmospheric hydrolysis unreacted CaO during wet magnetic separation. Based on the XRD results in figure 3a, we may conclude that magnetic separation may be applied for the separation of metallic constituents (Cu, Co and Fe) from the reduced material. The XRD patterns for the magnetic and non-magnetic fractions of the reduced mixture at 1073 K are shown in figure 3b. The main constituents in the magnetic fraction are Fe_3O_4 and $Cu_{2-x}S$ phases. The encapsulated $Cu_{2-x}S$ phase with Fe_3O_4 facilitates magnetic separation after reduction at 1073 K.

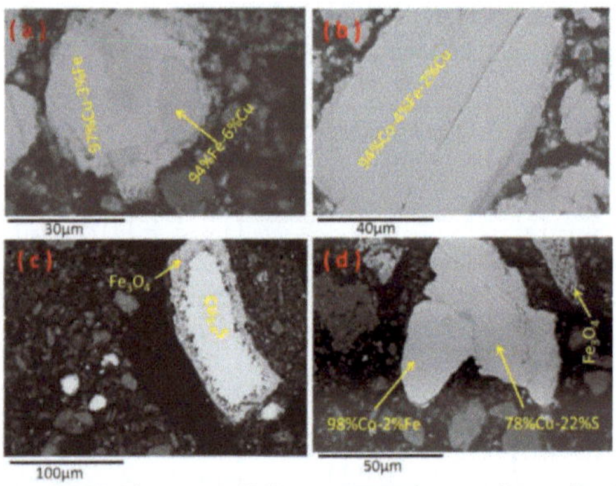

Figure 4 – Backscattered SEM images of Nkana concentrates after reduction in the presence of CaO and carbon black (MS:CaO:C = 1:2:3); (a) and (b) 1273 K and, (c) and (d) 1073 K

The XRF analysis of the magnetic and non-magnetic fractions is shown in table 2, from which it is clear that the magnetic fraction has about 92 wt.% Cu, Co and Fe elements. The main impurities in the magnetic fraction are Ca and S, which agree with the XRD results in figure 3a where CaS is the dominant phase. The SEM analysis showed that CaS in the magnetic fraction was partially sintered with the metallic/alloy phases. The XRF results in table 2 shows that a smaller fraction of Cu and Fe (about 9 wt.%) remained within the non-magnetic fraction and SEM-EDX analysis revealed that these were in the form of; (i) unreacted or residual mineral sulphide phases (FeS and Cu_5FeS_4) and (ii) very fine (< 5 µm) metallic/alloy phases that were sintered with the CaS phase.

Table 2 – XRF semi-quantitative analysis (wt. %) of the magnetic and non-magnetic fractions for the sample which was reduced at 1273 K with a molar ratio of MS:CaO:C = 1:2:3

Fraction	Reduction temperature	Cu	Fe	Co	S	Ca	Si	Al	Mg	Ti
Magnetic	1273 K	29.5	58.1	3.9	3.2	2.8	1.0	1.2	0.2	0.1
Non-magnetic	1273 K	3.8	4.2	0.1	17.3	47.2	17.3	4.6	4.8	0.5
Magnetic	1073 K	22.5	63.7	3.9	6.8	0.8	1.3	0.4	0.4	0.1
Non-magnetic	1073 K	5.2	6.5	0.1	18.1	36.5	20.3	4.8	4.8	0.4

On the other hand, there was no sintering between CaS and the metallic/alloy phases at 1073 K, as a result the concentration of CaS is low in the magnetic fraction. The main impurity in the magnetic fraction is sulphur, owing to the presence of the $Cu_{2-x}S$ phase. By comparison, the concentrations of Cu and Fe are slightly higher in the non-magnetic fraction for the sample which was reduced at 1073 K than at 1273 K. This is due to the fact that there was a larger proportion of the unreacted or residual mineral sulphide (FeS and Cu_5FeS_4) at 1073 K than at 1273 K.

3.2.1 Effect of Molar Ratio of MS:CaO:C The magnetic separation of the reduced sample was affected by the molar ratio of MS:CaO:C or MS:CaO and CaO:C. It was important to add CaO at twice the stoichiometric ratio, in order to achieve rapid and complete exchange reaction between mineral sulphides and CaO. However, by doubling ratio of MS:CaO (MS:CaO:C = 1:2:1) magnetic separation was adversely affected. For example, the magnetic fraction after reduction at 1273 K and molar ratio of MS:CaO:C = 1:2:1, had more than 15 wt.% CaS. The poor magnetic separation leaves high residual concentrations of CaS in the magnetic fraction. The reasons for high retention of CaS are: (i) dispersion of some metallic phases in CaS containing matte, as shown in figure 5a, and (ii) the wetting and sintering between the metallic phases and CaS. Since the MS–CaO exchange reaction occurs more rapidly than the reduction reaction at MS:CaO:C = 1:2:1 ratio, as a result the CaS formed reacts with metal oxides and produces a Ca_2CuFeO_3S liquid phase [13]. Metallisation of Cu and Fe occurs via reduction of the Ca_2CuFeO_3S liquid phase [13] which promotes sintering between the CaS and metallic phases. The sintering tendency between the metallic/alloy and CaS phases decreased when the ratio of C:CaO increased from more than 1:1.

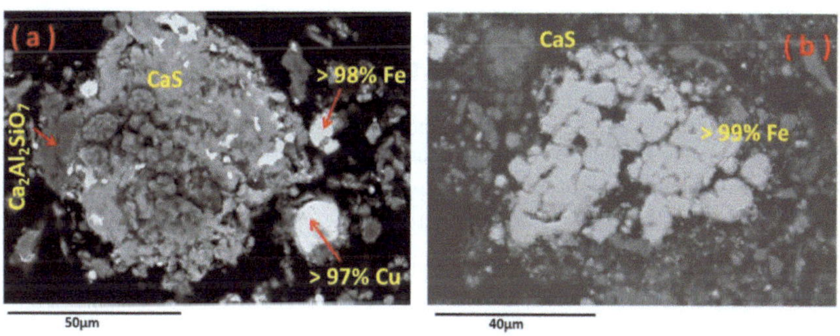

Figure 5 – The backscattered electron SEM images for the reduced samples at 1273 K, at molar ratios of: (a) MS:CaO:C = 1:2:1 and (b) MS:CaO:C = 1:2:2. The bright areas are the metallic phases

3.2.2 Effect of Reaction Time The SEM images for the partially reacted samples at 1273 K are shown in figure 6. The metallic phase mixture (Fe and Co) co-exists with $Cu_{2-x}S$ in the partially reacted $CuCo_2S_4$ and $CuFeS_2/Cu_5FeS_4$ mineral particles. The experimental results in figure 6 broadly agree with the thermodynamic prediction in figure 1 where we show that the metallic Co and Fe co-exist with Cu_2S [14]. The SEM-EDX analysis showed that the concentrations of Co and Fe elements may be as high as 99 wt.%, implying that a relatively rich fraction of magnetic alloy with Co and Fe may be separated magnetically after partial reduction. However, after such a short reducing interval, the analyzed particle size distribution of Co and Fe appears to be less than 30 μm, thereby implying that fine grinding might be necessary for liberating the metallic particles from unreacted Cu_2S, which might increased the overall processing cost.

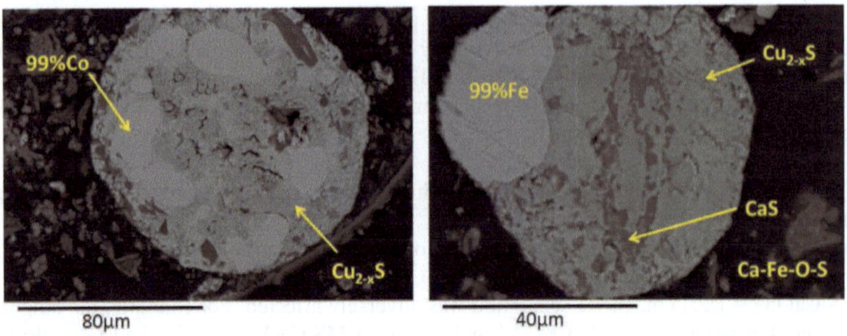

Figure 6 - Backscattered SEM images of the partially reduced samples, after 10 minutes at 1273 K, with carbon black at molar ratio of MS:CaO:C = 1:2:2

3.3 Energy Analysis

The theoretical energy requirement for carbothermic reduction of the Nkana sulphide concentrates in the presence of CaO was calculated using the HSC 5.1 software [15] by considering the following assumptions:

i. The reduction reaction 1 completes to 100 % by forming metallic phases (Cu, Co and Fe), CaS and CO gas.
ii. There is no reaction between CaO and the gangue minerals.
iii. The reactants (mineral sulphide + CaO + C) and the products (metallic phases, CaS, gangue minerals and CO gas) enter and leave the reaction chamber at 298 K and 1273 K, respectively.
iv. Moisture (H_2O) and CO_2 from the carbonates in the mineral concentrates react with carbon to form H_2 and CO gases, according to equations 5 and 6.

$$H_2O(g) + C(s) = H_2(g) + CO(g) \quad 5$$

$$\mathrm{CaMg(CO_3)_2 + 2C = CaO \bullet MgO + 4CO(g)} \quad 6$$

v. The off-gas contains only CO and H_2 gases, from reactions 1, 5 and 6.

The theoretical heat requirement was calculated as 0.6 kWh per kg of concentrates. The heat produced in burning off CO gas from reduction of mineral sulphide was calculated as -0.54 kWh such that the overall process requires a net energy of 0.06 kWh per kg of concentrates. By comparison, the theoretical heat requirement for smelting Cu-Co slag is about 0.8 kWh per kg of feed material.

4.0 Conclusions

1. Metallic phases containing Cu, Co and Fe were magnetically separated from the samples, after reduction at 1273 K as shown in figure 3a. The magnetic fraction is composed of 92 wt.% metallic phase (Cu, Co and Fe) (see table 2).
2. The molar ratio of carbon black should be at least equal to, more preferably higher than that of CaO at 1273 K, in order to minimise the sintering between the metallic and CaS phases, as shown in figures 5a and 5b.
3. The main components of magnetic fraction is metallic cobalt, $Cu_{2-x}S$, Fe_3O_4 and Fe_2O_3 after reduction at 1073 K (see figures 3b, 4c and 4d).

4. High purity Co (> 98 wt.% Co) is obtained after reduction at 1073 K due to the lack of metallisation of Cu and Fe, as shown in figure 4d.
5. Preferential metallisation of Co and Fe occurs by stopping the reactions after 10 minutes at 1273 K (see figure 6). However, the fine structures of Co and Fe formed appears quite difficult to be separate magnetically, as these particles of magnetic alloy cannot be liberated from CaS easily after grinding.

5.0 Acknowledgements

The authors acknowledge the financial support from the Institute of Materials, Minerals and Mining (IOM^3) in London and the Copperbelt University in Zambia.

6.0 References

1. Davenport, W.G.L., et al., *Extractive Metallurgy of Copper (4th Edition).* Chemical, Petrochemical & Process. 2002: Elsevier. 1-452.
2. Banda, W., *High Temperature Phase Equilibria in the Fe-Co-Cu-Si system Pertaining to slag cleaning*, in *Department of Process Engineering*2006, Stellenbosch: Stellenbosch.
3. Jones, R.T., et al., *Recovery of cobalt from slag in a DC arc furnace at Chambishi, Zambia.* Journal of The South African Institute of Mining and Metallurgy, 2002. **102**(Compendex): p. 5-9.
4. Banda, W., N.T. Beukes, and J.J. Eksteen, *Factors influencing base metal recovery from waste reverberatory furnace slags in a 50 kVA laboratory DC plasma arc furnace.* Journal of The South African Institute of Mining and Metallurgy, 2004. **104**(3): p. 201-207.
5. Hara, Y.R.S. *Reduction Mechanisms of Copper, Cobalt and Iron during Low Temperature Recovery from Mineral Sulphide Concentrates*. in *Characterization of Minerals, Metals, and Materials 2013*. 2013. John Wiley & Sons.
6. Hara, Y. and A. Jha, *Carbothermic reduction of Zambian sulphide concentrates in presence of lime.* Mineral Processing and Extractive Metallurgy, 2013. **122**(3): p. 146-156.
7. Jha, A., S.H. Tang, and A. Chrysanthou, *Phase equilibria in the metal-sulfur-oxygen system and selective reduction of metal oxides and sulfides .1. The carbothermic reduction and calcination of complex mineral sulfides.* Metallurgical and Materials Transactions B-Process Metallurgy and Materials Processing Science, 1996. **27**(5): p. 829-840.
8. Bale, C.W., et al., *FactSage thermochemical software and databases.* Calphad: Computer Coupling of Phase Diagrams and Thermochemistry, 2002. **26**(2): p. 189-228.
9. West, E.G., *Copper and its alloys*. 1982, Chichester: Ellis Horwood.
10. Cullity, B.D., *Introduction to Magnetic Materials*. 1972, London Addison-Wesley.
11. Jones, R.T. and A.C. Deneys, *Using a direct-current arc furnace to recover cobalt from slags.* JOM, 1998. **50**(10): p. 53-56.
12. Hara, Y.R.S. and A. Jha. *A Novel Low Energy Route for the Extraction of Copper and Cobalt Metals/Alloys from the Zambian Sulphide Concentrates*. in *Characterization of Minerals, Metals, and Materials*. 2012. John Wiley & Sons.
13. Hara, Y.S.R., *Mineral sulphide-lime reactions and effect of CaO/C mole ratio during carbothermic reduction of complex mineral sulphides.* International Journal of Minerals, Metallurgy, and Materials, 2014. **21**(1): p. 1-11.
14. Hara, Y.S.R., *Study of Reaction Mechanisms for Copper-Cobalt-Iron Sulfide Concentrates in the Presence of Lime and Carbon.* JOM, 2013: p. 1-11.
15. Roine, A., *HSC Chemistry 5.1*, 2002, Outokumpu Research Oy: Finland.

Energy Technology 2015: Carbon Dioxide Management and Other Technologies
Edited by: Animesh Jha, Cong Wang, Neale R. Neelameggham, Donna P. Guillen, Li Li, Cynthia K. Belt, Randolph Kirchain, Jeffrey S. Spangenberger, Frank Johnson, Andrew J. Gomes, Amit Pandey, and Peter Hosemann
TMS (The Minerals, Metals & Materials Society), 2015

LCA EVALUATION FOR DIFFERENT TREATMENT PROCESSES OF NICKEL LATERITE ORE

Shuai Wang[1,2], Hongxu Li*[1,2], Chao Li[1,2], Xiangxin Hao[1,2], Qinghua Bao , Lifeng Zhang[1,2],

1 School of Metallurgical and Ecological Engineering, University of Science and Technology 30# Xueyuan Road, Beijing, 100083, China
2 Beijing Key Lab of Green Recycling and Extraction of Metals, University of Science and Technology, 30# Xueyuan Road, Beijing, 100083,China
* Corresponding Author
Keywords: Nickel laterite ore; Resource consumption; LCA evaluation; Environmental impact.

Abstract

Lots of resources and energy were consumed from nickel laterite ores mining to nickel products. In addition, large amounts of waste gas, waste water and solid waste were emitted, so it is urgent to explore and utilize laterite nickel ores in more efficient ways. In this article, life cycle assessment method is used to evaluate the three kinds of nickel production processes, energy consumption and waste emissions in these processes were analyzed by using GaBi6 software. The results show that ammonia leaching method is the optimal process based on life cycle assessment. Global warming potential of ammonia leaching method is minimum, 18381(kg), 144.27% lower than that of electric furnace process. Acidification potential of ammonia leaching method is also more minimum comparing with blast furnace process, electrtc furmace process. Some optimization methods were proposed for these problems, including adjustment of production process and utilization of the wastes comprehensively.

Introduction

Scale of stainless steel production expanded rapidly in China that makes the demand of nickel metal has been greatly increased in recent years[1]. The nickel is widely in used in producing stainless steel, part of nickel used for electroplating and chemical industry, etc[2]. The huge demand of nickel leads to the big change of nickel ores mining as well as the nickel extraction process from lower grade resources followed with the degradation of environment. As the nickel sulfide ore reserves dwindling, laterite nickel ore mining and utilization has become increasingly important. However, because of the complexity of the laterite nickel ore, the current development and utilization of technology has a lot of options, although these methods have different characteristics and their respective advantages, but in the process energy consumption and environmental load is significantly different, So it is necessary to make the necessary evaluation of different process routes, for reference and selection. Based on the above consideration, the purpose of this paper is to evaluate whole environment of three different laterite process flowsheet from beginning of ore mining to their production considering the amount of material and consumed and the waste released. GaBi 6 software would be used as life cycle assessment tools and to make scenario analysis, the three typical process routes for nickel production as blast furnace process, electric furnace smelting reduction process and ammonia leaching method would be selected for comparison analysis and process environment evaluation.

Treatment processes of laterite nickel ore

From 1879, New Caledonia developed the first laterite treatment process based on the iron blast furnace technology[3], so many various techniques have been developed like: (1) Nickel iron smelting, (2) Nickel matte smelting, (3) Blast furnace/electric furnace smelting (pig nickel-iron), (4) High pressure acid leaching (HPAL), (5) Atmospheric pressure leaching (AL), (6) Extra-high pressure acid leaching (EPAL), (7) Heap leaching (HL), (8) Sulfate atmospheric leaching (SAL),and (9) Carol process[4]. However, not all of the above techniques in the different stages of development have been commercialized, especially some hydrometallurgy processes, three typical process routes were described as follows:

Electric furnace smelting reduction process

Nickel-iron production electric furnace process flow is shown in Fig.1 [5]. First of all, crushing the ore to 50~150mm, send the ore to drying kiln for drying, making the ore non-caked and have low pulverization rate. Then preheat and roast it at 700℃ in calcination rotary kiln, calcine was got and then add the calcine with volatile coal (10~30mm)into electric furnace, getting nickel-iron after reduction smelting at 1000℃[6]. Almost all of the nickel oxides and cobalt oxides were reduced into metals in the process of reduction smelting in electric furnace. By adding right amount of coke to adjust reduction of iron, finally the finished nickel-iron alloy will be produced after converting rough nickel and iron.

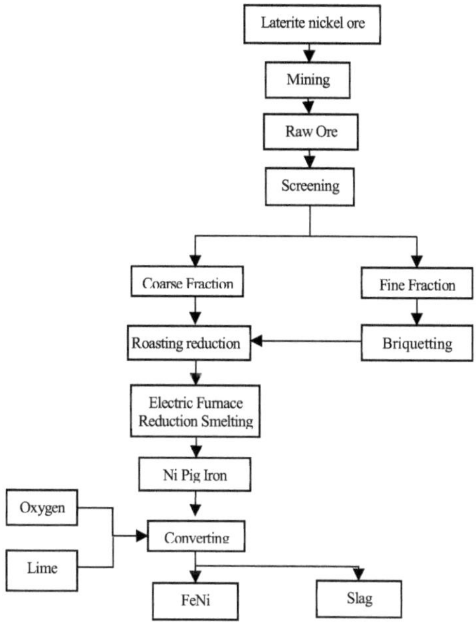

Fig.1. Nickel iron production electric furnace process flow

Blast furnace process

There is no essential difference between the traditional blast furnace iron making process and producing low nickel pig iron from laterite ore. The main process is crush and screening of nickel oxide ore, burdening-sintering, and smelting of sinter in the blast furnace. But as a result of composition and physical characteristics of nickel laterite ore are much different from general iron ore, the burdening of laterite ore sintering and blast furnace smelting have their own characteristics. Although blast furnace smelting process has advantages of small investment and low energy consumption, but its poor adaptability of minerals, the has the strict request for content of magnesium, it also can't handle powder ores. In addition, blast furnace process has strict requirements about burden[7].

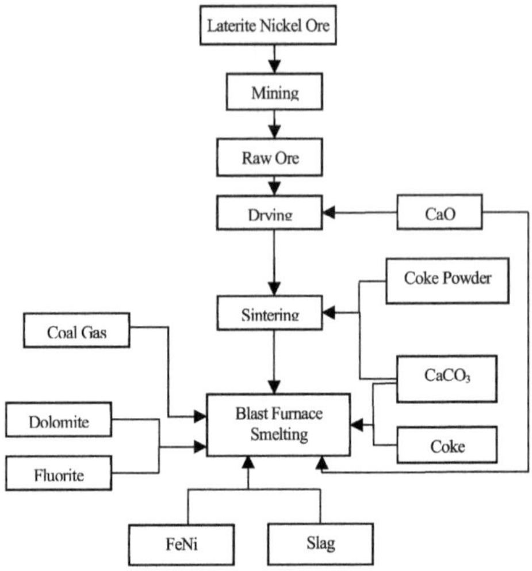

Fig.2. Nickel iron production blast furnace process flow

Ammonia leaching method

The laterite nickel ore was ammonia-leached on multistage atmospheric pressure after drying and reduction roasting. Nicaro factory is a representative in cuba. Reduction roasting is aimed to reduce the silicate nickel and nickel oxide into metals by maximum. The most of Fe_2O_3 was reduced into Fe_3O_4, only a minority of Fe_2O_3 was reduced into Fe by controlling reduction conditions at the same time. Ammonia leaching is the process in which nickel and cobalt roasting become nickel-ammonia complex and cobalt-ammonia complex, and then they got into the solution. Firstly, drying and grinding laterite ore, and then reduction roasting at 600~700℃, making nickel, cobalt and iron reduce into the alloy, the following is reflux ammonia leaching. The nickel and cobalt can form complex with ammonia, they got into leach liquor with other valuable metals. Removing iron from mother liquor after sulfide precipitation, through steamed ammonia, alkali type nickel sulfate was produced. Alkali type nickel sulfate was calcined into nickel oxide which can also be reduced into the nickel powder[8].

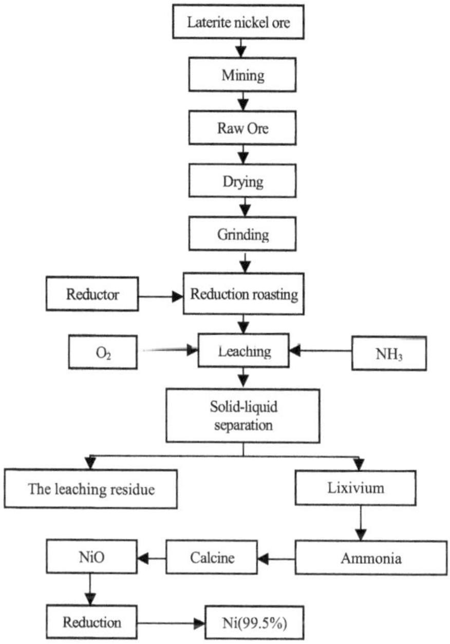

Fig.3. Ammonia leaching process flow

LCA process

Goal and scope

In order to study impact of the metallurgical methods of extract nickel metal on the environment, using GaBi 6 software to make life-cycle assessment of the three extract nickel processes from energy input, energy consumption, environmental impact and resource usage, comparing each other.

This article cites the example of electric furnace smelting reduction process, blast furnace process and ammonia leaching method. Due to the limitation of the experimental data, taking heavy and non-ferrous metal smelting design manual[9] data as the research object and combining with actual production conditions of domestic nickel metal smelters, using LCA method to quantitatively assess the environmental load of the process of producing nickel metal. The purpose is to comprehensively assess environmental load of nickel metal production process in our country and compare impact of different smelting processes on the environment, so as to seek the optimal process of the nickel smelting.

In this paper, the scope of the assessment system is from ore mining, ore dressing, grinding, production of material consumed to nickel metal extraction process, impact of waste emissions on the environment in the whole process were considered. Production of auxiliary raw materials includes lime production, the activated carbon production, power generation, diesel oil and coal production. Some input and output data of production process reference GaBi 6 software database and collected data. Electric furnace smelting reduction process includes mining, ore-dressing, drying, rotary kiln roasting and reduction smelting and refining process, auxiliary materials and energy production, such as lime, activated carbon and electricity.

Inventory Analysis

Table 1 Life cycle inventory of product 1t nickel

Process	Electric furnace smelting reduction process	Blast furnace process	Ammonia leaching method
Mining	Crude ore：100t Power：1053kwh	Wet ore：214 t Power：2135.7kw.h	Crude ore：112t Power：1613kwh
Drying	Dry ore：35.4t Power：3876.8kwh Coal：1939.4kg emissions：dust:0.01581t	Dry ore：75.76t Power：5248.87kwh CaO：7.6t Coal:3907.6kg	Ore:39.2t Power：3470kwh Coal：2007.7kg Emissions：dust
Rotary kiln roasting	Calcine：22.6t Power：4550kwh Powdered coal：125kg Limestone：316kg Water discharge： Cyanide wastewater Discharge to atmosphere：dust:0.42t, 260℃ Lime:5kg/t Ore:806kg/h, 900℃	-	Calcine：28.3t Power：6440kwh Reductor：980kg Discharge to atmosphere：SO_2、dust Degree of roasting reduction:63.8%
Sintering	-	Sinter:83.4t Power:7744.9kwh Limestone：156kg Coke powder：328.87kg Coal:2200kg	-
Electric furnace reduction smelting	Melted nickel iron：1.2t Powdered coal：0.13t Coke：352.7kg Na_2SO_4:1.82t Power:21000kwh Water discharge：wastewater Solid emissions：slag	-	-
Blast furnace smelting	-	ferro-nickel:1t Power15237.43kwh Coke：2tlimestone 331kg CaO:492kg Dolomite:234kg Fluorite:10t Coal:6040kg	-
Blowing	ferro-nickel：1t Na_2CO_3：16.8kg O_2：80m^3 CaO：216kg Power：12.26kwh Solid emissions：waste slag	-	-

Leaching	-	-	Lixivium：1.524t Power：35.6kwh Ammonia：1.2t Ammonium carbonate：0.62t Flocculant：113.2g O_2：1*10^6L Water：12100L CO_2：1.6t 大气排放：CO_2、NH_3 Water discharge： waste water Solid emissions：waste slag Pulp density：70%
Refining	-	-	nickel：1t Power：18.9kwh Water：456.87t

<u>LCA model</u>

The established LCA models of three nickel laterite processing technologies are shown as Fig.1, which demonstrates the process of electric furnace smelting reduction process, blast furnace process and ammonia leaching method. Blue arrows and red arrows in the figure represent the material flow and energy flow respectively.

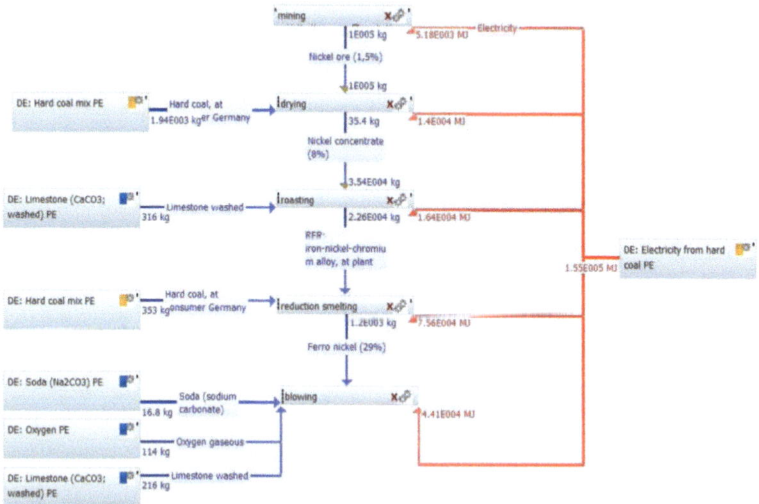

Fig.4. Electric furnace process of nickel iron production of life cycle diagram

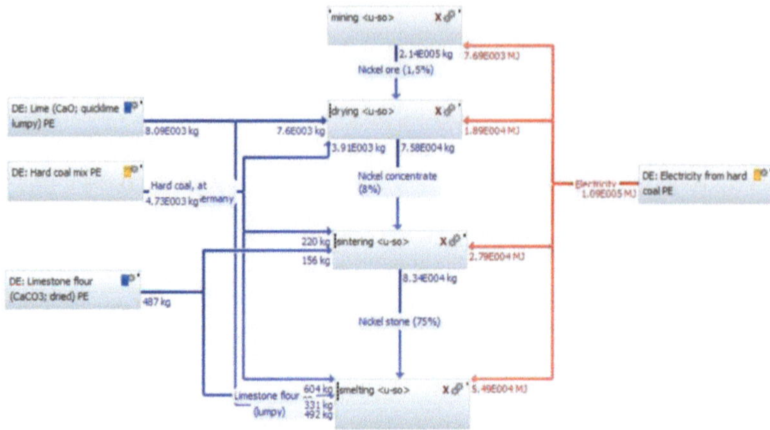

Fig.5. Blast furnace process of nickel iron production of life cycle diagram

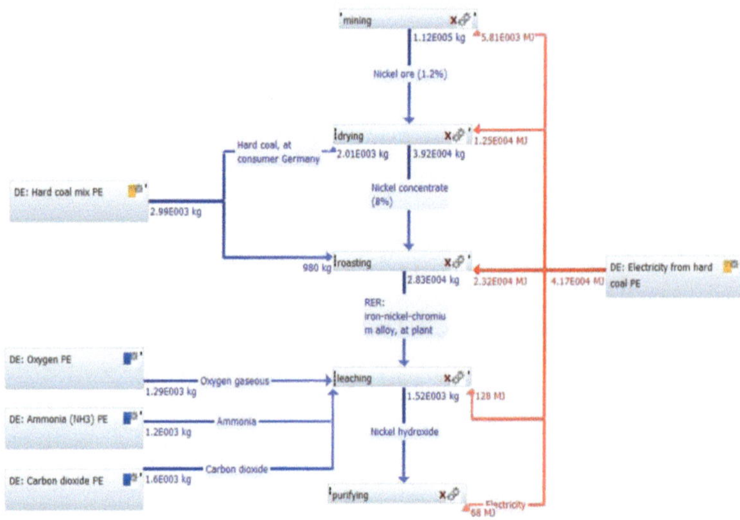

Fig.6. Ammonia leaching process of nickel production of life cycle diagram

<u>Life cycle assessment</u>

The article adopts CML 2001methods, the environmental impact of product life cycle was divided into three parts: Material and energy consumption, pollution (greenhouse effect, ozone depletion, acidification, human toxicity, and so on), and damage. There are ten indexes for the entire life cycle of the three processes are shown in Fig.10: resource consumption index(ADP), acidification potential(AP), eutrophication potential index(EP), the global warming index(GWP), ozone depletion potential index(ODP), the formation of ozone potential index(POCP), freshwater aquatic ecotoxicity pot. (FAETP), human toxicity potential(HTP), marine aquatic ecotoxicity pot.(MAETP), and terrestric ecotoxicity potential(TETP). LCA indexes for the entire life cycle of the three processes are shown in Fig.7, 8, 9. We can see the power consumption accounted for the vast majority of the environmental load in electric furnace smelting reduction process and pyrogenic smelting blast furnace process, but the hard coal consumed in the production process play a larger role in the process of blast furnace process.

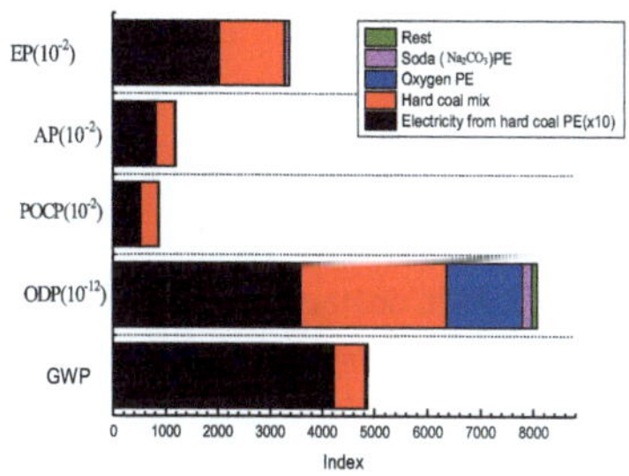

Fig.7. Electric furnace process

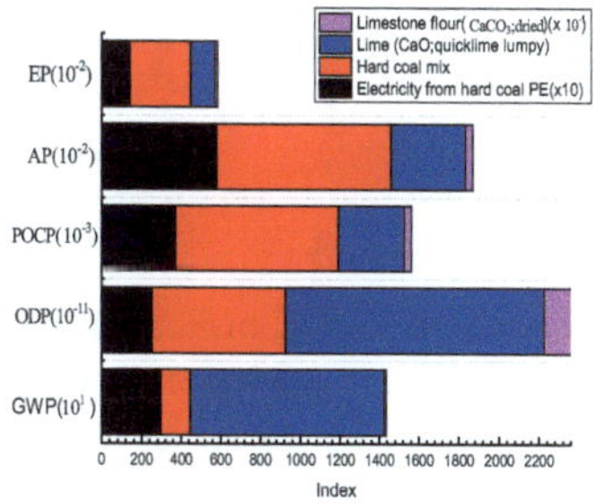

10^1 **Fig.8.** Blast furnace process

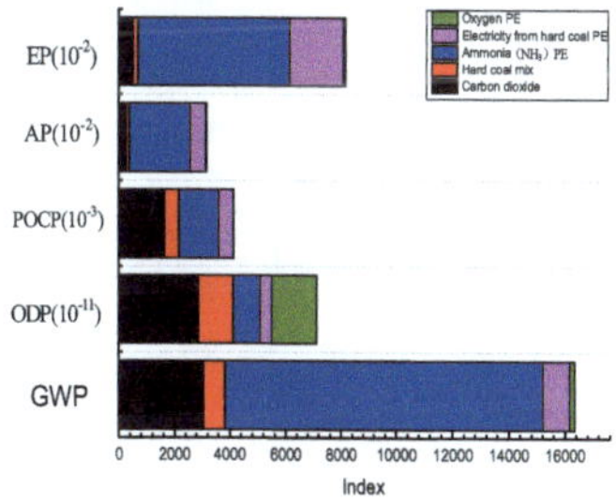

Fig.9. Ammonia leaching method

The relative value of environmental load is shown in **Fig.10**. It shows that, in pyrometallurgy, the largest index is POCP, flowing AP and GWP, owning to the huge consumption of power. At the same time, in the reduction roasting ammonia leaching, the largest index is ADP, following AP and GWP, the main reason is huge consumption of power in mining, drying and roasting gives rise to air pollution.

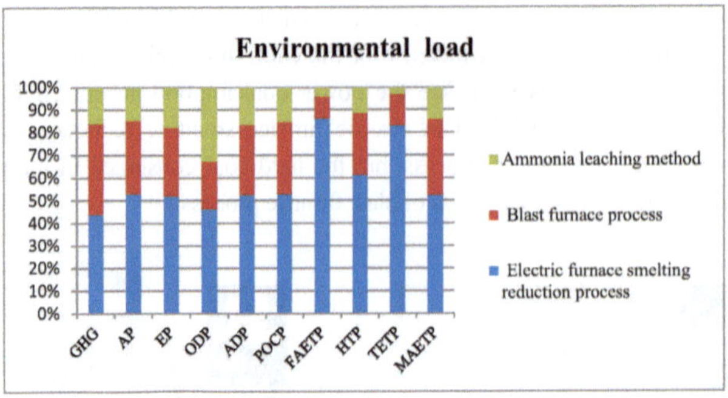

Fig.10. Environment load of treatment processes of laterite nickel ore

As is shown in the figure, environmental impact decreased in order of: ammonia leaching method>blast furnace process>electric furnace smelting reduction process. The specific numerical data of GWP and AP are indicated in **Fig.11.**

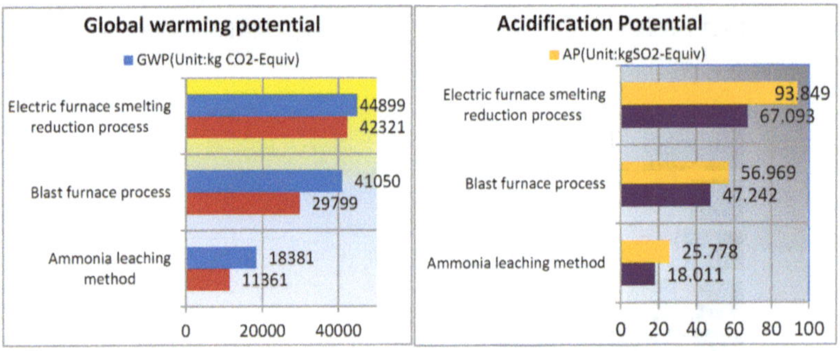

Fig.11. The global warming index and acidification potential index of three processes (Conversion to SO_2 and CO_2 equivalent)

The impact on the global warming index decreased in order of: electric furnace smelting reduction process>blast furnace process>ammonia leaching method. Electrical energy consumption accounts for the majority of environmental impact potential in whole life cycle. The coal consumption in blast furnace process and the usage of ammonia also make great contribute to environmental impact potential.

Optimization tips

In the three processes, electricity consumption results in most of the environmental impact, as a whole, the more electricity consumption, the greater the impact on the environment. Therefore, reducing the electricity consumption may become the direction of our future efforts, focusing on decreasing environmental impact in ferronickel production. Comparing with hydrometallurgy process, although the pyrometallurgy has the largest environmental impact, considering Chinese national conditions, it will not be eliminated in the short term.

Conclusion

In this paper, life cycle assessment was applied to nickel production, studying raw materials, energy consumption and pollutant emissions in the entire life cycle of laterite nickel ore treatment process, quantitatively analyzing various environmental implications, providing the basis for decision makers to choose production process and management system. In this paper, the following conclusions:

- The main factor of the environmental impact is energy consumption and material consumption through analyzing the calculating.
- The results of assessment show that the main environment impact of smelting process for ozone depletion, acidification, global warming. The main environmental influence of ammonia leaching process is marine ecological toxicity, global warming and ozone depletion.
- Ammonia leaching method is more environment-friendly than pyrogenic smelting process.

ACKNOWLEDGEMENT

The authors gratefully acknowledge the financial support of the National Science Foundation of key funds PRC for the research project [51234008], also the financial support of Beijing technical development project [00012132]

REFERENCES

[1] Wuzhuang Zhao. Substantial increase of nickel demands caused by rapid growth of stainless steel production deserves close attention 2007, 3:6.

[2] Jiadong Zhang. Safety evaluation of nickel resource in China 2013, 28-29.

[3] Xinghua Lan. Technology difficulty of recovering nickel from laterite 2007, 4:28.

[4] Nawshad Haque, Terry Norgate. Estimation of greenhouse gas emissions from ferroalloy production using life cycle assessment with particular reference to Australia. Journal of Cleaner Production;2013, 39:220-230.

[5] Bangsheng Zhang, Kaixi Jiang, Haibei Wang. Progress in laterite nickel ore smelting method in China. Nonferrous Metals Engineering & Research 2012, 33:16-19.

[6] Xiaowen Zhou, Jianchun Zhang, Xianping Luo. The Current Situation and Prospects of the Process and Technology of Nickel Extraction from Laterite-nickel ore 2008,(1):19-20

[7] Youping Zhang, Yusheng Zhou, Zhaoyi Li. Characteristics of laterite ore and pyrometallurgy process analysis 2007, (6):18.

[8] Xiaoming Li, Lin Tang, Shiliang Liu. Laterite nickel ore processing technology. Ferroalloy 2007, 195(4):24.

[9] Zhuo Sun, Jia Zhang, Derun Wang. Heavy and non-ferrous metal smelting design manual 1996,460-480.

Energy Technology 2015: Carbon Dioxide Management and Other Technologies
Edited by: Animesh Jha, Cong Wang, Neale R. Neelameggham, Donna P. Guillen, Li Li, Cynthia K. Belt, Randolph Kirchain, Jeffrey S. Spangenberger, Frank Johnson, Andrew J. Gomes, Amit Pandey, and Peter Hosemann
TMS (The Minerals, Metals & Materials Society), 2015

LOW TEMPERATURE SULPHIDIZATION OF Cu-Co SLAG IN THE PRESENCE OF CALCIUM SULPHIDE

Yotamu R.S. Hara and Animesh Jha

The Institute for Materials Research, Houldsworth Building
Clarendon Road, Leeds University, Leeds LS2 9JT (UK), England

Keywords: slag, sulphidization, recovery, calcium sulphide, cobalt, copper and matte

Abstract

The investigation focuses on sulphidization of Cu, Co and Fe from a 7wt%CuO-7wt%CoO-10wt%CaO-40wt% SiO_2-(30 wt%Fe,6 wt%Al)$_2O_3$ slag in the presence of calcium sulphide (CaS), over a temperature range of 1173 K – 1323 K. The metal oxides phases (CuO, CoO, FeO and Fe_3O_4) in the slag were converted into metal sulphide phases (Cu_2S, $CuFeS_2$, FeS, Cu_5FeS_4, Co_8S_9) via MO + CaS + SiO_2 = MS + $CaSiO_3$ reaction, where M represents metallic Cu, Co or Fe. Cu_2S and CoS formed preferentially at low molar ratio of MO/CaS and below 1173 K. The sulphidized slag was reduced in the presence of lime (CaO) and carbon at 1323 K by following the equilibrium: MS + CaO + C = M + CaS + CO(g). The reaction products were characterised by X-ray diffraction and scanning electron microscopy techniques. The proposed method of recovering metallic phases (Cu, Co and Fe) from a Cu-Co slag via sulphidization and reduction at lower reduction temperature than 1373 K offers an energy efficient process.

1.0 Introduction

Copper and cobalt are two major valuable constituents in the conventional copper smelting and converter slag [1-3]. These two metals are lost into the slag phase via oxidation and metal/matte entrapment [2]. The content of copper and cobalt in the copper smelting and converting slag may be as high as 2 wt.% and 0.8 wt.%, respectively [1]. According to London Metal Exchange in September, 2014, the price of copper and cobalt per tonne of metal is about $7000 and $33000, respectively. Comparing the slag composition of copper and cobalt with mineral concentrates, the slag becomes a viable source of these metals, which at present has not been exploited in Zambia, for example.

1.1 Recovery of Cu and Co from Slag

The recovery of copper and cobalt from the slag phase can be achieved via a combination of hydrometallurgical and pyrometallurgical processes [2, 3]. The hydrometallurgical treatment of Cu-Co slag mainly involves the following steps: (i) acid leaching in which metallic Cu and Co are dissolved, (ii) purification of the leach solution, and (iii) electro-winning of Cu and Co from the leach solution. However, most of the copper and cobalt are dissolved in the silicate and ferrite phases which makes leaching difficult, irrespective of the lixiviant employed [3]. The leaching of silicate yields silica gel which is hard to filter when compared with the leaching of ferrite which yields a leachant with increased level of iron in the leach solution [3, 4]. Overcoming these problems remain a problem in the hydrometallurgical extraction of Cu-Co slag, which otherwise would be the easiest route.

In pyrometallurgical processing, the Cu-Co slag is smelted with carbon (reductive smelting) at about 1823 K to obtain the molten iron rich Cu-Co-Fe alloy and slag phases [1

5], from which the fine powders of Cu-Co-Fe alloy are atomized. The Cu-Co-Fe alloy powders are then leached in autoclave to selectively solubilise Cu and Co and, followed by refining and electro-winning [5, 6].

In the combined pyrometallurgical - hydrometallurgical processing, the slag is roasted in the presence of pyrite (FeS_2) [7, 8], sulphuric acid (H_2SO_4) [3], ferric sulphate ($Fe_2(SO4)_2$) [9] or ammonium sulphate ($(NH_4)_2SO_4$) [10] with the aim of producing copper and cobalt sulphate. The roast calcine is leached in water or sulphuric acid to solubilize copper and cobalt sulphates. However, the recovery of cobalt is poor via this method [8, 9], due to the low reactivity of CoO in $CoFe_2O_4$ phase.

1.2 Sulphidization of Slag in the Presence of Calcium Sulphide (CaS)

Calcium sulphide (CaS) is the reaction product of the carbothermic reduction of the mineral sulphide concentrates in the presence of CaO as shown in equation 1. At present, CaS has no industrial value and its disposal posses threat to environment due to the risk of evolution of harmful hydrogen sulphide (H_2S) gas in the presence of moisture. It is for this reason that in the present investigation we focus on the utilisation of CaS for the treatment of the Cu-Co slag by following the equilibrium in equation 2 which has not been investigated or reported in literature. The values of Gibbs energy change for the sulphidization of Cu_2O, CoO and FeO in the presence of CaS and SiO_2 are compared in figure 1a from which it is evident that the reactions are thermodynamically feasible above 773 K. The sulphidization of the metal oxides may also be achieved in the presence of carbon, as shown in figure 1b. The values of Gibbs energy change for the sulphidization of metal oxides with CaS and SiO_2 are more negative in the presence than in the absence of carbon.

$$MS + CaO(s) + C = M + CaS(s) + CO(g) \quad 1$$

$$MO + CaS + SiO_2 = MS + CaSiO_3 \quad 2$$

Reaction 2 may proceed via; (i) ion exchange reaction between metal oxide and CaS (reaction 3) and (ii) formation of calcium silicate ($CaSiO_3$) (equation 4).

$$MO + CaS = MS + CaO \quad 3$$

$$SiO_2 + CaO = CaSiO_3 \quad 4$$

The Gibbs energy analysis shows that copper oxides (Cu_2O and CuO) may be sulphidized in the absence of SiO_2, according to reaction 4 [11]. However, for sulphidation of FeO and CoO SiO_2 is required.

It is clear from equation 2 and figure 1 that the sulphidization of Cu-Co slag in the presence of CaS will yield metal sulphide (Cu_2S, CoS and FeS) and calcium silicate ($CaSiO_3$) rich phase mixture. The sulphidized slag may be then treated via following pyrometallurgical techniques, to obtain either metal sulphide or metallic phase mixture:

i. Carbothermic reduction in the presence of CaO (see equation 1) and magnetic separation may yield the metallic phases (Cu, Co and Fe) from the non-magnetic fraction of CaS and $CaSiO_3$ phases. The non-magnetic fraction containing CaS may be recycled back into the process.
ii. Holding the sulphidized slag say at 1523 K, may allow metal sulphide (matte) / slag ($CaSiO_3$ rich phase) phase separation which is similar to a conventional copper matte

smelting. The matte (metal sulphide) phase may be then processed in the treatment of Cu-Co-Fe mineral sulphide concentrates.

iii. Froth flotation has been applied to concentrate the mineral sulphides from ores and this technique may be applied to recover the mineral sulphide phases from the sulphidized slag.

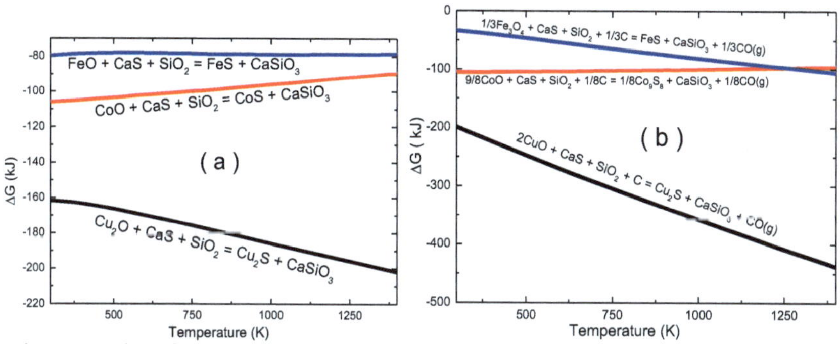

Figure 1 – The values of Gibbs energy changes for the sulphidization of metal oxides in the presence of CaS, SiO_2; (a) in the absence of carbon and (b) in the presence of carbon, computed from the HSC software 5.1 database [12]

2.0 Experimental

The Cu-Co slag sample was synthesised in the laboratory by melting the mixture of metal oxide (CuO, CoO, Fe_2O_3, CaO, Al_2O_3 and SiO_2) powders in accordance with the composition in table 1, which is comparable with a typical Nkana dump slag in Zambia. For synthesis the sample was melted in air for 2 hours at 1523 K. The transition metal oxides were added in high concentrations, well above the concentrations in normal slag for the easy of analysis and understanding of the metal recovery process [13]. The details of the slag sample preparation are found in our previous studies [13].

Table 1 – Composition (wt. %) of the synthetic slag used in the investigation

Constituent	CuO	CoO	Fe_2O_3	CaO	Al_2O_3	SiO_2
Weight (%)	7	7	30	10	6	40

From the XRD pattern in figure 2, it is evident that there are three phases present in the synthetic slag: (i) the two dominant crystalline phases are $Co_{0.68}Fe_{2.38}O_4$ and SiO_2, and (ii) the amorphous glassy phase, designated by a broad peak between 20° and 35°.

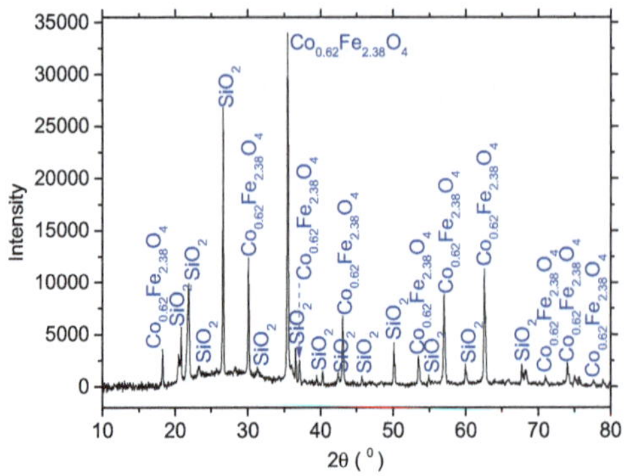

Figure 2 – XRD pattern of the slag, synthesised in air (21 % O_2) at 1523 K

2.1 Sulphidization of the Slag

Calcium sulphide was synthesised by reducing > 99 wt.% pure anhydrous calcium sulphate ($CaSO_4$) with activated charcoal by following the stoichiometry in equation 5. The mixture was held for 2 hours at 1323 K. The powder diffraction analysis of the reduced sulphate showed the presence of CaS.

$$CaSO_4 + 2C = CaS + 2CO_2(g) \quad 5$$

For the sulphidization of the slag in the presence of CaS, 2 g of the slag was ground to less than 150 μm and mixed with CaS, according to equation 2, with and without carbon (graphite). The mixture of slag and CaS were transferred inside an alumina crucible, which was then placed inside the graphite clay (salamander) crucible. Graphite blocks were placed in the gap between the alumina and salamander crucibles for minimizing oxidation of the samples during reduction with residual oxygen in the furnace atmosphere. Each reaction was investigated by maintaining a preselected isothermal temperature in an elevating hearth furnace, inside which the flow of argon gas at 2 litres min^{-1} was maintained

2.2 Carbothermic Reduction of the Sulphidized Slag

The reduction reactions were analysed using a thermogravimetric analysis (TGA) balance and the weight of the sample was monitored at the interval of 10 seconds from which the weight loss occurring as a result of CO gas evolution was recorded for determining the extent of overall reactions completed.

2.3 Characterisation

After reaction, each sample was ground in a mortar and pestle and, analysed using an X-ray powder diffraction machines (Phillips X'pert machine, with a copper target, $\lambda = 1.5418$ Å). Each powder sample was scanned in the 2θ range of 5° and 70° [14]. For scanning electron microscopy (SEM) examination, a portion of the sample was mounted in a mixture of 2 parts epoxy resin and 1 part epoxy harder and, cured for 12 hours at 298 K. After resin had hardened, the sample was then gently ground and polished. The polished surfaces were coated with an 8 nm platinum layer to prevent charging of the samples during SEM analysis, using secondary electron and backscattered electron (BSE) imaging techniques.

3.0 Results and Discussion

3.1 Sulphidization of Slag in the Presence of CaS and Graphite

The XRD patterns of the sulphidized slag samples are shown in figures 3a and 3b from which it is apparent that the metal oxides (Cu-O, Co-O and Fe-O) in the slag reacted with CaS and yielded a mixture of metal sulphide phases namely $CuFeS_2$, $CuFe_2S_3$ and FeS. The sulphide phase mixture co-exists with the calcium silicate phases ($CaSiO_3$ and Ca_2SiO_4) and $Ca_2Al_2SiO_7$, as predicted from the Gibbs energy calculations in figures 1a and 1b. The XRD peaks for the cobalt sulphide (CoS) cannot be observed in figures 3a and 3b because it dissolves, mainly in the FeS phase and partly in the $CuFeS_2$ and $CuFe_2S_3$ matrices. Based on the XRD results in figures 3a and 3b, it can be concluded that the Cu-Co slag sulphidized in the presence of CaS and yielded matte (metal sulphide) and slag ($CaSiO_3$, Ca_2SiO_4 and $Ca_2Al_2SiO_7$), as predicted in equation 2 and, figures 1a and 1b.

From the powder diffraction analysis in figures 3a and 3b, the following conclusions maybe drawn. Although there is an excess of carbon present, the Fe_3O_4 does not reduce readily at 1173 K in the presence of carbon. As expected in the absence of carbon, the residual phase concentration of Fe_3O_4 is higher than in the presence of carbon. This is also apparent from the powder diffraction analysis of the sample heat treated at 1323 K in the presence and absence of carbon at 1323 K. The increased diffraction intensity of FeS with temperature points out that the quantity of FeS bearing sulphides (CoS / Cu_2S) increases with temperature, as expected from the computational equilibrium data below:

$$Fe_3O_4 + 3CaS + 3SiO_2 = 3FeS + 3CaSiO_3 + 0.5O_2(g);\ \Delta G_{1323K} = -72\ kJ \qquad 7$$

$$Fe_3O_4 + 3CaS + 3SiO_2 + C = 3FeS + 3CaSiO_3 + CO(g);\ \Delta G_{1323K} = -300\ kJ \qquad 8$$

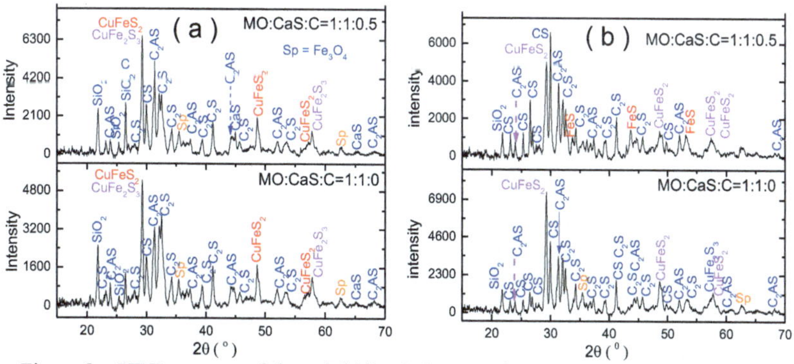

Figure 3 – XRD patterns of the sulphidized slag samples at; (a) 1173 K and (b) 1323 K, where CS = $CaSiO_3$, C2S = Ca_2SiO_4 and C_2AS=$Ca_2Al_2SiO_7$

In figure 4, with increasing metallisation the relative diffraction intensities of sulphide constituents decrease, as expected from the chemical equilibrium in equations 7 and 8. The metallic/alloy phases (Fe and Co-Fe) co-exist with CaS because not all of CaS was consumed by the sulphidization reaction. Metallisation occurs via equation 9, at the higher stoichiometric ratio of carbon. There is no metallic Cu in figure 4 even though the reduction of Cu_2O via equation 9 is thermodynamically more feasible than that of CoO and FeO. The absence of metallic Cu confirms that Cu_2S is preferentially formed via equation 10.

$$MO + C = M + CO(g) \tag{9}$$

$$2Cu + FeS = Cu_2S + Fe \tag{10}$$

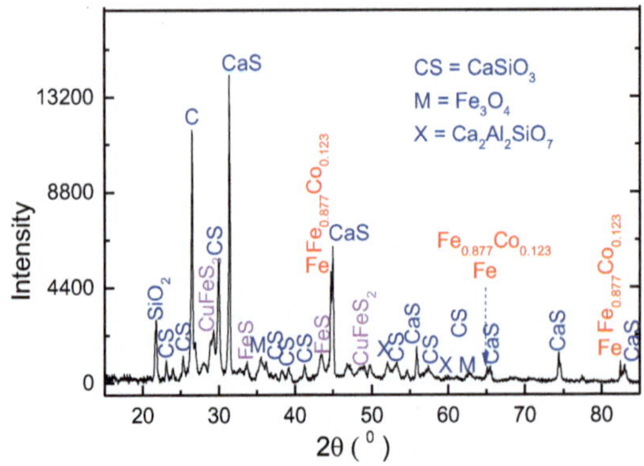

Figure 4 – XRD pattern for the slag sample which was sulphidized in the presence of CaS, with graphite at 1173 K, mole ratio of MS:CaS:C = 1:1:5

The representative SEM images of the sulphidized slag at 1173 K are shown in figure 5. The metal sulphide (matte) appears bright under the backscattered electron image from which it is apparent that the matte has separated out from the slag formed insitu, due to mutual phase immiscibility. Several areas of the sulphidized slag samples were analysed by the SEM-EDX technique in order to characterise the phases present that cannot otherwise be identified by the X-ray powder diffraction. The analysed composition of phases using semi-quantitative SEM-EDX technique is inserted in figure 5 from which it is apparent that CoS dissolved in an FeS matrix during matrix.

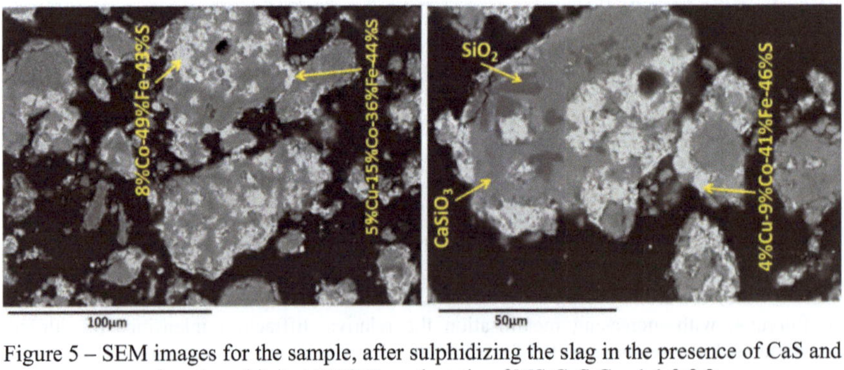

Figure 5 – SEM images for the sample, after sulphidizing the slag in the presence of CaS and carbon (graphite) at 1173 K, mole ratio of MS:CaS:C = 1:1.3:2.2

3.2 Carbothermic Reduction of the Sulphidized Slag

The plots of the % reduction versus time curves are shown in figure 6a from which it is apparent that the reduction reactions completed in about 65 minutes and 15 minutes at 1173 K and 1323 K, respectively. By comparison, the direct reduction of the slag with carbon, without sulphidization, did not reach completion even after 2 hours at 1323 K [13]. The rapid reduction of the sulphidized slag occurs in this case due to the presence of the liquid phase [15-17]. The XRD patterns of the reduced sulphidized slag are shown in figure 6b from which the absence of metal sulphide is apparent. The metallic/alloy phase mixture containing Cu, Co-Fe and Fe formed during carbothermic reduction of the sulphidized slag, in the presence of CaO. In particular, the Co-Fe alloy formed as a result of high solubility of metallic Co in the austenite iron matrix. The metallic phases co-exist with CaS, excess CaO and a mixture of calcium silicates ($CaSiO_3$ and Ca_2SiO_4) and $Ca_2Al_2SiO_7$.

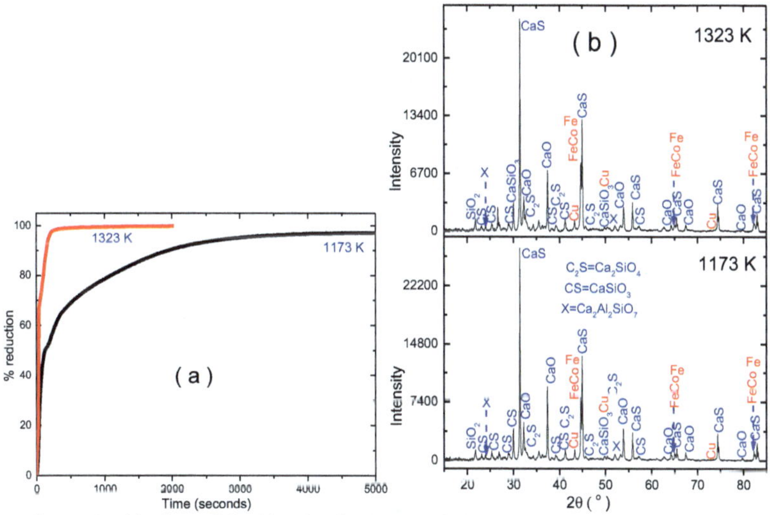

Figure 6 – (a) plots of the % reduction versus times curves and (b) XRD patterns, for the carbothermic reduction of the sulphidized slag in the presence of CaO (MS:CaO:C = 1:2:3)

The SEM images of the reduced sulphidized slag are shown in figure 7 in which the metallic/alloy constituents appear bright in a backscattered electron image. The average particle size of metallic fractions after reaction is larger than those obtained by direct reduction of slag with carbon. This is due to the formation of a liquid phase [15], during reduction of the metal sulphide above 1073 K.

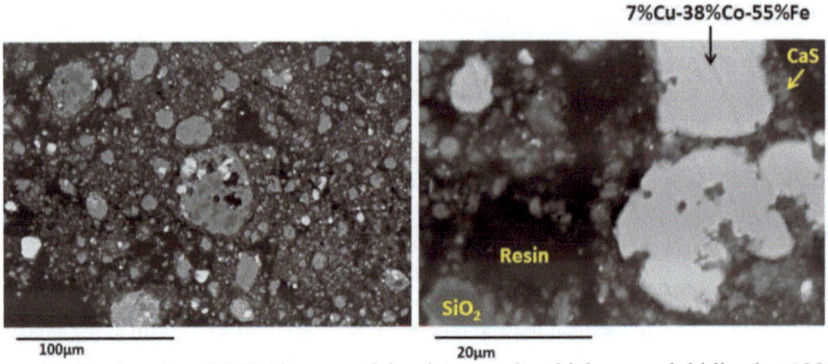

Figure 7 – Backscattered SEM images of the slag sample which was sulphidized at 1323 K (MO:CaS:C = 1:1.5:2) and reduced with carbon black at 1323 K (MS:CaO:C = 1:2:3)

3.3 Energy Analysis

The theoretical energy requirement for the production of Cu, Co and Fe from slag via sulphidization – carbothermic reduction was calculated using HSC 5.1 software [12] and the following assumptions were made:
Sulphidization – (i) Cu_2S, CoS and FeS are the only mineral sulphides in the sulphidized slag, (ii) alumina is in the form of $Ca_2Al_2SiO_7$ phase and (iii) the reactants and products enter and leave the reaction chamber at 298 K and 1323 K, respectively.
Carbothermic reduction – (i) CO gas is the only gas generated during reduction of mineral sulphides and (ii) the reactants and products enter and leave the reaction chamber at 1323 K.

The theoretical energy balance for sulphidization of slag and carbothermic reduction was calculated as 0.45 kWh and 0.15 kWh, respectively, per kg of slag, thus giving the total theoretical energy requirement of 0.6 kWh per kg of slag. By comparison, the theoretical energy requirement for smelting the slag at 1873 K is 0.8 kWh per kg of slag. Based on the above analysis, it can be concluded that the theoretical energy requirement for treating Cu-Co slag via sulphidization – carbothermic reduction at 1323 K is about 75 % to that of smelting the slag at 1873 K.

4.0 Conclusions

1. The metal oxides (Cu-O, Co-O and Fe-O) present in the slag transforms into a mixture of metal sulphide ($CuFeS_2$, Cu_5FeS_4, Cu_2S and FeS) via sulphidization reaction in the presence of CaS, at 1173 K and 1323 K, as shown in figures 3 - 5. The observed sulphidization maximises in the presence of carbon (graphite) at 1323 K.
2. The mixture of metallic constituents (Fe and Co-Fe) and metal sulphide ($CuFeS_2$, Cu_5FeS_4 and FeS) co-exist with CaS, during partial metallisation at high mole ratio of graphite (MO:CaS:C = 1:1:5), as shown in figure 4. This is because graphite is an ineffective reducing agent when compared with active carbon.
3. The extent of sulphidization of Fe_3O_4 with and without carbon is much lower than that of oxides of the copper and cobalt as predicted in the Gibbs energy data in figures 1a and 1b.
4. The carbothermic reduction of the sulphidized slag completes in about 65 minutes and 15 minutes at 1173 K and 1323 K, respectively (see figure 6a) which were also ascertained from the analysis of metallisation using XRD and SEM techniques.

5.0 Acknowledgements

The authors acknowledge the Institute of Materials, Minerals and Mining (IOM[3]) in London and the Copperbelt University in Zambia for the financial support.

6.0 References

1. Jones, R.T., et al., *Recovery of cobalt from slag in a DC arc furnace at Chambishi, Zambia.* Journal of The South African Institute of Mining and Metallurgy, 2002. **102**(Compendex): p. 5-9.
2. Davenport, W.G.L., et al., *Extractive Metallurgy of Copper (4th Edition).* Chemical, Petrochemical & Process. 2002: Elsevier. 1-452.
3. Mututubanya, A., *Cobalt recovery from old Nkana Copper Slag via solid state carbothermic reduction and sulphation*, in *Metallurgy*. 2013, University of Zambia: Lusaka.
4. Deng, T. and Y. Ling, *Processing of copper converter slag for metal reclamation. Part I: extraction and recovery of copper and cobalt.* Waste Management & Research, 2007. **25**(5): p. 440-448.
5. Banda, W., *High Temperature Phase Equilibria in the Fe-Co-Cu-Si system Pertaining to slag cleaning*, in *Department of Process Engineering*. 2006, Stellenbosch: Stellenbosch.
6. Jones, R.T. and A.C. Deneys, *Using a direct-current arc furnace to recover cobalt from slags.* JOM, 1998. **50**(10): p. 53-56.
7. Matusewicz, R. and E. Mounsey, *Using ausmelt technology for the recovery of cobalt from smelter slags.* JOM, 1998. **50**(10): p. 51-52.
8. Tümen, F. and N. Bailey, *Recovery of metal values from copper smelter slags by roasting with pyrite.* Hydrometallurgy, 1990. **25**(3): p. 317-328.
9. Altundoğan, H.S. and F. Tümen, *Metal recovery from copper converter slag by roasting with ferric sulphate.* Hydrometallurgy, 1997. **44**(1–2): p. 261-267.
10. Sukla, L.B., S.C. Panda, and P.K. Jena, *Recovery of cobalt, nickel and copper from converter slag through roasting with ammonium sulphate and sulphuric acid.* Hydrometallurgy, 1986. **16**(2): p. 153-165.
11. Maweja, K., T. Mukongo, and I. Mutombo, *Cleaning of a copper matte smelting slag from a water-jacket furnace by direct reduction of heavy metals.* Journal of Hazardous Materials, 2009. **164**(2–3): p. 856-862.
12. Roine, A., *HSC Chemistry 5.1.* 2002, Outokumpu Research Oy: Finland.
13. Hara, Y.S.R. and A. Jha, *Kinetic and Thermodynamic Analysis of the Reduction of Oxides of Cu and Co in a SiO2-CaO-(Al, Fe)2O3 Slag*, in *Celebrating the Megascale*. 2014, John Wiley & Sons, Inc. p. 553-562.
14. Hara, Y. and A. Jha, *Carbothermic reduction of Zambian sulphide concentrates in presence of lime.* Mineral Processing and Extractive Metallurgy, 2013. **122**(3): p. 146-156.
15. Jha, A., *Carbothermic reduction of sulphide minerals*, in *Department of Metallurgy and Materials Science*. 1984, Imperial College London.
16. Hara, Y.S.R., *Study of Reaction Mechanisms for Copper-Cobalt-Iron Sulfide Concentrates in the Presence of Lime and Carbon.* JOM, 2013: p. 1-11.
17. Hara, Y.S.R., *Mineral sulphide-lime reactions and effect of CaO/C mole ratio during carbothermic reduction of complex mineral sulphides.* International Journal of Minerals, Metallurgy, and Materials, 2014. **21**(1): p. 1-11.

Energy Technology 2015: Carbon Dioxide Management and Other Technologies
Edited by: Animesh Jha, Cong Wang, Neale R. Neelameggham, Donna P. Guillen, Li Li, Cynthia K. Belt, Randolph Kirchain, Jeffrey S. Spangenberger, Frank Johnson, Andrew J. Gomes, Amit Pandey, and Peter Hosemann
TMS (The Minerals, Metals & Materials Society), 2015

A KINETIC ANALYSIS OF ACID LEACHING OF NIOBIUM AND ZIRCONIUM FROM TITANIA WASTE RESIDUE STREAM: AN ENERGY EFFICIENT METHODOLOGY FOR THE RECLAMATION OF METAL VALUES

Terence Makanyire[1], Animesh Jha[1], Stephen Sutcliffe[2]

[1]The University of Leeds
School of Chemical and Process Engineering
Clarendon road, Ls2 9JT, UK
[2]Greatham Works, Tees Road, Hartlepool, TS25 2DD, UK

Keywords: Niobium, zirconium, extraction, leaching, kinetics, rutile

Abstract

Residues from rutile chlorination plants often contain relatively high concentrations of critical metals, essential for energy devices. Ironically, the conventional method of extraction of niobium and zirconium is quite energy demanding and detrimental to the environment. In this investigation, the kinetics of dissolution of niobium and zirconium from titania waste in hydrochloric acid solutions are investigated. The thermodynamic stability of minerals demands the use of chlorination, carbothermic reduction or alkaline fusion for breakdown of their mineral concentrate and upgrading before leaching in acid, usually hydrofluoric acid. Reclamation of niobium and zirconium values from titania waste presents an opportunity for a low energy process to be utilised. The effects of parameters leaching temperature (25 – 90 °C), acid concentration (0.5 – 2.5M), stirring speed and solid-to-liquid ratio were determined in the experiments. Leachates were analysed by ICP OES, using which the models for the leaching processes and activation energies were determined.

Introduction

Zirconium and niobium have exceptional resistance to corrosion and chemical attacks, making them good materials for the nuclear, aerospace and electronics industries as well as for making materials for building chemical process equipment. Due to their unreactive nature, the extraction of Nb_2O_5 and ZrO_2 from mineral ores is usually carried out in many steps usually employing energy intensive processes and aggressive acids such as hydrofluoric acid. Depending on the titaniferous feedstock, the neutralised chloride waste from the titanium dioxide manufacturing plants may contain as much as 3 wt% niobium oxide and 6 wt% zirconium oxide, which makes the waste an economically viable resource for these two metal oxides.

Niobium pentoxide (Nb_2O_5) can only be dissolved by fusion with alkaline fluxes such as NaOH or acidic fluxes such as $KHSO_4$ or in hydrofluoric acid (HF) [1]. Conventional processes for niobium recovery usually involve direct leach from ores using 70 – 80 wt% hydrofluoric acid with nitric or sulphuric acid [7, 8] at elevated temperatures, with residence times in the region of ten hours to obtain optimum dissolution of the niobium oxide. Hydrofluoric acid leaching forms the complex niobium heptafluoride H_2NbOF_5. Leaching is normally followed by solvent extraction to separate most of the niobium form the gangue. In many cases, however, the concentration of niobium in the mineral ore is usually low [3] which makes direct leaching uneconomical. Alternatively, the mineral concentrate is fused with alkali, usually NaOH at temperatures as high as 800 °C followed by water leaching to form an insoluble niobate which can be solubilised by hydrochloric acid. This approach reduces the loss of the highly volatile and hazardous HF into the environment. Alkali fusion method, however, has the disadvantage of requiring 6 to 8 times the reaction requirement in alkali consumption [4]. Some researchers [5] and [6] have proposed a sub-molten salt leaching approach to niobium recovery but this method employs a high alkali-to-ore ratio which can be as high as 7:1 and is too energy intensive. Yang et al., (2014), leached 98% niobium using 10 M sulphuric acid and heating for 2 hours at 200 °C under oxygen pressure and found the leaching to be largely dependent on leaching temperature, following a surface reaction control kinetic model.

Zircon ($ZrSiO_4$) is the most abundant zirconium ore and main source for commercial production of zirconium, its compounds and alloys [9]. It is usually obtained as a co-product or by-product of mining and processing of heavy mineral sands like ilmenite and rutile [9]. Due to its high stability ($dG_{1400\ K} = -1498.1\ kJ\ mol^{-1}$) [2], decomposition of zircon requires use of an elevated temperature and aggressive chemicals. The widely used decomposition process involves caustic fusion of the zircon using an excess of sodium hydroxide at 650 °C or at 1000 °C where sodium carbonate is used [9].

$$ZrSiO_4 + 4NaOH = Na_2ZrO_3 + Na_2SiO_3 + 2H_2O \qquad \text{Eq. 1}$$

Sodium zirconate is formed as shown by Eq. 1 together with sodium silicate and all reaction products are then slurried in water to dissolve the Na_2SiO_3 and hydrolyse Na_2ZrO_3, forming $Zr(OH)_4$. The hydroxide can then be filtered and calcined to ZrO_2 or treated with mineral acids for further processing to other compounds. Most of the zirconium processing technologies are related to nuclear energy production hence most of the knowledge is limited to a few companies.

Due to the limited information available in the literature on the recovery of niobium and zirconium from industrial wastes, the objective of this study is to investigate the kinetics of zirconium and niobium dissolution in hydrochloric acid using neutralised titania waste as a secondary raw material. This study will also contribute to a better understanding of dissolution of niobium and zirconium not only from neutralised titania waste but also from a wide range of industrial wastes which contain the metals.

Experimental

Materials

The neutralised titania waste as wet cake was obtained from a titanium dioxide plant employing a typical range of ore feedstocks. Analytical grade HCl was used for dissolutions and dilutions were done using deionised water. All glassware was cleaned with dilute nitric acid and rinsed with distilled water several times before use.

Leaching experiment

A batch of 5 kg of neutralised titania cake was oven dried to constant weight (approximately 2.5 kg) at 80 °C, ground and sieved through a 100 micron sieve. Acid leaching experiments were conducted batch-wise in a closed 1000 mL Pyrex reaction flask. Agitation was provided by a magnetic stirrer that enabled adequate dispersion of the particles without evaporation loss of the solution and heating by a thermostatically controlled hot plate. In each run, 20 g of the dried titania cake was added to 400 mL HCl of certain temperature and molality as required. For determining the effect of solid-to-liquid ratio, the volume of acid was adjusted before adding 20 g dried cake. At selected time intervals, 5 mL of slurry was withdrawn and filtered for analysis.

Characterization

Niobium and zirconium content of the sampled solutions was determined by inductively coupled plasma optical emission spectrometry (ICP OES), using which the dissolution models were developed. The elemental composition of the cake (2.4 wt% as Nb_2O_5, 5.5 wt% as ZrO_2 and the rest being mainly TiO_2 and Fe_2O_3) was determined by X-ray fluorescence spectrometry (XRF).

Results and discussion

Effect of leaching temperature

The leaching was carried out in the temperature range 25 – 90 °C for an initial hydrochloric acid concentration of 1.5 M. Figure 1 presents the influence of temperature on the dissolution of zirconium and niobium. The results indicate that both zirconium and niobium extraction gradually increase with increasing temperature in the range 25 – 70 °C. When the leaching temperature is increased to 90 °C, both zirconium and niobium extraction decreased and this may be due to homogeneous precipitation of the hydrous oxides of zirconium and niobium [10 – 12]. The change in temperature from 25 °C to 70 °C raises zirconium extraction from 34 to 48% and niobium extraction from 29 to 46% after one 60 minutes. Increasing the leaching temperature further to 90 °C lowers the extraction to 37% and 36% for zirconium and niobium respectively. A similar phenomenon has been observed on other transition metals such as nickel which precipitates as an oxalate at temperatures above 50 °C [14] and titanium, which precipitate as hydrous titanium dioxide at temperatures above 70 °C [13].

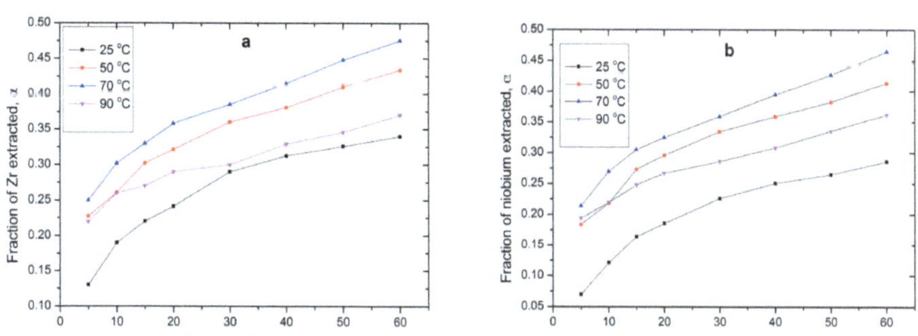

Figure 1. The effect of temperatures on the fractions of zirconium (a) and niobium (b) extracted in a 1.5 M HCl acid medium.

Effect of acid concentration

The effect of hydrochloric acid concentration on the extraction of zirconium and niobium has been studied within a range of 0.5 to 2.5 M at 70 °C. From Figure 2, it can be seen that the acid concentration has a significant influence on both zirconium and niobium extraction. Increasing the acid concentration from 0.5 M to 2.5 M raises the extraction from 4.5 to more than 61% for zirconium and from 3.6 to over 60% for niobium. This is because increasing the concentration of reactants increases the rate across the interfacial boundary layer and hence the rate of leaching.

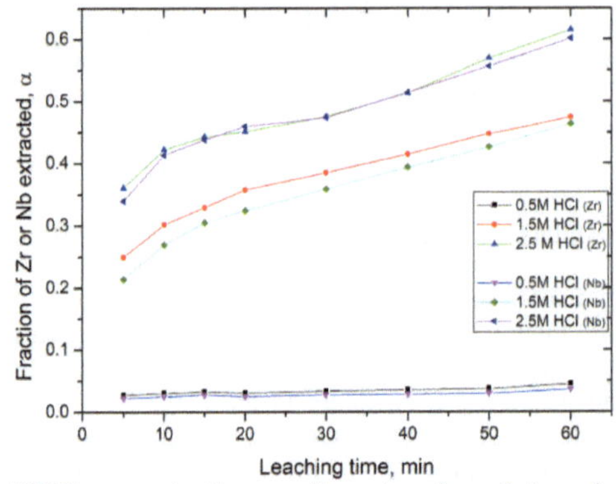

Figure 2. The effect of HCl concentrations on the extraction of zirconium and niobium at 70 °C.

Effects of solid-to-liquid ratio and stirring speed

The effect of solid-to-liquid ratio in the range 1/20 – 1/5 has been studied at 70 °C, an initial acid concentration of 2.5 M and a leaching time of 60 minutes. Figure 3 presents (a) the influence of solid-to-liquid ratio and (b) the effect of stirring speed on the extraction of zirconium and niobium.

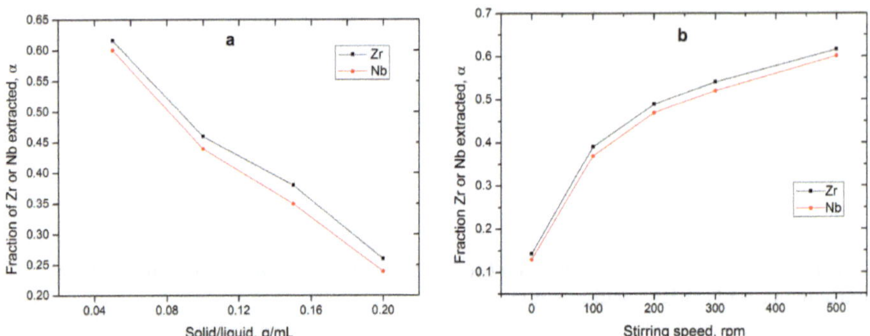

Figure 3. The effect of solid-to-liquid ratio on extraction of zirconium and niobium at 70 °C and an initial HCl concentration of 2.5 M.

The results indicate that the extraction decreases significantly as the solid-to-liquid ratio is increased. This may be because when the solid-to-liquid ratio is high, the association of the zirconium and niobium with HCl is greatly reduced. The extraction of niobium and zirconium ions increases with increasing stirring speed as shown in Figure 3 (b). This is a typical behavior of a diffusion limited chemical reaction [15] where increasing the stirring speed increases the interfacial area and hence the mass transfer rate.

Kinetic analysis

The rate-limiting step is the slowest step in the leaching process and it controls the overall leaching kinetics. The rate-limiting step can either be [16]:

1. Diffusion in the liquid film surrounding the solid particles (film diffusion).
2. Diffusion within the particles – in pores or through the solid phase itself (particle diffusion).
3. Chemical reaction at the surface of the particle (surface reaction).

It is therefore critical to establish a quantitative measurement of the leaching kinetics and mechanism. The experimental data were correlated to diffusion and surface reaction controlled kinetics models for solid-liquid reactions, connecting the fraction reacted (α) and time (t). A particle diffusion controlled reaction should give a straight line with a rate constant (k) as the gradient when $1-3(1-\alpha)^{\frac{2}{3}}$ is plotted against leaching time (t) and if the leaching is controlled by surface reaction, a plot of $1-(1-\alpha)^{\frac{1}{3}}$ against time should give a straight line with rate constant, k as the gradient [17]. The kinetic model developed by Dickinson and Heal (1999) [18]:

$$\left((1-\alpha)^{-\frac{1}{3}}-1\right)+\frac{1}{3}\ln(1-\alpha)=kt \qquad \text{Eq. 2}$$

describe a scenario where both the interface transfer and diffusion across the leached layer affect the rate of the leaching reactions. A plot of $\left((1-\alpha)^{-\frac{1}{3}}-1\right)+\frac{1}{3}\ln(1-\alpha)$ against t should therefore give a straight line with gradient k if the leaching reaction is controlled by both interfacial and particle diffusion.

Tables 1 and 2 compare the rate constant data for zirconium and niobium at various temperatures, obtained from three different linear regression analyses for the models mentioned above. It is observed from both Tables 1 and 2 that the error is least in the case of Dickinson and Heal's model [17] meaning that both the interface transfer and diffusion across the leached layer affect the rate of the leaching reactions. In view of this, Eq. 2 was used for calculating the Arrhenius plots for determination of activation energies.

Table 1: A comparison of linear regression analysis of rate constant at various temperatures obtained by curve fitting to different rate equations.

Temperature	$1-3(1-\alpha)^{\frac{2}{3}}$		$1-(1-\alpha)^{\frac{1}{3}}$		$\left((1-\alpha)^{-\frac{1}{3}}-1\right)+1/3\ln(1-\alpha)$	
(°C)	k (s^{-1})	Error	k ($^{s-1}$)	Error	k (s^{-1})	Error
25	7.12×10^{-4}	5.42×10^{-5}	1.41×10^{-3}	1.72×10^{-2}	1.74×10^{-4}	8.64×10^{-6}
50	1.06×10^{-3}	4.73×10^{-5}	1.57×10^{-3}	1.23×10^{-4}	2.73×10^{-4}	8.49×10^{-6}
70	1.27×10^{-3}	5.23×10^{-5}	1.71×10^{-3}	1.26×10^{-4}	3.47×10^{-4}	1.09×10^{-5}
90	6.18×10^{-4}	3.52×10^{-5}	1.03×10^{-3}	8.09×10^{-5}	1.49×10^{-4}	7.80×10^{-6}

Table 2. A comparison of linear regression analysis of rate constant at various temperatures obtained by curve fitting to different rate equations.

Temperature (°C)	$1-3(1-\alpha)^{\frac{2}{3}}$		$1-(1-\alpha)^{\frac{1}{3}}$		$\left((1-\alpha)^{-\frac{1}{3}}-1\right)+1/3\ln(1-\alpha)$	
	k (s^{-1})	Error	k ($^{s-1}$)	Error	k (s^{-1})	Error
25	5.32 x 10^{-4}	3.10 x 10^{-5}	1.38 x 10^{-3}	1.71 x 10^{-4}	1.24 x 10^{-4}	5.17 x 10^{-6}
50	1.02 x 10^{-3}	5.53 x 10^{-5}	1.67 x 10^{-3}	1.59 x 10^{-4}	2.52 x 10^{-4}	1.02 x 10^{-5}
70	1.28 x 10^{-3}	4.54 x 10^{-5}	1.84 x 10^{-3}	1.21 x 10^{-4}	3.39 x 10^{-4}	1.31 x 10^{-5}
90	6.66 x 10^{-4}	2.58 x 10^{-5}	1.18 x 10^{-3}	7.13 x 10^{-5}	1.57 x 10^{-4}	6.09 x 10^{-6}

Figures 4 and 5 show (a) the plot of $\left((1-\alpha)^{-\frac{1}{3}}-1\right)+\frac{1}{3}\ln(1-\alpha)$ versus leaching time and (b) the Arrhenius plots in terms of the leaching of zirconium and niobium respectively. To calculate the apparent activation energy (E_a), the Arrhenius equation (Eq. (3)) is employed:

$$\ln k = \ln A - \left(\frac{E_a}{RT}\right) \qquad \text{Eq. 3}$$

where k is the apparent rate constant, A is the pre-exponential factor, R is the universal gas constant and T is the absolute temperature. The plot of ln k against $1/T$ should be a straight line with a slope of $-Ea/R$ and the intercept of ln A.

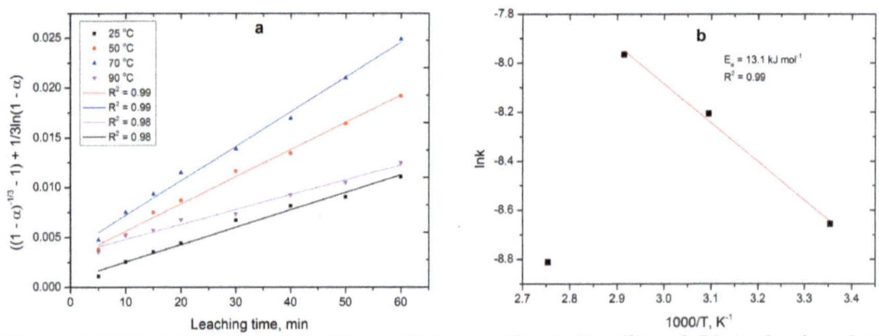

Figure 4. (a)A plot of the data in Figure 1(a) according to Eq. (2) and (b) Arrhenius plot for the leaching of zirconium.

The apparent activation energies were determined from the slopes of temperature ranges 25 – 70 °C to be 13.1 ± 1 kJ mol^{-1} and 21.3 ± 3 kJ mol^{-1} for zirconium and niobium respectively. These values are consistent with values expected of a diffusion controlled reaction [19]. Apparent rate constants from leaching at 90 °C were not used in the calculation of activation energies due to possible precipitation of zirconium and niobium oxides/oxychlorides during leaching, which underestimates the apparent rate constant for the reaction.

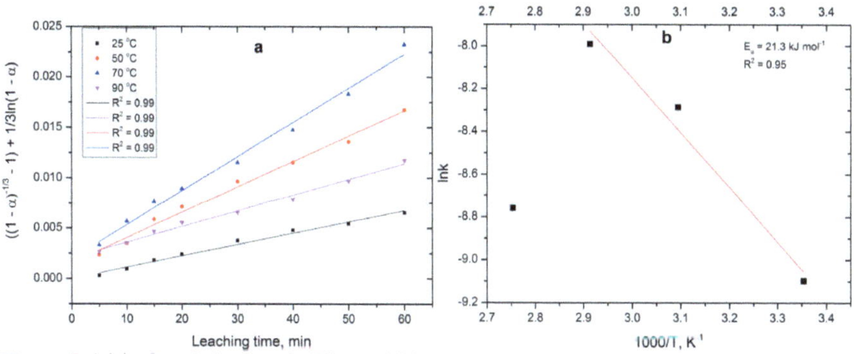

Figure 5. (a)A plot of the data in Figure 1(b) according to Eq. (2) and (b) Arrhenius plot for the leaching of niobium.

Conclusions

From the results obtained in this investigation, the following conclusions may be drawn:

1. An increase in temperature from 25 to 70 °C raises the extraction of both zirconium and niobium. However on further increasing the temperature to 90 °C leads to a decrease in both zirconium and niobium extraction possibly, due to the hydrothermal precipitation as either oxides or oxy-chlorides.
2. The leaching process follows the kinetic model $\left((1-\alpha)^{-\frac{1}{3}}-1\right)+\frac{1}{3}\ln(1-\alpha)=kt.$
3. The concentration of acid has a significant influence on the extraction of both zirconium and niobium. Increasing concentration from 0.5 M to 2.5 M leads to an increase in recovery of both zirconium and niobium from around 4% to 60%.
4. Both the interface transfer and diffusion across the leached product layer affect the overall rate of reactions with activation energies of 13.1 ± 1 kJ mol^{-1} and 21.3 ± 3 kJ mol^{-1} for zirconium and niobium, respectively.
5. As solid-to-liquid ratio increases, the extraction of both zirconium and niobium decreased significantly.

Acknowledgements

The authors wish to thank the Engineering and Physical Sciences Research Council and Huntsman Pigments for the financial support. The help from Huntsman Pigments' analytical team with analysis of samples is greatly appreciated.

References

[1] Ullmann, F., Gerhartz, W., Yamamoto, Y., Campbell, F., Pfefferkorn, R., Rounsaville, J. and Ullmann, F. (1985). *Ullmann's encyclopedia of industrial chemistry*. 1st ed. Weinheim, Federal Republic of Germany: VCH.

[2] Schuiling, R., Vergouwen, L. and Rijst, H. (1976). Gibbs energies of formation of zircon ($ZrSiO_4$), thorite ($ThSiO_4$), and phenacite (Be_2SiO_4).

[3] Yang, X., Zhang, J., Fang, X. and Qiu, T. (2014). Kinetics of pressure leaching of niobium ore by sulfuric acid. *International Journal of Refractory Metals and Hard Materials*, 45, pp.218--222.

[4] Guo W., Wang Z. (2009). *Modern tantalum and niobium metallurgy*. Beijing: Metallurgical Industry Press.

[5] Zhou, H., Zheng, S., Zhang, Y. and Yi, D. (2005). A kinetic study of the leaching of a low-grade niobium--tantalum ore by concentrated KOH solution. *Hydrometallurgy*, 80(3), pp.170--178.

[6] Wang, X., Zheng, S., Xu, H. and Zhang, Y. (2009). Leaching of niobium and tantalum from a low-grade ore using a KOH roast--water leach system. *Hydrometallurgy*, 98(3), pp.219--223.

[7] Kirk, R., Othmer, D., Kroschwitz, J. and Howe-Grant, M. (1996). *Encyclopedia of chemical technology*. 1st ed.

[8] Habashi, F. (1997). *Handbook of extractive metallurgy*. 1st ed. Weinheim: Wiley-VCH.

[9] da Silva, R., Dutra, A. and Afonso, J. (2012). Alkali fusion followed by a two-step leaching of a Brazilian zircon concentrate. *Hydrometallurgy*, 117, pp.93--100.

[10] Da Silva, G., Da Silva, M. and Caetano, T. (2002). Preparation and characterization of hydrous zirconium oxide formed by homogeneous precipitation. *Materials Research*, 5(2), pp.149--153.

[11] Akhtar, K., Haq, I. and Hira, U. (2013). Synthesis and Characterization of Uniform Zirconia Particles by Homogeneous Precipitation Method. *High Temperature Materials and Processes*, 32(4), pp.391--395.

[12] Rodrigues, L. and da Silva, M. (2010). Thermodynamic and kinetic investigations of phosphate adsorption onto hydrous niobium oxide prepared by homogeneous solution method. *Desalination*, 263(1), pp.29--35.

[13] Pfaff, G. (2008). *Special effect pigments*. 1st ed. Hannover: Vincentz Network.

[14] Szymczycha-Madeja, A. (2011). Kinetics of Mo, Ni, V and Al leaching from a spent hydrodesulphurization catalyst in a solution containing oxalic acid and hydrogen peroxide. *Journal of hazardous materials*, 186(2), pp.2157--2161.

[15] Kokosa, J., Przyjazny, A. and Jeannot, M. (2009). *Solvent microextraction*. 1st ed. Hoboken, N.J.: Wiley.

[16] Paul, E., Atiemo-Obeng, V. and Kresta, S. (2004). *Handbook of industrial mixing*. 1st ed. Hoboken, N.J.: Wiley-Interscience.

[17] Gavhane, K. (2008). *Chemical Reaction Engineering II*. 1st ed. Mumbai: Arihant.

[18] Havlík, T. (2008). *Hydrometallurgy*. 1st ed. Cambridge: Cambridge International Science Pub.

[19] Han, K. (2002). *Fundamentals of Aqueous Metallurgy*. 1st ed. Littleton: SME.

Energy Technology 2015: Carbon Dioxide Management and Other Technologies
Edited by: Animesh Jha, Cong Wang, Neale R. Neelameggham, Donna P. Guillen, Li Li, Cynthia K. Belt, Randolph Kirchain, Jeffrey S. Spangenberger, Frank Johnson, Andrew J. Gomes, Amit Pandey, and Peter Hosemann
TMS (The Minerals, Metals & Materials Society), 2015

THE OPTIMIZATION FORMULA DESIGN OF $Cu_xZn_{1-x}O$ INFRARED RADIATION MATERIAL AND COATING SLURRY

Yuhao Ding[1, 2], Hao Bai[1, 2,*], Chao Lian[2], Wei Wei[2], Wenquan Liu[3]

[1]State Key Laboratory of Advanced Metallurgy, University of Science and Technology 30# Xueyuan Road, Beijing 100083, China
[2]School of Metallurgical and Ecological Engineering, University of Science and Technology 30# Xueyuan Road, Beijing 100083, China
[3]China Metallurgical Industry Planning and Research Institute
Corresponding author: baihao@metall.ustb.edu.cn

Keywords: infrared radiation; solid solution; emissivity; optimization model; molar content.

Abstract

Infrared radiation material(IRM) is main content for coating slurry applied on the inner surface of industrial furnaces, which improve heat transfer and efficiency of furnaces can rise by 5%~30%. In this paper, the $Cu_xZn_{1-x}O$ infrared radiation material that can greatly improve the materials' emissivity and the coating slurry were prepared. X-ray diffraction analysis, differential thermal analysis and scanning electron microscope-energy dispersive spectrum analysis indicate that the IRM was prepared successfully. The emissivity of three wavebands was measured. In order to obtain higher emissivity, optimization models were proposed and with calculation by Matlab software, the optimal molar content of CuO can be obtained as 3.6% in IRM and the optimal formula for coating slurry which includes IRM, dispersant, thickener and adhesive was obtained. The emissivity of the coating under the optimal formula was measured to be 0.965 and 0.958 in two main waveband which is in agreement with optimized calculation.

Introduction

In China, about 70% of the total energy is consumed in industry[1-2] and 25%~40% of the energy is used for combustion in industrial furnaces. Since the energy utilization efficiency of furnaces is generally low, it is important to take measures to improve their performance. Infrared radiation coating applied on the surface of inner wall in furnaces has been widely used[3].The infrared radiation coating can enhance emissivity of the furnaces' inner wall by which radiation transfer can be strengthened between the heated materials and the furnaces' wall[4]. The application of infrared radiation coating can raise the efficiency by 5%~30%[5].
Researchers have studied the coating's energy-saving mechanism from many aspects.,

since it is found "doping" improves emissivity especially in the wavelength range of 1-5μm, one explanation is of free carrier absorption mechanism whereby an electron within a band absorbs radiation by transferring from a low-energy level to an empty high-energy level[6], d-d transitions in octahedral coordination[7], distortion of the crystal lattice. For semiconductor material free carrier absorption mechanism generally dominates the emissivity enhancement in the wavelength region of 1-5μm. Zinc oxide (ZnO), an important II–VI semiconductor with a band gap of 3.3eV and a large exciton binding energy of 60meV, is the hexagonal wurtzite structure[8-10]. Recently, many elements such as Al, Mg, Co, Ga, Ce and Cu[11-13] have been doped or alloyed into ZnO samples and good properties have been obtained. Cu is a prominent luminescence activator in II–VI compounds, which is considered that Cu dopant could modify the luminescence of ZnO crystals through creating localized impurity levels. In addition, Cu has many physical and chemical properties that are similar to those of Zn. Cu doping has been reported to be able to change the microstructure and optical properties of ZnO thin samples[13].

In this paper, $Cu_xZn_{1-x}O$ series solid solution were prepared, Cu doped ZnO crystals can greatly improve the infrared radiation properties of ZnO crystals. Based on this idea, IRM and infrared radiation coatings were prepared in laboratory, and then the emissivity was measured. According to the experimental results, optimal mathematical models were applied to determine the optimum formula of IRM and infrared radiation coatings slurry.

Preparation of CuO-ZnO materials

According to the formulas of preparing $Cu_xZn_{1-x}O$ materials and the analytical method of performance mentioned in the paper of Chao Lian[14] etc, the samples were prepared and the solid solution reaction was confirmed. Thus the emissivity of 3~5μm and 8~14μm waveband were measured at 773K and the result was shown in Fig.1.

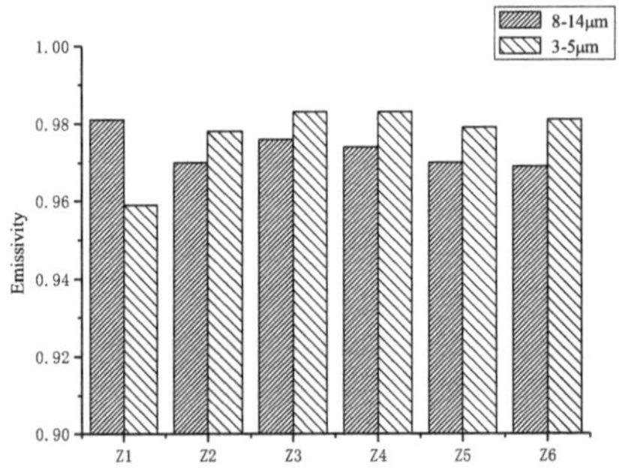

Fig.1 The observed values of each formula's emissivity

And the maximum emissivity in two waveband are all more than 0.98. In the article by Chao Lian[14] et al, the validity of the emissivity strengthening effect of solid solution has been described. But owing to complicated mechanism of emissivity strengthening effect of solid solution, the correlation between emissivity strengthening and degree of solid solution is not linear. Thus, in order to obtain the optimized emissivity performance of $Cu_xZn_{1-x}O$ and coating made of $Cu_xZn_{1-x}O$, it is necessary to optimize the formulae of both $Cu_xZn_{1-x}O$ material and the coating produced from the material.

Optimization of the CuO-ZnO materials

According to Planck's law, the curve E_λ-λ of the gray body's surface whose emissivity is 0.9 can be obtained at 773K, shown in Fig.2. The area below the curve indicates emissive power in 1~22 μm waveband, which is divided into five parts, and the proportion of each part can be obtained by integral computing, the result shown in Table.1.

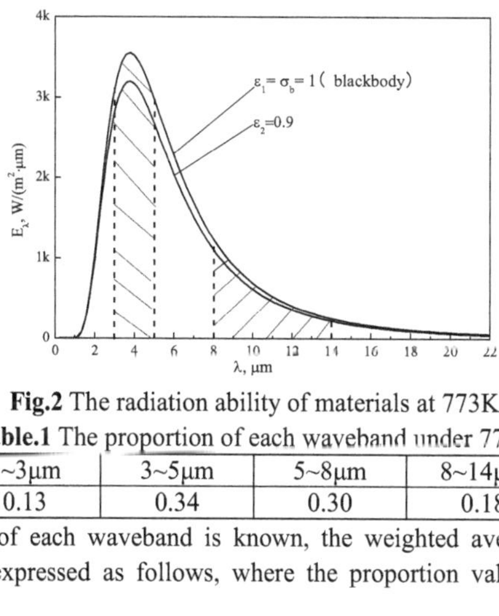

Fig.2 The radiation ability of materials at 773K

Table.1 The proportion of each waveband under 773K

waveband	1~3μm	3~5μm	5~8μm	8~14μm	14~22μm
proportion	0.13	0.34	0.30	0.18	0.05

If the emissivity of each waveband is known, the weighted average emissivity in 1~22μm can be expressed as follows, where the proportion value of each part is weight factor:

$$\bar{\varepsilon} = 0.13\varepsilon_1 + 0.34\varepsilon_2 + 0.30\varepsilon_3 + 0.18\varepsilon_4 + 0.05\varepsilon_5 \qquad (1)$$

For each sample of $Cu_xZn_{1-x}O$ material based on different formulas, there would be different emissivity ε_i in every waveband, which leads to different values of $\bar{\varepsilon}$. Thus, there must be some relation between the $\bar{\varepsilon}$ and the contents in each formula. The optimization method is to obtain optimal formula by solving extreme value problem for maximum of $\bar{\varepsilon}$.

Since the Dual-band radiation emissivity measuring meter was used, the emissivity was measured only in the waveband of 3~5μm and 8~14μm. Thus the Equ.1 was simplified into the form shown as follows.

$$\bar{\varepsilon} = 0.34\varepsilon_2 + 0.18\varepsilon_4 \qquad (2)$$

Since the Dual-band radiation emissivity measuring meter was used, the emissivity was measured only in the waveband of 3~5μm and 8~14μm. Thus the Equ.1 was simplified into the form shown as follows.

$$\bar{\varepsilon}=0.34\varepsilon_2+0.18\varepsilon_4 \quad (2)$$

The mole fraction of Cu which was expressed as x was the independent variable of function of each waveband's emissivity. The optimization of formula was carried out according to the data shown in Fig.1. The precision can meet the demand when the fitted equation is a five-order poly-nominal function. The results of fitting are shown as follows:

$$\varepsilon_2=0.906+0.074\times x-0.024\times x^2+3.30\times 10^{-3}\times x^3-2.01\times 10^{-4}\times x^4+4.48\times 10^{-6}\times x^5$$

$$\varepsilon_4=0.739+0.195\times x-0.0544\times x^2+6.828\times 10^{-3}\times x^3-3.941\times 10^{-4}\times x^4+8.524\times 10^{-6}\times x^5$$

With the Matlab software, extreme computing with objective function base on Equ.2 made the optimal formula of spinal materials being 3.52% of CuO and 96.48% of ZnO. The sample prepared base on the optimal formula was tested and the emissivity values are shown in Table.2, where the fitted values are also given for comparison. As shown in Table.2, the fitted and experimental results are agreed with each other well.

Table.2 The experimental values and fitted values of CuO-ZnO materials

waveband	3~5μm	8~14μm
Fitted values	0.985	0.977
Experimental values	0.965	0.958
Deviation	2.1%	2%

Preparation and Optimization of coating slurry

The $Cu_xZn_{1-x}O$ series infrared radiation coating slurry are composed of $Cu_xZn_{1-x}O$ materials, adhesive, dispersant and thickener [15]. These additives would affect the emissivity of the coating, but how and to what degree they affect the emissivity is unknown. Thus, in order to obtain the optimal formula of the coating slurry for the best emissivity, it is necessary to study the relation of the component of the slurry and the emissivity.

Preparation of coating slurry

Phosphate, betonies and sodium hexametaphosphate were selected as adhesive, thickener and dispersant respectively. The formula design and optimization were through mixture regression design method, which include extreme vertices design, boundary surface centroid design and overall centroid design, where the content is expressed by mass percentage. If the number of factors is p in the mixture regression design, the content of the *pth* factor is expressed as x_p. Then the criteria of the design can be concluded as follows:

$$\begin{cases} 0 \le x_i \le 1(\mathrm{i}=1,2,3,...,\mathrm{p}) \\ x_1+x_2+x_3+...+x_p=1 \end{cases}$$

Each component changes within a certain range, which was determined according to previous research and experiments. The ranges of the components of the coating materials are shown as follows:

$$\begin{cases} \text{CuO-ZnO material}(x_1): 0.3 \le x_1 \le 0.6 \\ \text{sodium hexametaphosphate}(x_2): 0.05 \le x_2 \le 0.15 \\ \text{bentonite}(x_3): 0.15 \le x_3 \le 0.2 \\ \text{phosphate}(x_4): 0.3 \le x_4 \le 0.5 \end{cases}$$

Table.3 Formula design

Number		x_1	x_2	x_3	x_4
Extreme Vertices Design	1	0.50	0.05	0.15	0.30
	2	0.30	0.05	0.15	0.50
	3	0.45	0.05	0.20	0.30
	4	0.40	0.15	0.15	0.30
	5	0.35	0.15	0.20	0.30
	6	0.30	0.05	0.20	0.45
	7	0.30	0.15	0.15	0.40
	8	0.30	0.15	0.20	0.35
Boundary Surface Centroid Design	9	0.36	0.10	0.18	0.36
	10	0.30	0.10	0.18	0.42
	11	0.39	0.05	0.18	0.38
	12	0.34	0.15	0.18	0.33
	13	0.38	0.10	0.15	0.37
	14	0.35	0.10	0.20	0.35
Overall Centroid Design	15	0.42	0.10	0.18	0.30

According to different rules of the design methods, some formulas can be obtained as shown in Table.3. Based on these formulas, 15 kinds of coating slurry were made by mixing, water adding and stirring. After coated on the mullite refractory brick and dried, these samples were measured for emissivity.

Optimization of coatings

The regression equation with quadratic form was used for the fitting computing for the data of mixture design experiments. The equation can be expressed as follows, where x_i indicates one component in the formula and b_i and b_{ij} undetermined factors.

$$\hat{\varepsilon} = \sum_{i=1}^{4} b_i x_i + \sum_{i<j} b_{ij} x_{i} \qquad (3)$$

Thus, with the SPSS statistical software, the regression equation of 3~5μm can be obtained as Equ.4. The comparison between fitted and experimental values in this waveband is shown in Fig.4.

$$\varepsilon_2 = -24.797x_2 - 11.556x_3 + 1.188x_4 + 38.435x_1x_2 + 20.246x_1x_3 - 3.362x_1x_4 + 32.685x_2x_3 + 31.484x_2x_4 + 19.317x_3x_4 - 0.197 \quad (4)$$

Table.4 the emissivity of the formulas

Number	x_1	x_2	x_3	x_4	ε_2	ε_4
1	0.50	0.05	0.15	0.30	0.754	0.874
2	0.30	0.05	0.15	0.50	0.913	0.878
3	0.45	0.05	0.20	0.30	0.801	0.871
4	0.40	0.15	0.15	0.30	0.814	0.841
5	0.35	0.15	0.20	0.30	0.698	0.848
6	0.30	0.05	0.20	0.45	0.911	0.955
7	0.30	0.15	0.15	0.40	0.833	0.859
8	0.30	0.15	0.20	0.35	0.774	0.824
9	0.36	0.10	0.18	0.36	0.919	0.831
10	0.30	0.10	0.18	0.42	0.860	0.917
11	0.39	0.05	0.18	0.38	0.772	0.864
12	0.50	0.05	0.15	0.30	0.915	0.898
13	0.30	0.05	0.15	0.50	0.910	0.930
14	0.45	0.05	0.20	0.30	0.930	0.906
15	0.40	0.15	0.15	0.30	0.910	0.904

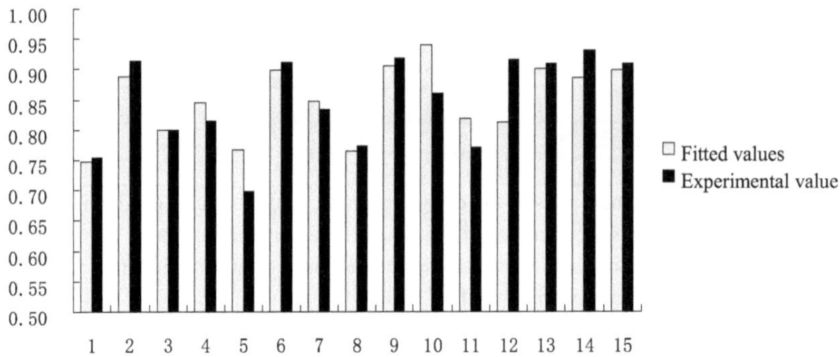

Fig.4 Comparison of fitting results and the experimental values in 3~5μm

Similarly, the regression equation of 8~14μm waveband can be obtained like Equ.5. The comparison between fitted and experimental values is shown in Fig.5.

$$\varepsilon_4 = -9.861x_2 + 4.536x_3 - 0.981x_4 + 13.345x_1x_2 - 10.868x_1x_3 - 2.228x_1x_4 - 4.614x_2x_3 + 11.901x_2x_4 - 3.384x_3x_4 + 1.803 \quad (5)$$

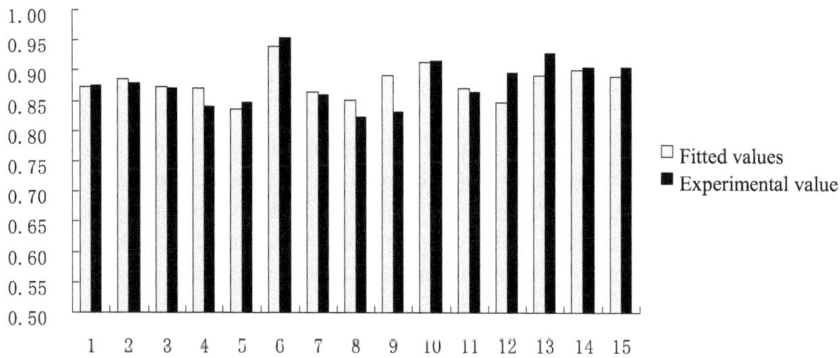

Fig.5 comparison of fitting results and the measured values in 8~14μm

To optimize the formula, combining Equ.4 and Equ.5 into Equ.2, and final regression equation was obtained as Equ.6.

$$\varepsilon=0.34\varepsilon_2+0.18\varepsilon_4 \tag{6}$$

$$=-10.20596x_2-3.11256x_3+0.22734x_4+15.47x_1x_2+4.9274x_1x_3$$
$$-1.54412x_1x_4+10.28238x_2x_3+12.84674x_2x_4+5.95866x_3x_4+0.25756$$

The optimal solution can be obtained with Matlab, which is x_1=30%, x_2=8.58%, x_3=17.13%, x_4=44.29%. The optimal fitted values are calculated to be 0.947 and 0.916 in 3~5μm and 8~14μm waveband respectively. In order to verify the accuracy of the model, the coating slurry was prepared and the emissivity was measured at773K.The result is shown in Table.5.

Table.5 the fitting values and experimental values

Waveband	3~5μm	8~14μm
Fitted values	0.947	0.916
Experimental values	0.939	0.913
Deviation	0.85%	0.33%

As shown in Table 5, the deviations between fitted and measured values in 3~5μm and 8~14μm band at 773K are 0.85% and 0.33%. This indicates that the model can accurately predict the emissivity for a certain formula.

Table.6 the emissivity of optimal formula of base materials and coating slurry

Waveband	3~5μm	8~14μm
Optimal materials	0.965	0.958
Optimal coating	0.939	0.913

In summary, the emissivity of optimal formula of the IRM and coating slurry were compared in 3~5μm band and 8~14μm band. The result is shown in Table.6, which shows that the preparation of coatings was successful based on the optimization method, and this offers helpful information for preparing CuO-ZnO based high emissivity coatings in the future.

Conclusions

(1).By using optimization model, optimal formula which is 3.52% of CuO and 96.48% of ZnO was obtained for preparation of $Cu_xZn_{1-x}O$ infrared radiation material. At 773K, the emissivity of the optimal formulation are 0.965 and 0.958 in 3~5μm band and 8~14μm band.

(2). Based on the $Cu_xZn_{1-x}O$ material, with the mixture regression design method, high emissivity coating slurry was prepared and the emissivity of the coating was measured. By further optimizing the emissivity of the coating slurry, the optimal formula, which contains 30.00% of $Cu_xZn_{1-x}O$ material, 8.58% of sodium hexametaphosphate, 17.13% of bentonite, 44.29% of phosphate, was obtained. The deviation between the fitted values and the experimental values of the coating was small, which means the optimization method is reasonable.

References

[1] F. D. Shi. The Cause of the Change of Chinese Industrial Energy Consumption. System's Engineering, 2008, 26(4): 55-56.

[2] Y. Y. Jiang, H. L. Wang. Research on the relation between industrial energy consumption and industrial economy growth of China[J]. Statistics and Decision, 2009(4): 114-116.

[3] L. P Ma, D Zhang. The research and development of a new infrared radiation ceramics coating technology [J]. Energy saving technology. 1998(2): 15-16.

[4] Jianxian Zhang, Yongjun Zou. High-emissivity Coating and Its Application[J]. Infrared Technology, 2007, 29(8):491-494

[5] J. C Zhou, J. K Chen, P. L Tu. High Quality Infrared Coating and It's Application in Resistance Furnace[J]. Heat treatment of metals. 1992(8): 40-45.

[6] N.M. Ravindra, R.K. Singh, "Transient Thermal Processing Techniques in Electronic Materials", TMS Soc., Warrendale, PA. Aug. 1996, 159–164.

[7] S.Singh, M.S.R.Rao,"Optical and electrical resistivity studies of isovalent and aliovalent 3d transition metal ion doped ZnO", Phys. Rev.B., 80(2009) 045210-045220.

[8] J. Alaria et al., "No ferromagnetism in Mn doped ZnO semiconductors", Chemical Physics Letters, 415 (2005) 337–341.

[9] M.H.F. Sluiter et al., "First principles based design and experimental evidence for a ZnO-based ferromagnet at room temperature", Phys. Rev. Lett., 94(2005). P. 187204-1.

[10] J.J.Ding et al.,"Effect of substrate and annealing on the structural and optical properties of ZnO:Al sampless", Journal of Physics and Chemistry of Solids, 71 (2010) 346–350

[11] W.I. Park et al., "Photoluminescent Properties of ZnO/Zn0.8Mg0.2O Nanorod Single-Quantum-Well Structures", J. Phys. Chem. B, 108 (2004) 15457–15460.

[12] Z.Q.Ma et al., "Electrical properties of Na/Mg co-doped ZnO thin sampless", Thin Solid Sampless, 515 (2007) 8611–8614.

[13] Z.Zhang et al., "Cu-Doped ZnO Nanoneedles and Nanonails: Morphological Evolution and Physical Properties", J. Phys. Chem. C, 112 (2008), pp. 9579–9585.

[14] Chao Lian et al. Infrared radiation properties of CuO-ZnO-based sintered material prepared for energy-saving coating. The Minerals, Metals & Materials Socity. 2013

[15] Yulong Zhang, Shiguo Du. Design and examples of coating formula[M]. Beijing: China Textile Press, 2010.

Energy Technology 2015: Carbon Dioxide Management and Other Technologies
Edited by: Animesh Jha, Cong Wang, Neale R. Neelameggham, Donna P. Guillen, Li Li, Cynthia K. Belt, Randolph Kirchain, Jeffrey S. Spangenberger, Frank Johnson, Andrew J. Gomes, Amit Pandey, and Peter Hosemann
TMS (The Minerals, Metals & Materials Society), 2015

THE ROLE OF AUSTENITIZING ROUTINES OF PIPE STEELS DURING CCS

Anja Pfennig[1], Johann Schiz[1], Axel Kranzmann[2]

[1] HTW University of Applied Sciences Berlin,
Wilhelminenhofstraße 75 A, Gebäude C, 12459 Berlin
anja.pfennig@htw-berlin.de

[2] BAM Federal Institute of Materials Research and Testing,
Unter den Eichen 87, 12205 Berlin

Keywords: corrosion, CCS, carbon storage, aquifer, austenitizing

Abstract

Properties of pipe steels for CCS technology require resistance against the corrosive environment of a potential CCS-site (heat, pressure, salinity of the aquifer, CO_2-partial pressure). The influence of austenitizing in heat treatment routines of two different injection pipe steels (1.4034, X46Cr13 and 1.4021, X20Cr13) was evaluated. Steel coupons were austenitized at different temperatures (900- 1050 °C) for different lengths of time (30-90 min) before quenching and annealing prior to long term corrosion experiments (60°C, 100 bar, artificial brine close to a CCS-site in the Northern German Basin, Germany). In general, fewer pits are found on X46Cr13. Comparing steels with 13% chromium each the higher carbon content of X46Cr13 (0.46% C) results in a lower number of pits compared to X20Cr13 (0.20% C). It is found that neither the carbon content of the steels nor austenitizing temperature has much influence, but local corrosion behaviour is most susceptible towards austenitizing time.

Introduction

Engineering a geological on-shore saline aquifer CCS-site (CCS Carbon Capture and Storage [1]) corrosion of injection pipe steels may become an issue when emission gasses, e.g. from combustion processes of power plants, are compressed into deep geological layers [2-5]. CO_2-corrosion is sensitively dependent on alloy composition, contamination of alloy and media, environmental conditions like temperature, CO_2 partial pressure, flow conditions and protective corrosion scales [3,6-10]. Because the solubility of $FeCO_3$ in water is low (p_{Ksp} = 10.54 at 25 °C [11] generally steels, that are exposed to CO_2-environment, precipitate slow growing surface layers mainly comprised of $FeCO_3$ (siderite) [2], which is also found in pits of locally corroded samples [8,9]:

	$CO_2\,(g) + H_2O\,(l)$	$\rightarrow$	$H^+ + HCO_3^-\,(aq)$	(1)
cathodic:	$2\,HCO_3^- + 2\,e^-$	$\rightarrow$	$2\,CO_3^{2-} + H_2$	(2)
anodic:	Fe	$\rightarrow$	$Fe^{2+} + 2e^-$	(3.1)
	$Fe_2+ + CO_3^{2-}$	$\rightarrow$	$FeCO_3$	(3.2)
	$Fe_2+ + 2\,HCO_3-$	$\rightarrow$	$Fe(HCO_3)_2$	(4.1)
	$Fe(HCO_3)_2$	$\rightarrow$	$FeCO_3 + CO_2 + H_2O$	(4.2)

Possibly the initial formation of $Fe(OH)_2$ [3,12] leads to an increase of the local pH near the hydroxide film resulting in the precipitation of an internal and external ferrous carbonate film, Han et. al [12]. Localized corrosion may then start especially at grain boundaries when the ferrous hydroxide film is locally damaged. The ferrous carbonate film is exposed to the brine, dissolves and is locally depassivated. As a consequence the carbonate film detaches (Fig. 1).

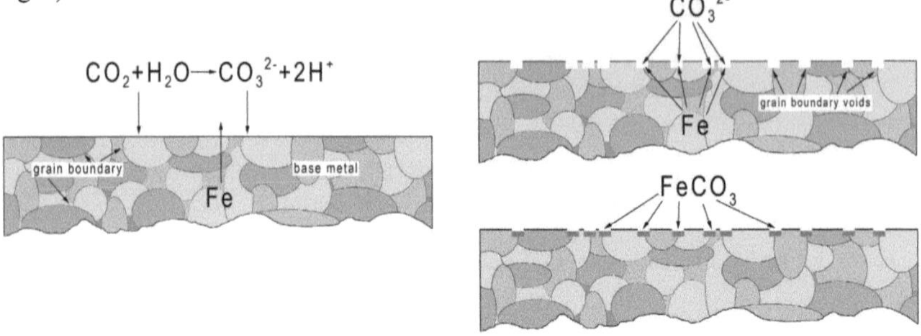

Fig. 1. Schematic drawing of pit precipitation on injection steels initiated at grain boundaries

The influence of heat treatment, that is: temperature and time of austenitizing, cooling rate as well as temperature and time of annealing, has been shown by various authors. The presence and amount of retained austenite as a microstructural component resulting from the heat treatments applied has a beneficial effect on the pitting corrosion resistance of 13%-chromium steels (13CrNiMo) [13]. A higher Ni and Cr content in the heat treated steels improve the corrosion resistance [14,15]. Hou et al. introduce a method of empirically calculating the influence of alloying elements in heat treated steels [16]. Cvijović and G. Radenković showed that the corrosion resistance of duplex steels with chromium contents even as high as 22-27% varied with solidification mode and annealing condition [17]. In general raising the annealing temperature lowers the pitting potential of lean duplex stainless steels [17-19]. The lowest potentials, corresponding to the transition from metastable to stable pitting, are observed for annealing at 900 °C while a maximum improvement of corrosion stability can be achieved by annealing at 1200 °C [17]. The better corrosion resistance of martensitic stainless steels with 13% Cr at higher austenitizing temperature (980-1050 °C) is related to the dissolution of carbides [19-21]. The precipitation of Cr-rich $M_{23}C_6$ and M_7C_3 carbides reduced the resistance of passive film and pitting corrosion [19] and has high impact on mechanical properties due to secondary hardening [20]. The influence of heat treatment on the microstructure and mechanical properties is well known [19,20,21]. However for C-Mn (carbon) steels in a H_2S-containing NaCl solution the martensitic microstructure has the highest corrosion rate up to two orders of magnitude higher than ferritic or ferritic-bainitic microstructures due to the fact that martensitic grain boundaries are more reactive [22].

Experimental

The steel qualities for laboratory experiments X20Cr13 (1.4021, AISI 420 J) and X46Cr13 (1.4034, AISI 420 C) are used as injection pipe. Exposure tests in CO_2-saturated aquifer brine were carried out using samples made of thermally treated specimen of the steels with 8 mm thickness, 20 mm width, 50 mm length. Austenitizing prior to exposure was done at 950 °C, 1000 °C and 1050 °C for 30 min, 60 min and 90 min. Following usual heat treatment protocols the specimen were then annealed at 650 °C for 30 min to gain martensitic microstructure with sufficient hardness and toughness. A hole of 3.9 mm diameter was used for sample positioning. Samples of each base metal were positioned within the vapour phase and within the liquid phase. The brine (as known to be similar to the Stuttgart Aquifer [23]: Ca^{2+}: 1760 mg/L, K^{2+}: 430 mg/L, Mg^{2+}: 1270 mg/L, Na^{2+}: 90,100 mg/L, Cl^{-}: 143,300 mg/L, SO_4^{2-}: 3600 mg/L, HCO_3^{-}: 40 mg/L) was synthesized in a strictly orderly way to avoid precipitation of salts and carbonates. Flow control (3 NL/h) of the technical CO_2 (purity 99,995 vol.-%)) into the brine at ambient pressure was done by a capillary meter GDX600_man by QCAL Messtechnik GmbH, Munic. The exposure of the samples between 700 h to 8000 h was disposed in autoclaves according to the conditions at the geological site at 60 °C at 100 bar and for comparison as well at ambient pressure – each material in a separated reaction vessel [8].

Before corrosion tests the surfaces of the steels were activated by grinding with SiC-Paper down to 120 µm under water. After the corrosion tests, the samples were cut partly for scale analysis with the corrosion layer and prepared partly for kinetic analysis after the scale was etched. Descaling of the samples was performed by exposure to 37% HCl. Then parts of the samples were embedded in a cold resin (Epoxicure, Buehler), cut and polished first with SiC-Paper from 180 µm to 1200 µm under water and then finished with diamond paste 6 µm and 1 µm.

Different light optical and electron microscopy techniques were performed on specimens to investigate the layer structures and morphology of the samples. X-ray diffraction was carried out in a URD-6 (Seifert-FPM) with CoKα-radiation with an automatic slit adjustment, step 0.03° and count 5 sec. Phase analysis was performed by matching peak positions automatically with PDF-2 (2005) powder patterns. Mainly structures that were likely to precipitate from the steels were chosen of the ICSD and refined to fit the raw-data-files using POWDERCELL 2.4 [24] and AUTOQUAN ® by Seifert FPM. To characterise the pitting corrosion, 3-D-images were realized by the double optical system Microprof TTV by FRT. Kinetics of the corrosion were determined by the corrosion rates which were calculated via mass change of the samples before and after corrosion testing according to DIN 50 905 part 1-4. The measurement of the layer thicknesses and residual pipe wall thicknesses was performed by the line scan method using the semi-automatic analyzing program Analysis Docu ax-4 by Aquinto. The scale thickness was measured for each exposure time taking a set of 10 to 20 micrographs. Then a set of 100 line scans was placed over the scale residing in the average scale thickness of the sample.

Results and Discussion

The inhomogeneous corrosion layer specimen surfaces after 700 h, 2000 h and 4000 h of exposure to CO_2-saturated saline aquifer water on consists of either a non consistent carbonate layer or carbon precipitates covering pits precipitated on the samples surface. As a measure of quality the hardness was analyzed prior to exposure revealing that longer austenitzing times lead to lower hardness and higher austenitizing temperatures will result in lower hardness (Fig. 2).

To evaluate the influence of the heat treatment the samples were examined via light optical methods to predict the amount of counted pits and the pit depths. Kinetics was obtained via weight loss according to DIN 50 905 after exposure to the CO_2-saturated aquifer water (Fig. 3 and Fig. 4)

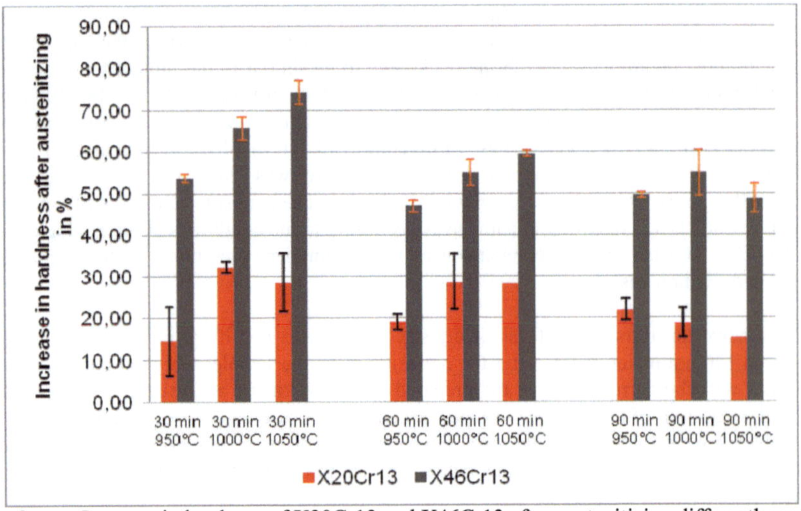

Fig. 2: Increase in hardness of X20Cr13 and X46Cr13 after austenitizing differently and annealing at 650 °C for 30 min.

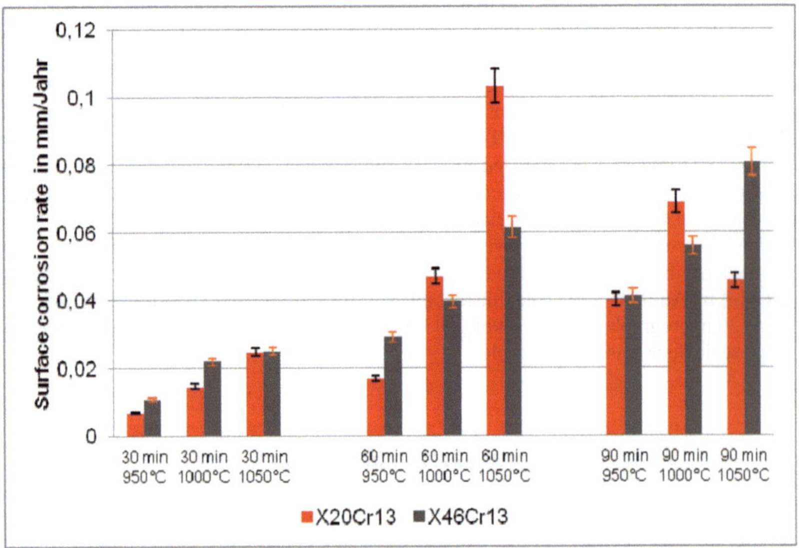

Fig. 3: Surface corrosion rate of X20Cr13 and X46Cr13 after austenitizing differently and annealing at 650 °C for 30 min. prior to exposure to CO_2-saturated saline aquifer water for 700 h.

In general the lower the austenitizing time and the lower the austenitizing temperature the better is the corrosion resistance. The austenitizing time has the greatest influence on the surface corrosion rate (Fig. 3), which scales with the elongation of austenitizing, but is neglectable when evaluating local corrosion phenomena [25]. Specimens heated to 950 °C and annealed for 30 minutes exhibit the lowest corrosion rates. The highest corrosion rate is found for X20Cr13 austenitized at 1050 °C for 60 minutes (Fig. 3). After 700 h the corrosion rates have the highest values, but as time proceeds the rates decrease.

Comparing local corrosion of X20Cr13 and X46Cr13 Fig. 4 shows that X20Cr13 shows a better pit corrosion resistance. X20Cr13 exhibits an average of 3508814 pits per m^2, whereas the X46Cr13 averages 9622 pits per m^2. Nevertheless the average pit on X46Cr13 (diameter average = 249 μm) is 5 times larger as the average pit on the X20Cr13 (49 μm).

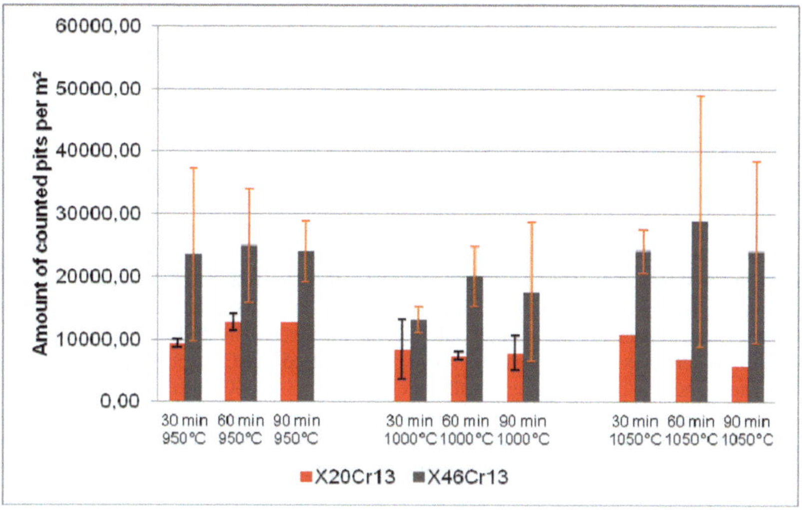

Fig. 4: Amount of counted pits of X20Cr13 and X46Cr13 after austenitizing differently and annealing at 650 °C for 30 min. prior to exposure to CO_2-saturated saline aquifer water for 700 h.

Figure 5 shows results from combined DoE-Analysis performed after Klein [26]. Although dependent parameters (such as both carbon content or autenitizing time) may not be evaluated the results are given, proving that the corrosion resistance towards surface corrosion is only dependent on the austenitizing time, but neither the austenitizing temperature nor the carbon content of the base material [9]. The fact that the carbon content exhibits the biggest impact on the local corrosion behaviour is shown in the results of the DoE analysis but since none of the parameters are within the liability regime the results here prove a tendency but do not show significant dependence of carbon content on the local corrosion behaviour.

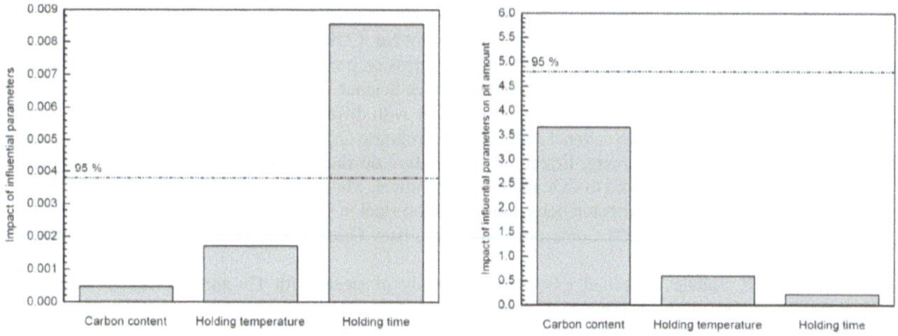

Fig. 5: The impact of the experimental parameters on the corrosion rate after 4000 h of exposure time combined with confidence intervals.

In the beginning the corrosion rates are significantly higher at higher austenitizing temperatures and longer holding times, but after 4000 h the corrosion rates are of not much difference. Most likely the corse grain structure is corrosively attacked at the grain boundaries, but once a sufficient corrosion layer has precipitated the mutual diffusion of ionic species towards the metal surface and into the surface is significantly slowed down with increasing thickness of the corrosion layer. In addition the grain growth process is limited at shorter austenitizing temperature giving the equilibration as little time as possible with additionally short austenitizing time. If grain growth is slow the small grain sizes within the microstructure will result in lower corrosion rates.

Conclusions

During intermissions of the injection of CO_2 -supposedly in its liquid or supercritical phase- the water level may rise into the injection pipe. This will then lead to the precipitation of corrosion products and formation of pits. The shorter austenitizing times and lower austenitizing temperatures exhibit the lowest corrosion rates. Still the austenitizing time has no significant impact on the number of pit and pit sizes. The following facts could be summarized acting as an aid to heat treatment of steels used in CCS-technology:

- The higher the carbon content of the steels (X20Cr13<C46Cr13) the less amount of pits are initiated.
- Shorter austenitizing at lower austenitizing temperature results in better corrosion resistance. Especially the austenitizing time has high impact.

Acknowledgement

This work was supported by the FNK (Fachkonferenz für wissenschaftlichliche Nachwuchskräfte) of the Applied University of Berlin, HTW and by IMPACT (EU-Project EFRE 20072013 2/21).

Literature

[1] D.C. Thomas, Carbon Dioxide Capture for Storage in Deep Geologic Formations – Results from CO_2 Capture Project, Volume 1: Capture and Separation of Carbon Dioxide form Combustion Sources, CO_2 Capture Project, Elsevier Ltd UK 2005, ISBN 0080445748

[2] S. Nešić, Key issues related to modelling of internal corrosion of oil and gas pipelines – A review, Corrosion Science 49 (2007) 4308–4338

[3] L.J. Mu, W.Z. Zhao, Investigation on Carbon Dioxide Corrosion Behaviors of 13Cr Stainless Steel in Simulated Strum Water, Corrosion Science, Manuscript No. CORSCI-D-09-00353 (2009) 1-24

[4] Z.D. Cui, S.L. Wu, S.L. Zhu, X.J. Yang, Study on corrosion properties of pipelines in simulated produced water saturated with supercritical CO_2, Applied Surface Science 252 (2006) 2368-2374

[5] Pfennig, A., Kranzmann, A., Reliability of pipe steels with different amounts of C and Cr during onshore carbon dioxide injection, International Journal of Greenhouse Gas Control 5 (2011) 757–769

[6] H. Zhang, Y.L Zhao, Z.D. Jiang, Effects of temperature on the corrosion behaviour of 13Cr martensitic stainless steel during exposure to CO_2 and Cl^- environment, Material Letters 59 (2005) 3370-3374

[7] Y.-S. Choi and S. Nešić, Corrosion behaviour of carbon steel in supercritical CO_2-water environments, No. 09256, NACE Corrosion 2008 Conference and Expo, New Orleans, Louisiana, USA, March 16th – 20th, 2008

[8] A. Pfennig, R. Bäßler, Effect of CO_2 on the stability of steels with 1% and 13% Cr in saline water, Corrosion Science, Vol. 51, Issue 4 (2009) 931-940,

[9] A. Pfennig, P. Zastrow, A. Kranzmann, Influence of heat treatment on the corrosion behaviour of stainless steels during CO2-sequestration into saline aquifer, International Journal of Green House Gas Control 15 (2013) 213–224

[10] D.S. Carvalho, C.J.B. Joia, O.R. Mattos, Corrosion rate of iron and iron-chromium alloys in CO_2-medium, Corrosion Science 47 (2005) 2974-2986
[11] B.R. Linter, G.T Burstein, Reactions of pipeline steels in carbon dioxide solutions, Corrosion Science 41 (1999) 117-139
[12] J. Han, Y. Yang, S. Nešić, B. N. Brown, Roles of passivation and galvanic effects in localized CO_2 corrosion of mild steel, Paper No. 08332, NACE Corrosion 2008, New Orleans, Louisiana, USA, March 16th – 20th, 2008
[13] J. Banaś, U. Lelek-Borkowska, B. Mazurkiewicz, W. Solarski, Effect of CO_2 and H_2S on the composition and stability of passive film on iron alloy in geothermal water, Electrochimica Acta 52 (2007) 5704–5714.
[14] P.D. Bilmes, C.L. Llorente, C.M. Méndez, C.A. Gervasi, Microstructure, heat treatment and pitting corrosion of 13CrNiMo plate and weld metals, Corrosion Science, ISSN: 0010-938X, Vol: 51, Issue: 4, (2009) 876-882
[15] Ş. Bülbül, Y Sun, Corrosion behaviours of high Cr-Ni cast steels in the HCl solution, Journal of Alloys and Compounds 598 [2010] 143-147
[16] B. Hou, Y. Li, Y. Li, J. Zhang, Effect of alloy elements on the anti-corrosion properties of low alloy steel, Bull. Mater. Sci, Vol. 23, No. 3 (2000) 189-192
[17] Z. Cvijović and G. Radenković, Microstructure and pitting corrosion resistance of annealed duplex stainless stell, Corrosion Science 48 (2006) 3887-3906
[18] L. Zhang, W. Zhang, Y. Jiang, B. Deng, D. Sun, J. Li, Influence of annealing treatment on the corrosion resistance of lean duplex stainless steel 2101
[19] Y.-S. Choi, J.-G. Kim, Y.-S. Park, J.-Y. Park, Austenitzing treatment influence on the electrochemical corrosion behaviour of 0.3C-14Cr-3Mo martensitic stainless steel, Materials Letters 61 (2007) 244-247
[20] A. N. Isfahany, H. Saghafian, G. Borhani, The effect of heat treatment on mechanical properties and corrosion behaviour of AISI420 martensitic stainless steel, Journal of Alloys and Compounds 509 [2011] 3931-3936
[21] J.-Y. Park, Y.-S. Park, The effects of heat-treatment parameters on corrosion resistance and phase transformation of 14Cr-3Mo martensitic stainless steel, Materials Science and Engineering A 449-451 [2007] 1131-1134
[22] D. Dyja, Z. Stradomski, A. Pirek, Microstructural and fracture analysis of aged cast duplex steel, Strength of Materials, Vol. 40, No. 1 [2008] 122-125
[23] A. Förster, B. Norden, K. Zinck-Jørgensen, P. Frykman, J. Kulenkampff, E. Spangenberg, J. Erzinger, M. Zimmer, J. Kopp, G. Borm, C. Juhlin, C. Cosma, S. Hurter, 2006, Baseline characterization of the CO2SINK geological storage site at Ketzin, Germany: Environmental Geosciences, V. 13, No. 3 (September 2006), pp. 145-161.
[24] SW. Kraus and G. Nolze, POWDER CELL – a program for the representation and manipulation of crystal structures and calculation of the resulting X-ray powder patterns, J. Appl. Cryst. (1996), 29, 301-303
[25] J. Schiz, Einfluss der Austenitisierungsparameter auf die Korrosionsbeständigkeit von Stählen in CCS Umgebung, Bachelor-Thesis zur Erlangung des Grades B.Sc. im Fachbereich Maschinenbau an der HTW Berlin, 2011
[26] Klein, B. Versuchsplanung DoE. Kassel : Oldenbourg Verlag, 2004. 3-486-27387-6.

Energy Technology 2015: Carbon Dioxide Management and Other Technologies
Edited by: Animesh Jha, Cong Wang, Neale R. Neelameggham, Donna P. Guillen, Li Li, Cynthia K. Belt, Randolph Kirchain, Jeffrey S. Spangenberger, Frank Johnson, Andrew J. Gomes, Amit Pandey, and Peter Hosemann
TMS (The Minerals, Metals & Materials Society), 2015

EVALUATION OF SURFACE TENSION FOR THE $NaNO_3$-KNO_3-$Ca(NO_3)_2$ SYSTEM AND ITS SUB-SYSTEM

Jifang Xu [1], Jingjing Zhao [1], Jixu Wang [2], Jianchao Li [3], Kang Wan [1]

[1] Shagang School of Iron and Steel, Soochow University, Suzhou 215021, Jiangsu P. R. China;
[2] Shanghai Key Laboratory of Modern Metallurgy & Material Processing, Shanghai University, Shanghai 200072, P. R. China;
[3] Material Department, Vocational and Industry Institute of Hebei, Shijiazhuang 050091, Hebei P. R. China

Keywords: Surface tension, Nitrate salts, Butler equation, Model, Solar energy applications

Abstract

Molten Nitrates salts have been used as thermal energy storage media for solar energy applications. Solar salt $NaNO_3$-KNO_3-$Ca(NO_3)_2$ (trade name HitecXL) is one of the most popular thermal energy storage medium. And the surface tension of nitrate salts is very useful for optimizing of the energy storage process. On the basis of Butler's equation, the thermodynamic model for determining the surface tension of molten ionic mixtures was extended to $NaNO_3$-KNO_3-$Ca(NO_3)_2$ system and its sub-system. The composition dependence of the surface tension in binary and ternary systems with different temperature were evaluated with the present model using information based on the surface tension and molar volume of the pure nitrates components, and on the anionic and cationic radii. Compared with the calculated values of surface tension as well as the data available in the literature, the results showed that the present model is working well.

Introduction

Among various technologies of renewable energy for sustainable development, solar energy is becoming one of the most promising sources of energy. In the past decade, the solar thermal power plant has been considered as the most potential way to massively utilize the solar energy in the future [1]. Molten nitrates having excellent characteristic of low volatility and viscosity, large high electrical conductivity, heat capacity and chemical stability, can offer huge potential to be used as the heat transfer fluid directly as well as the thermal storage medium. Molten nitrates represent the most likely candidates because of their exceptional thermal properties and low cost. Molten salts consisting of sodium nitrate and potassium nitrate (60-40 wt.%, *Solar Salt*) and $NaNO_2$-$NaNO_3$-KNO_3 (commercially called *Hitec heat transfer salt*) have been used successfully in solar thermal power. The ternary mixtures of and $NaNO_3$-KNO_3-$Ca(NO_3)_2$ (HitecXL) is one of the most popular thermal energy storage medium. The use of this salt in parabolic trough plants could lead to an increase in concentrating solar power plant efficiency by working at temperatures higher than those for conventional heat transfer fluid [2]. Obviously, the knowledge of the physical properties of molten salts is very important not only to the industrial application, but also for the fundamental understanding of the structure.

For the purposed of develop and synthesize of molten salt mixtures satisfying the requirement, several phase diagrams for the $NaNO_3$-KNO_3-$Ca(NO_3)_2$ system and its sub-system have been published and the physical properties of molten salts were measured [3-6]. Since many phenomena are closely related to surface tension of molten salt. It may directly affect the efficiency of heat transmission and exchange as thermal energy storage media, therefore, influencing the thermal efficiency of the relevant processes. Owing to the inherent problems and difficulties associated with measurements at high temperature and corrosion experimental environment especially for the multi-component systems, the calculation becomes more and more important for acquiring physicochemical data of molten salt. In order to find a suitable molten salt with excellent properties, it is important to clarify the relation between the composition and properties for molten salt. To our knowledge, the information on the properties of molten nitrate salts (particularly over a wide temperature and composition range) are very lack and do not fully satisfy the needs of the industry. Given this, only a few binary systems and the discrete experimental data of ternary system are reported. Therefore, it is necessary to develop a reliable model to make full use of the available data and extract as much information as we can.

In order to estimate surface tension for multi-component metallic and ionic systems, *Tanaka et al.* [7] developed a predicting model for determining the surface tension of molten multi-component systems, thermodynamically-based on Butler's equation [8] and on the procedure proposed by *Speiser et al.* [9]. In the present work, based on the surface tension and molar volume of the pure components, and on the ionic radii of the components in the system, the surface tension of $Ca(NO_3)_2$-KNO_3-$NaNO_3$ system and its sub-system were evaluated at different temperature and over a wide composition range.

Model For Estimation Of The Surface Tension Of Molten Nitrate Salt

As we known, the molten ionic mixtures easily have surface relaxation, and their ionic structures depend upon the ratio of the radius of cation to that of anion. Therefore, we should consider the ratio of the radius of cation to that of anion to evaluate ionic structure and physico-chemical properties of ionic mixtures. Butler deduced a surface tension calculation equation based on the concept of chemical potential and surface energy. On account of the fundamental consideration mentioned above. *Tanaka et al.* modified the model by using the ratio of the cation to anion radius as a model parameter. The present model reproduces the following characteristics of the composition dependence of surface tension in molten alkaline–halide ionic mixtures precisely [7]. This improved thermodynamic model is characterized by the simplicity with which it can be extended to multi-component systems, particularly since the calculations are conducted only with respect to information based on the surface tension and molar volume of the pure components, and on the ionic radii of the components in the system. In order to orient the readers, a brief description of the thermodynamic model for determining the surface tension is given below. Detailed discussions about this method can be found in reference.

The ANO_3-BNO_3 binary system

The surface tension (σ) of the ANO_3-BNO_3 molten nitrates is calculated as follows:

$$\sigma = \sigma_{ANO_3}^{Pure} + \frac{RT}{A_{ANO_3}} \ln \frac{M_{ANO_3}^{Surf}}{M_{ANO_3}^{Bulk}} \quad (1)$$

$$\sigma = \sigma_{BNO_3}^{Pure} + \frac{RT}{A_{BNO_3}} \ln \frac{M_{BNO_3}^{Surf}}{M_{BNO_3}^{Bulk}} \quad (2)$$

Where

$$M_i^P = \left(\frac{R_i}{R_{NO_3^-}} \cdot N_i^P\right) / \left(\sum_{i=A,B} \frac{R_i}{R_{NO_3^-}} \cdot N_i^P \right) \quad (3)$$

The ANO_3-BNO_3-CNO_3 ternary system

The surface tension (σ) of the ANO_3-BNO_3-CNO_3 molten nitrates is calculated as follows:

$$\sigma=\sigma_{ANO_3}^{Pure}+\frac{RT}{A_{ANO_3}}\ln\frac{M_{ANO_3}^{Surf}}{M_{ANO_3}^{Bulk}} \quad (4)$$

$$\sigma=\sigma_{BNO_3}^{Pure}+\frac{RT}{A_{BNO_3}}\ln\frac{M_{BNO_3}^{Surf}}{M_{BNO_3}^{Bulk}} \quad (5)$$

$$\sigma=\sigma_{CNO_3}^{Pure}+\frac{RT}{A_{CNO_3}}\ln\frac{M_{CNO_3}^{Surf}}{M_{CNO_3}^{Bulk}} \quad (6)$$

Where

$$M_i^P=(\frac{R_i}{R_{NO_3^-}}\cdot N_i^P)/\left(\sum_{i=A,B,C}\frac{R_i}{R_{NO_3^-}}\cdot N_i^P\right) \quad (7)$$

In the binary and ternary system, subscripts i is refer to A, B and C cations, which refers to the Na^+, K^+, Ca^{2+} ions in the present work. Superscripts P='Surf' and 'Bulk' indicate the surface and bulk, respectively. R is the gas constant, T is the absolute temperature, and σ^{Pure} is the surface tension of pure molten nitrates, which is treated as a model parameter. $A_i=N_0^{1/3}\cdot V_i^{2/3}$ corresponds to the molar surface area in a monolayer of pure molten component i (N_0: Avogadro's number, V_i: molar volume of the pure molten component i). N_i^P is the mole fraction of the component i in phase P (P=Surf or Bulk). R_A , R_B and R_C correspond to the radii of cations A, B and C. the ionic radii data were taken from the data compiled by *Shannon* [10], while the molar volumes of the pure nitrates, as recommended by *Janz* [11], were used in the present calculations. These values are listed in Table 1 and 2, respectively. The equations for determining the temperature dependences of surface tension are listed in Table 3. It should point out that the accuracy of experimental data is then very important. However, the available literature on the subject shows that there are large discrepancies among different authors in different laboratories. In addition, the accuracy statements are nonexistent or ill defined. The density of calcium nitrate has not been determined experimentally, presumably because that pure molten salt has limited thermal stability, so its molar volume cannot be calculated directly. Mixtures of $Ca(NO_3)_2$ with alkali nitrate salts in which the latter are the majority constituent are stable at elevated temperature and , in particular, the density of mixtures of KNO_3-$Ca(NO_3)_2$ melts have been reported, and the molar volume of calcium nitrate in the molten state were calculated by these density data [12]. Similarly, the relationship between temperature and the surface tension of pure calcium nitrate is not found in the public reference, the reason is mentioned above. Only one surface tension data for calcium nitrate was founded (about 101.5×10^{-3} N/m at 833K). Therefore, its surface tension was calculated by the surface tension of $NaNO_3$-$Ca(NO_3)_2$ melts [6], and the temperature dependence of surface tension for calcium nitrate were also shown in Table 3. When these parameters are available, it is straightforward to construct a model for calculating the surface tension of the $Ca(NO_3)_2$-KNO_3-$NaNO_3$ system and its sub-system.

Table 1 Ionic radii of cations and nitrate ion (Å)

Na^+	K^+	Ca^{2+}	NO_3^-
0.95	1.33	0.99	1.21

Table 2 Molar volume of pure component (m^3/mol)

Nitrate	Temperature (K) dependence of molar volume (m^3/mol)	
$NaNO_3$	$(38.675+0.02032\times10^{-2}\times T)\times10^{-6}$	Ref. 11
KNO_3	$(45.776+0.02413\times10^{-2}\times T)\times10^{-6}$	Ref. 11
$Ca(NO_3)_2$	$(70.378+0.02565\times10^{-2}\times T)\times10^{-6}$	Ref. 12

Table 3 Temperature dependence of the surface tensions of pure components

Nitrate	Temperature (K) dependence of surface tension (N/m)	
$NaNO_3$	$(155.40+0.061\times T)\times 10^{-3}$	Ref. 11
KNO_3	$(148.48+0.064\times T)\times 10^{-3}$	Ref. 11
$Ca(NO_3)_2$	$(146.20+0.054\times T)\times 10^{-3}$	Calculated

Applications To Estimate The Surface tension Of Molten Nitrate Salt

The surface tension of the binary nitrate system

The importance of the knowledge of accurate data for thermo physical properties is self-evident. However, the practical determination of these properties at high temperatures is difficult. On the other hand, when data are available, the differences among different laboratories are sometimes dramatic [13]. As an example, Figure 1 shows the surface tension of pure KNO_3 at the $NaNO_3$-KNO_3 and KNO_3-$Ca(NO_3)_2$ binary system. The differences among different sets of data can be as high as 2.3%. For the surface tension of pure $NaNO_3$ the situation is similar. Figure 1 shows the differences can amount to 3.2% among the $NaNO_3$-KNO_3 and $NaNO_3$-$Ca(NO_3)_2$ binary system at elevated temperatures. The accuracy and the difficulty of system for experimental, the surface tension data for pure constituent surface tension on $NaNO_3$-KNO_3 system were used to calculate.

Figure 1 and Figure 2 shows the calculated results of $NaNO_3$-KNO_3, $NaNO_3$-$Ca(NO_3)_2$ and KNO_3-$Ca(NO_3)_2$ binary system at the 630K and 670K, respectively. It can be seen that the calculated data for the $NaNO_3$-KNO_3 binary system were in good agreement with the experimental results at the elevated temperature. While the discrepancies data of pure KNO_3 and pure $NaNO_3$ among different system, the calculated data for the $NaNO_3$-$Ca(NO_3)_2$ binary system above the experimental data from literature, the relative average error at 630K are about 2.5% and 3.5%, respectively. The comparison between the predications and the experimental results show that the general trend is the same in two binary system, the surface tension increases with increasing the $Ca(NO_3)_2$ content. The present observations indicated that it is possible to get a reasonable idea of estimate the surface tension for the binary nitrate system by the present model.

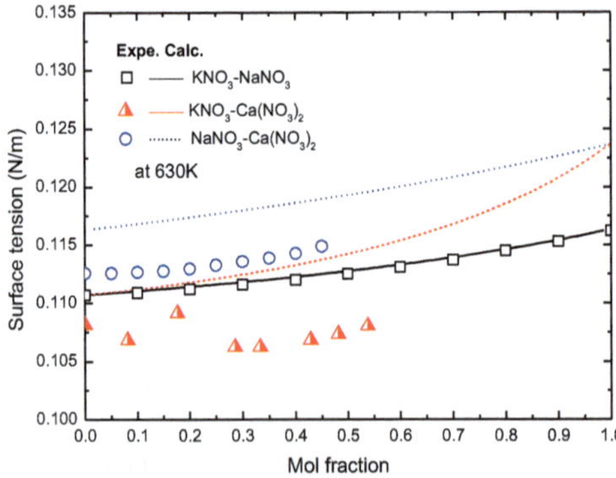

Figure 1 Surface tension of the selected binary nitrate system at 630K

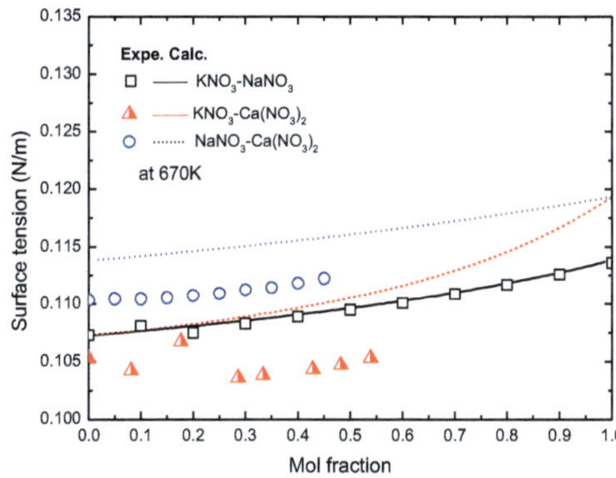

Figure 2 Surface tension of the selected binary nitrate system at 670K

The surface tension of the $NaNO_3$-KNO_3-$Ca(NO_3)_2$ ternary nitrate system

It should point out that the eutectic point composition in $NaNO_3$-KNO_3-$Ca(NO_3)_2$ system is $17NaNO_3$-$52KNO_3$-$30Ca(NO_3)_2$ (mol%), when the $Ca(NO_3)_2$ content in this system is higher, the melting point of molten salt will up to 700K, which is not suitable for both thermal energy storage media, and heat transfer fluid in solar thermal power [2]. Therefore, the experimental data with the content of $Ca(NO_3)_2$ below 75 mol% were reported [6]. By utilizing the pure constituent nitrate salts data reported in literature, the present model has been applied to estimate surface tension of $NaNO_3$-KNO_3-$Ca(NO_3)_2$ system at 630K and 670K. The comparison between the predications and the experimental results are presented in Figure 3 and Figure 4. The 673K liquidus isotherms were also shown in the figures [14]. The present work reflected that the surface tension of this system increases with the increasing content of $Ca(NO_3)_2$ and KNO_3. The relationship between surface tension and the temperature is the same as that of the experimental obtained. Thus, the overall trend that the effects of component content on the surface tension is in agreement with the experimental observation. Our calculation in the present work clearly indicates that the present model is working well. And more data about this system were provided for the application molten nitrate salts system as the thermal storage medium.

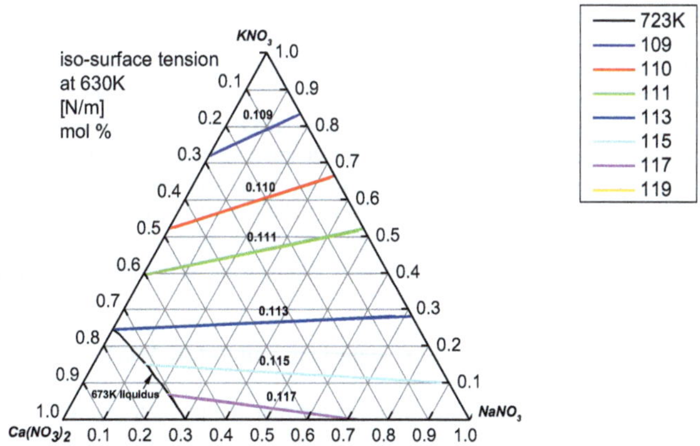

Figure 3 Surface tension of the $NaNO_3$-KNO_3-$Ca(NO_3)_2$ ternary system at 630K

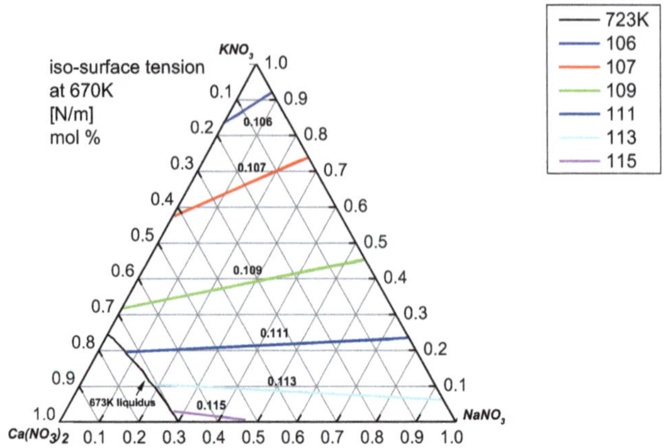

Figure 4 Surface tension of the $NaNO_3$-KNO_3-$Ca(NO_3)_2$ ternary system at 670K

Future Work

The cost of constituent nitrate salts is critically important since it will directly influence the overall capital investment of thermal energy storage systems. The commercial price of the individual nitrate constituent is in general quite volatile in the global market and tends to go up sharply in the recent years. Based on the market survey, the cost of the individual constituent nitrate can be ranked as the following sequence (in order from least expensive to most expensive): $Ca(NO_3)_2 \rightarrow NaNO_3 \rightarrow KNO_3 \rightarrow LiNO_3 \rightarrow CsNO_3$ [15]. Obviously, lithium nitrate, as one constituent of the multi-component mixtures under development, is more expensive, and hence the addition of lithium nitrate will considerably increase the overall cost of the storage media, although it just holds a small percentage in mixtures. Molten salts consisting of calcium nitrate are relatively inexpensive, and it showed a significant cost advantage as heat transfer fluids and storage media. A good strategy of blending the constituents of nitrate salt mixtures will reduce the cost further, although the per unit cost of a single component is generally reduced when produced in larger quantities and at lower purity levels. The cost may be reduced by optimizing the mixture to limit the amount of most expensive constituent, such as cesium and lithium, while still maintaining acceptable physical properties. An optimal strategy will be studied to minimize the cost whilst optimizing the chemical stability and physical properties of the mixture.

This paper presents the surface tension of the novel heat transfer fluid with a broad operating range that may be suitable for applications in solar thermal power. Future work includes an economic optimization of the $NaNO_3$-KNO_3-$Ca(NO_3)_2$ composition as well as full physical-property and thermo-property characterization, such as viscosity, melting behavior, density, thermo conductivity, high temperature stability and so on. The model will be applied to evaluate the surface tension of $NaNO_2$-$NaNO_3$-KNO_3 ternary system, $LiNO_3$-$NaNO_3$-KNO_3 ternary system, quaternary nitrate mixtures and higher order nitrate salts. The present model described in this paper can be used to develop advanced thermal materials for other heat transfer and thermal storage applications.

Conclusions And Future Work

The molten nitrate salt is an important thermal energy storage media, and has widely used in solar plants. Solar salt $NaNO_3$-KNO_3-$Ca(NO_3)_2$ (trade name *HitecXL*) is one of the most popular thermal energy storage medium. And the surface tension of nitrate salts is very useful for optimizing of the energy storage process. In order to find a suitable molten salt with excellent properties, it is important to clarify the relation between the composition and properties for molten salt. In this paper, on the basis of Butler's equation, the thermodynamic model for determining the surface tension of molten ionic mixtures was extended to $NaNO_3$-KNO_3-$Ca(NO_3)_2$ system and its sub-system. The composition dependence of the surface tension in binary and ternary systems with different temperature were evaluated with the present model using information based on the surface tension and molar volume of the pure nitrates components, and on the anionic and cationic radii. Compared with the calculated values of surface tension as well as the data available in the literature, the results showed that the present model is working well, and more data were provided for the application molten nitrate salts system as the thermal storage medium.

Acknowledgments

Thanks to the Natural Science Foundation of Jiangsu Province of China (BK20130308) for their kind financial support. Also wish to acknowledge support by the Innovation Fund for Students of Soochow University (2013xj044).

Reference

[1] M. Thirugnanasambandam, S. Iniyan and R. Goic, "A review of solar thermal technologies" *Renewable and Sustainable Energy Reviews*, 14(2010), 312-322.
[2] J.C. Gomez, N. Calvet, A.K. Starace and G.C. Glatzmaier, "$Ca(NO_3)_2$-$NaNO_3$-KNO_3 molten salt mixtures for direct thermal energy storage systems in parabolic trough plants" *Journal of Solar Energy Engineering*, 135(2013), 021016-1-8.
[3] A.W.C. Menzies and N.N. Dutt, "The liquidus surface of the ternary system composed of the nitrates of Potassium, Sodium, and Calcium" *J. Am. Chem. Soc.*, 33(1911). 1366-1375.
[4] P. I. Protsenko and A.G. Bergman, "Binary system KNO_3-$Ca(NO_3)_2$" *Zh. Obshch. Khim.*, 20(1950), 1365-1375.
[5] P. I. Protsenko and A.G. Bergman, "Binary system $Ca(NO_3)_2$-$NaNO_3$" *Russ. J. Gen. Chem. (English Transl.)*, 20(1950), 1421-1431.
[6] G.J. Janz, U. Krebs, H.F. Siegenthaler and R.P.T. Tomkins, "Molten salts: volume 3, Nitrates, Nitrites, and mixtures electrical conductance, density, viscosity and surface tension data" *J. Phys. Chem. Ref. Data*, 1(1972), 581-746.
[7] T. Tanaka, T. Kitamura and I.A. Back, "Evaluation of surface tension of molten ionic mixtures" *ISIJ Int.*, 46 (2006), 400-406.
[8] J.A.V. Butler, "The thermodynamics of the surfaces of solutions", *Proc. R. Soc. A*, 135(1932), 348-375.
[9] R. Speiser, D.R. Poirier and K. Yeum, "Surface tension of binary liquid alloys", *Scr. Metall.*, 21 (1987), 687-692.
[10] R.D. Shannon, "Revised effective ionic radii and systematic studies of interatomic distances in halides and chalcogenides", *Acta Crystallogr. A.*, A32 (1976), 751-767.
[11] G.J. Janz, *Molten Salt Handbook*, (New York: Academic Press, 1967).
[12] W.R. Bradshaw, "Effect of composition on the density of multi-component molten nitrate salts", (Report SAND2009-8221, Sandia National Laboratories, 2009).
[13] T. Bauer, D. Laing and R. Tamme, "Characterization of sodium nitrate as phase change material", *Int J Thermophys*, 33(2011), 91-104.
[14] E.M. Levin, H.F. McMurdie and F.P. Hall, *Phase Diagrams for Ceramists* (Columbus: The American Ceramic Society, 1956).
[15] C.Y. Zhao and Z.G. Wu "Thermal property characterization of a low melting-temperature ternary nitrate salt mixture for thermal energy storage systems", *Solar Energy Materials & Solar Cells*, 95 (2011), 3341-3346.

Energy Technology 2015
Carbon Dioxide Management and Other Technologies

ENERGY TECHNOLOGIES AND CARBON DIOXIDE MANAGEMENT SYMPOSIUM 2015

Solar Energy

Session Chairs:

Li Li
Shulan Wang

Energy Technology 2015: Carbon Dioxide Management and Other Technologies
Edited by: Animesh Jha, Cong Wang, Neale R. Neelameggham, Donna P. Guillen, Li Li, Cynthia K. Belt, Randolph Kirchain, Jeffrey S. Spangenberger, Frank Johnson, Andrew J. Gomes, Amit Pandey, and Peter Hosemann
TMS (The Minerals, Metals & Materials Society), 2015

Development of High Flux Solar Simulators for Solar Thermal Research

Ben M. Ekman, Geoffrey Brooks and M. Akbar Rhamdhani

Faculty of Engineering and Industrial Science
Swinburne University of Technology
Hawthorn, VIC 3122, Australia

Keywords: solar energy, solar simulator, concentrated heat flux

Abstract

High flux solar simulators, used to produce controlled high temperature experiments are a valuable tool for the research and development of high temperature material processes. As an alternative to a direct external solar concentrator where the sun's radiation is concentrated using a parabolic dish, an indoor solar simulator uses an array of high intensity discharge lamps attached to ellipsoidal mirrors to focus their light at a secondary focal point where temperatures in excess of 2000°C can be reached. To mimic, as closely as possible, the spectrum of the sun, the novel high flux solar simulator design uses metal halide lamps and not follow the conventionally practise of using xenon lamps. The $42kW_e$ simulator consisting of seven $6kW_e$ metal halide lamps delivered a peak thermal flux of approaching $1mW/m^2$ to the secondary focal plane of a closely coupled ellipsoidal reflector. A comparison of different designs and their performance is also presented in this paper.

Introduction

The application and design of solar simulators falls into two main classifications, non-concentrating uniformly distributed light used in the testing of photovoltaic (PV) cells and solar hot water collectors and high flux concentrators used to generate high temperatures exceeding 1000°C used for a variety of research applications including material processing, thermo-chemical reactions, the production of solar fuels and in simulated solar thermal power generation. The radiation heat source needs to resemble as closely as possible the spectrum of the sun and in the main has been a high intensity arc lamp coupled to an optical reflector resulting in the delivery of an intense beam of concentrated radiation with wavelengths in the UV, visible and infrared regions of the spectrum.

As early as 1973 researchers [1] inserted a tungsten-halogen heat source positioned at one focus of a suitably shaped ellipsoidal reflector to heat small objects placed at the secondary focal point. High flux solar simulators first began to appear in the early 1990's in the USA using Xenon lamps [2]. Argon arc lamps and high powered xenon lamps were used in 2003 to 2007 at ETH Zurich [3, 4]. By 2012 high intensity solar simulators were used in research centres [5, 6, 7, 8] that were capable of generating intense fluxes and concentrations of over 4000 suns all using Xenon lamps that were close coupled to a precision ellipsoidal reflector. Today only a handful continue to operate throughout the world that are capable of generating highly concentrated radiation fluxes that correspond to stagnation temperatures of over 2000K and up to 3700K. Optical configurations based on parabolic-shaped mirrors are commercially available for large-scale collection and concentration of solar energy for the generation of electrical power. The total amount of radiated power collected by any of these systems is proportional to the projected area of the mirrors. Their arrangement depends mainly on the concentrating system selected and the latitude of the site [9]. The common configurations used include linear concentrators such as parabolic troughs and Fresnel concentrators, or point concentrators such as the central solar towers and parabolic dishes. An alternative reliable research tool is required that is capable of providing an artificial source of concentrated energy with a spectral distribution as close as possible to that of natural sun light. A high flux solar simulator will create the constant conditions required for controlled high temperature experimentation. At the same time the design objectives must be to simulate the parameters that would be encountered by high concentration solar systems capable of processing materials on commercially viable scale including the spectral distribution and a large focal hot spot of the reflected sun image. Metal halide (MH) lamps are unique in that their spectral distribution closely replicates sunlight and their larger arc length results in a larger and less intense region of high flux.

This paper describes the design and fabrication of a high concentration solar simulator using an array of seven high powered Metal Halide lamps each coupled to a precision ellipsoidal reflector.

Design

Factors that were taken into consideration during the design stage included costs, the emission spectrum of the lamp, the lamp efficiency, lamp cooling, reflector size, shape and quality, reflector surface coating, the positioning of the lamp within the reflector, reflector support and safety. From a comparison of the lamps used in high intensity solar simulators, the xenon lamp has clearly been the preferred choice for two main reasons, the radiation spectrum and the size of the arc light source. The metal halide lamp emits the majority of the energy within the visible light wavelength of 350 um to 700 um sharply falling off in the UV region and a more gradual fall in the near and far infrared. By contrast the xenon lamp emission spectrum has a flat low energy distribution in the vision light region with a number of intense spikes in the infrared [10]. In fact the visible light emission constitutes only about 25 percent of the total light output, with most of the energy falling into the infrared spectral region. Approximately 70 percent of the xenon arc lamp output occurs at wavelengths longer than 700 nanometers, while less than 5 percent of the output consists of wavelengths less than 400 nanometers. Similarly with the Argon arc lamp which has low energy emission in the visible light region of 380nm to 700nm and high energy emission spikes in the 700nm to 1000nm range. By contrast the metal halide lamp spectral distribution is very similar to that of natural sunlight. The distribution profile is remarkably similar to that of sunlight except for a spike at 900nm. The high energy component of the radiation is contained within the visible range of the spectrum from 400nm to 800nm. The comparison is clearly seen when overlayed as shown in Figure 1a.

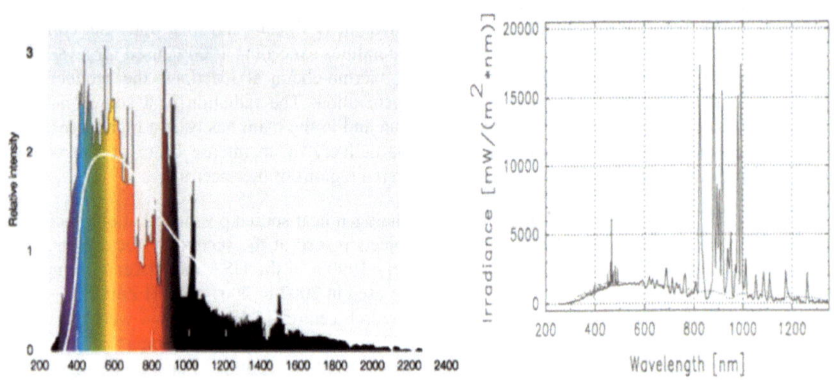

(a) (b)

Figure 1. (a)Spectral radiant intensity distribution of the metal halide lamp. The white line is the intensity distribution of the sun.(b) Spectral radiant intensity distribution of the xenon lamp [10].

The metal halide lamp in fact is the only artificial light source that emits a radiation spectrum that closely mimics that of natural sunlight. The intense infrared energy spikes that a xenon lamp emits, as seen in Figure 1b, may require either forced air cooling for low wattage lamps or water cooling for higher powered lamps. In addition the reflectors are more prone to damage and may require forced air cooling to their surface. With metal halide lamps approximately, 90% of the electrical energy supplied to the lamp is converted to and radiated out as energy, and the remaining 10% is lost through ohmic effects to the foils and electrodes. About 75% of the power consumed is radiated by the discharge arc itself. The radiation output is split up into 10% UV, 45% visible and 20% infrared radiation. Around 15% of the energy is emitted by the electrode and the bulb which can reach temperatures of more than 900°C [10] and as the arc size within the lamp increase, the radiative energy transfer from the arc to the target is reduced [3, 11]. This negatively affects the magnitude and distribution of the radiative flux in the target plane. For this reason even though the metal halide lamp emits a spectral distribution that more closely replicates sunlight, xenon lamps have been chosen over metal halide. However the luminous efficacy, which is described as how efficiently a lamp converts electrical energy into visible light, has in the past been very limiting with little choice but to choose a xenon lamp, this is no longer the case [10]. Xenon lamps that have a power range of 4 kW to 6 kW have a luminous flux range of 155,000 lm to 280,000 lm and an efficacy of 39 lm/W to 47 lm/W. The power equivalent in the metal halide lamp has a luminous flux range of 380,000 lm to 600,000 lm and an efficacy of 95 lm/W to 100 lm/W making the metal halide lamp more efficient in converting electrical power to light in this size range. In addition xenon lamps operate under very high pressure and are dangerous in the event of any exploding bulb. Metal halide on the other hand operates at much lower pressures and have a protective outer bulb for added protection. There is a direct relationship between the electrode gap, lamp voltage, operating pressure and luminous efficacy of a discharge lamp [10] and as a general rule lamps with small electrode gaps generally have low efficiencies.

Our design uses an array of commonly focused lamps, each comprising of a truncated ellipsoidal reflector closely coupled to a metal halide high intensity discharge lamp. An array of seven, 6000 W lamp/reflector modules, are arranged in a circular pattern. Although the arc gap is significantly larger in a metal halide lamp, researchers using xenon lamps have found [11] that the temperatures produced by these high intensity fluxes generated by short arc lamps resulted in a shorter life for the reflectors and because the energy at the target is concentrated into a small spot, are difficult to contain thus creating strong thermal gradients resulting in material thermal stress. Efforts have been made by some to defocus the rays and produce a preferred more uniform flux density distribution [11]. The metal halide lamps used in our design, because of the longer arc length, will have a more uniform flux density distribution without the need for post defocusing equipment. The lamp chosen as the radiation source is the OSRAM HMI 6000 W/SE, Hydrargyrum Medium-arc Iodide metal halide lamp. The lamp dimensions and the technical details are shown in Figures 2. As shown in Figure 3, the metal halide lamp consists of an inner and outer quartz glass envelope. The rounded 4mm diameter ends of the cathode and anode are separated by a 23 mm gap. Although larger metal halide lamps are commercially available these would require auxiliary cooling which would add significant complexity to the design and increase the overall cost of the simulator.

Each of the seven lamps were closely coupled to an ellipsoidal reflector. Having a multiple array of lamp/reflector units provides the flexibility required to allow varying power intensities at the target depending on the test circumstances. In order to concentrate the light emitted by the artificial light source, we utilise the fact that rays originating from a point source can be collected in their entirety on to a target point by placing the source and target points on the foci of an ideal highly reflective ellipsoid of revolution. In order to concentrate the light emitted by the artificial light source, we utilise the fact that rays originating from a point source can be collected in their entirety on to a target point by placing the source and target points on the foci of an ideal highly reflective ellipsoid of revolution. Unlike other geometric reflector shapes which have a single focal point such as parabolic or spherical reflectors where collimated light may be focused, the ellipsoidal reflector has two focal points. The ellipsoid, takes light that is generated at the primary focal point F1 and reflects it to F2 the secondary focal point. Precision made elliptical mirrors are therefore a highly efficient means of transferring radiation from the source to a target point. However since the source is not a point source but has a finite volume, the image at the secondary focal point is magnified to a degree depending on the source dimensions, the focal length as well as the eccentricity of the reflector. In addition reflectors with higher eccentricity result in the arc location being closer to the reflector surface with potentially damaging consequences due to the higher thermal loads on the surface of the reflector [11]. Further, the focal length of the reflector is to be minimised to avoid magnification of specular errors, the orientation of the lamp is to be confined to within vertical ± 135^0, as per manufacturers specifications, the distance between reflectors is to be minimised but within practical operational limits and the truncation diameter is to be maximised in order to capture the maximum amount of light. The effect of the truncation angle and the tilt angle are less pronounced [3]. One design challenge faced by this project was to mount each lamp at the focal point of the ellipsoidal reflector and to include a degree of focus adjustment.

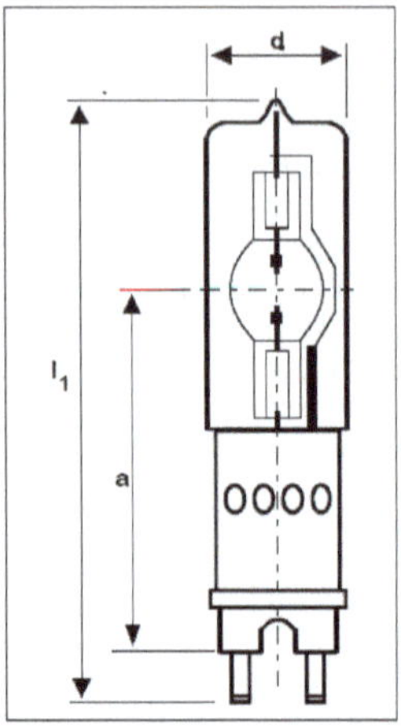

Technical data

Order reference		HMI® 6000 W/SE
Rated lamp wattage	W	6,000
Rated lamp voltage	V	123
Rated lamp current (~)	A	55
Ignition voltage (cold / hot)	kV_s	10 / 40, max. 60
Luminous flux	lm	600 000
Color temperature	K	6 000
Color rendering index CRI		> 90
Electrode gap e	mm	23
Lamp length (overall) l_1	mm	360
Bulb diameter d	mm	75
LCL a	mm	210

Figure 2. HMI 6000 W/SE XS Metal halide lamp [10].

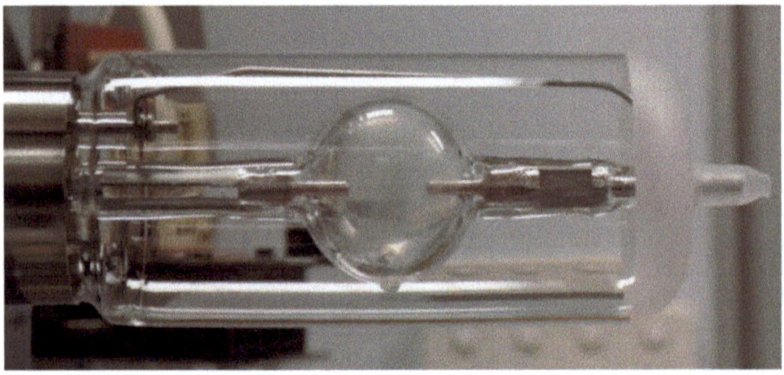

Figure 3. Photo of the installed $6kW_e$ metal halide lamp showing the inner and outer quartz glass envelopes and the electrode gap .

At the same time each reflector module must be fully adjustable to allow the beams to converge at the focal plane. Each lamp/reflector module must be supported by the frame and allow heat to dissipate. Considering a point light source located at the primary focal point, the portion of light emitted by the lamp but not reflected by the elliptical reflector was calculated to be 37%, this is made up of rear opening losses of 19% plus front aperture opening losses of 18%. The portion of light emitted by the lamp which makes contact with the surface of the reflector is 63%. A custom designed reflector with a smaller rear opening and a longer truncation length would improve on this however, bespoke designs incur significant costs and although produce improved results were cost prohibitive. The largest available elliptical reflector design produced was used for this simulator with a truncated diameter of 510 mm, a focal length of 1585 mm, a rear opening diameter of 140 mm, a reflector height of 267mm and an eccentricity of 0.92. Elliptical reflectors can be made in a number of ways, most commonly via the spinning of aluminium. A more accurate method is by the electro forming of nickel onto a precisely machined and polished mandrel to form the highly accurate ellipsoidal shape followed with surface coatings which results in different coefficients of reflectivity and wear resistance depending on the coating material. In this case the reflectors were made by electro forming of nickel and coated by electro vacuum deposition of aluminium followed by a surface coating of silicon oxide. The reflectivity of the surface as a function of wavelength is shown in Figure 4. With the reflector losses encountered the percentage reflectance curve in the 400 to 700 nm where most of the energy of the metal halide lamp occurs, averages 85% to 95%. With only 63% of the light emitted by the lamp making contact with the reflector, the result is that only 57% to 60% of the light from each lamp will reach the secondary focal point region at the target receiver. The remainder of the light will be scattered. By comparison the xenon lamp generates 70% of its energy in the over 700 nm range where the reflectivity performance is 10% less at an average of 80%.

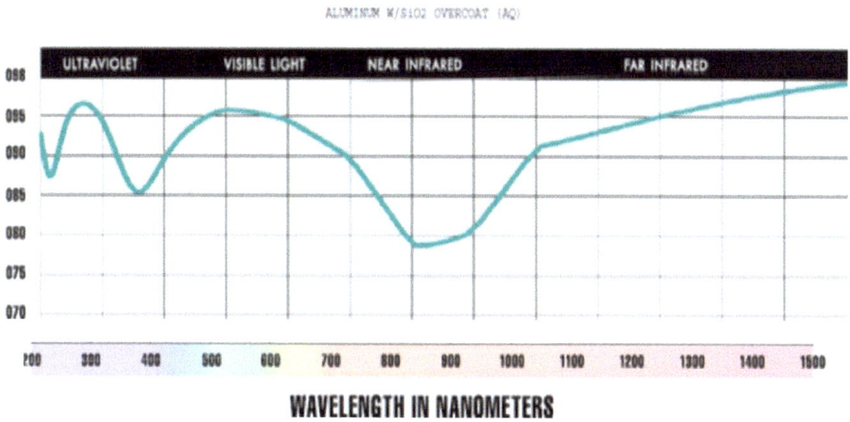

Figure 4. % Reflectivity v's wavelength chart for polished aluminium with silicon oxide coating [12].

A point source of light located at the primary focal point will be reflected by the elliptical reflector to the secondary focal point however all arc light sources have a dimension and the greater the arc dimension, the greater the distance from the primary focal point (located at the centre of the arc) to the end of the arc. The resultant geometric flux at the target plane will be related to the light angle (from γ to α) which is dependent on the size of the light source and the focal length of the ellipsoidal reflector. Light at the secondary focal plane that falls out of the size of the receiving aperture would be additional lost scatter and result in further reduction to the concentration efficiency. For a lamp containing a 9 mm arc gap, angles γ and α were 4.5° and 2° repectively. Focal lengths of 1585 mm and 3000 mm,were used to determine the size of the arc image magnification shown in the graph in Figure 5. Although focal length has a significant effect on the size of image at the secondary focal plane and a shorter focal length is preferred, it is the arc length that has the most significant effect. A receiver aperature diameter of less than 330 mm will result in a significant amount of radiation spillage and radiation shielding is required to protect the surrounding receiver hardware from damaging heat fluxes. A shorter focal length is limited as this creates other problems including a reduced working space and greater tilt angles that are restricted by the lamp manufacture as well as the increased difficulty of heat dissipation. A compound parabolic concentrator (CPC) with an inlet diameter of 350 mm, with an acceptance angle of 31°, matching the rim angle of the simulator, would enhance the flux concentration by a factor of 2 [3].

Another important aspect that must not be overlooked is that the solar simulator is an in-door research tool that is to be used as an alternative to a large parabilic dish where the suns rays are collected and directed to the focal point and the high temperatures are contained within a reactor. Initially these are small scale prototype reactor experiments however one of the goals would be to prove the viability for commercial size operations. Although there are currently no commercial high temperature solar reactors that are operational and processing material on a large scale, the solar simulator must mimic the paramaters of such commercial facility. In this cenario the suns rays are not parralel but have a finite angular size of the suns disc of about 33 minutes of an arc or about 9.6 mrad.

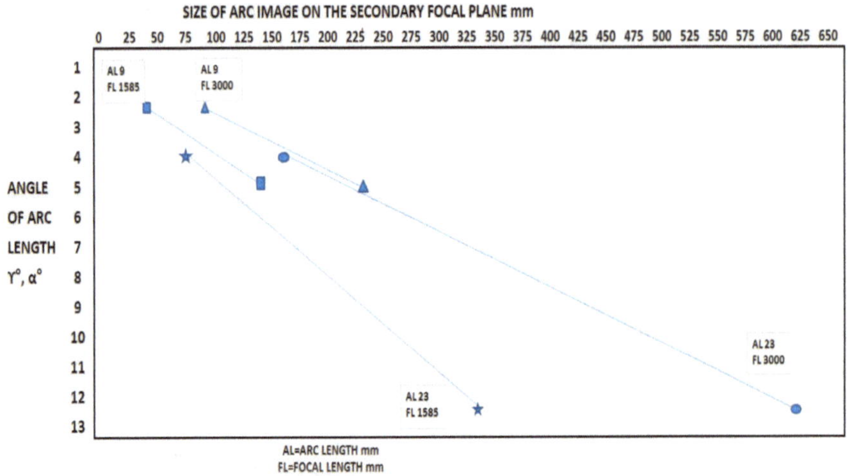

Figure 5. Size of arc image at the secondary focal plane for an arc length of 9 mm and 23 mm and a focal length of 1585mm and 3000mm.

Another important aspect that must not be overlooked is that the solar simulator is an in-door research tool that is to be used as an alternative to a large parabilic dish where the suns rays are collected and directed to the focal point and the high temperatures are contained within a reactor. Initially these are small scale prototype reactor experiments however one of the goals would be to prove the viability for commercial size operations. Although there are currently no commercial high temperature solar reactors that are operational and processing material on a large scale, the solar simulator must mimic the paramaters of such commercial facility. In this cenario the suns rays are not parralel but have a finite angular size of the suns disc of about 33 minutes of an arc or about 9.6 mrad. For a focal length of 20 m the sun will have a beam spread of approximately 800mm at the focal plane. This is demonstrated at the PROMES research facility in Odeillo France where the worlds largest 1830 m^2 parabolic mirror with an 18 m focal length concentrates 1MW of solar power onto a 800mm diameter spot to produce about 10 megawatt/m^2 at its focus [13]. Irrespective of the intensity of the hot spot which is relative to the helliost collection mirror area, the long focal length will produce a large hot spot diameter. Commercial solar reactors of the future will similarly require large focal lengths for operational constraints and to enable the processing of commercial volumes of materials. A large diameter focal spot produced by solar simulators using metal halide lamps are more representative of the concentrated sun required for industrial reactor designs. Table I lists the parameters of the lamp/reflector array chosen and how these compare to other published high flux solar simulator installations.

Arc Modelling and Optical Characterisation.

Illumination systems depend greatly on the source characteristics. The design of a solar simulator requires the accurate modelling of the source parameters in order to have the fabricated system agree with the design. For an arc source, often a spherical or a generic cylinder can be used to model the emission however arc sources tend to be deformed or bowed so the generic source model is a poor representation of the actual arc source. In order to effectively model the simulator design with optical design software, the most crucial parameter is the source model.

Table I Paramerters of the lamp and reflector array. [3, 6, 7]

HIGH FLUX **SOLAR SIMULATORS** - in date order

SIMULATOR	DATE	LAMP TYPE	POWER KW/LAMP	NUMBER OF LAMPS	TOTAL ELECTRIC POWER KW	DELIVERED RADIATIVE POWER KW	PEAK FLUX KW/M2	STAGNATION TEMPERATURE K	REFLECTOR DIAMETER d mm	REFLECTOR HEIGHT h mm	TRUNCATION ANGLE α^t	FOCAL LENGTH mm	ECCENTRICITY e	RIM ANGLE Φ_{rim}
ETH ZURICH-PETRASCH	2007	XENON	15	10	150	50	11000 Av 6800	3700	950	379	70	3000	0.87	27,41
UNI OF MINNESOTA-KRUEGER	2011	XENON	6.5	7	45.5	7.5	3700	2740	750	346	59.1	2032	0.89	37.7
UNI OF FLORIDA-ERICKSON	2012	XENON	6	7	42	NR	4230	2942	800	456	50	1800	(0.87)	26
EKMAN - SWINBURNE UNI.	2014	METAL HALIDE	6	7	42	15.3	1122	2100	508	283	50	1585	0.918	31

NR - NOT REPORTED (calculated)

Many lamp designs have existing ray files that have been determined either by the manufacturer or researchers which accurately model the light source of the arc however there were no ray files available for the 6000 Watt metal halide lamp used. One method to model the arc and generate a ray set or ray file is to capture the image of the arc with the asymmetry fully accounted. The difficulty is to obtain a true image of the arc. Figure 6(a) is a photo of the lamp using a DIN 13 shade glass filter with what appear to be a large spherical shaped arc. Figure 6(b) shows the arc under a variable neutral density filter that permits the cylindrically shaped arc to be visible and distinct from the over exposed glow surrounding it. The volume of the 23 mm long, 4 mm diameter cylindrical image of the arc source under ND filtered light is significantly smaller than that of a 23 mm diameter spherical source. The centre of the arc is positioned at the focus of the ellipsoidal reflector and ray tracing is performed. The lamp/reflector design was entered into FRED optical software including the orientation of all 7 lamps together with their material properties.

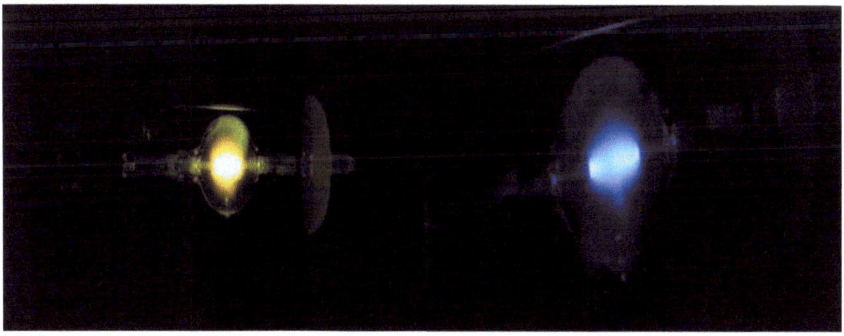

(a) (b)

Figure 6. (a) Photo of a metal halide lamp using a Din 13 shade filter. The arc is not visible but appears to be spherical in shape. (b) Photo of the metal halide lamp taken under a variable ND filter. The arc is now visible and cylindrical in shape.

A selection is made of the number of rays to be traced through the optical mechanical system whilst encountering various optical interactions. While the spectral definition of the light source used has a significant effect on the accuracy of the simulation of the system, accuracy also increases with the number of rays traced however larger ray numbers result in longer processing times. The results of ray tracing for a single reflector are shown in Figure 7(a).

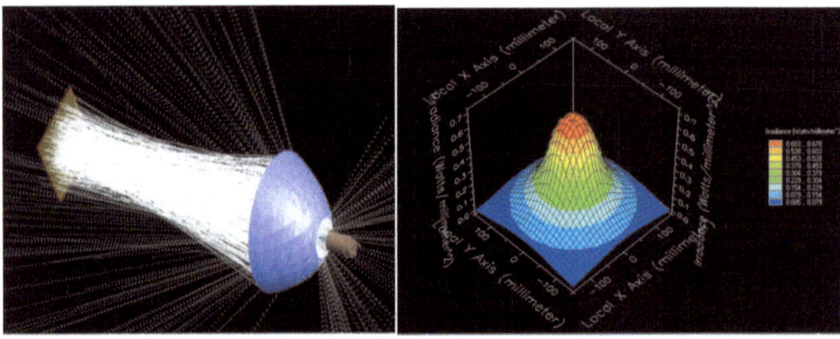

(a) (b)

Figure 7. (a) Results of ray tracing for a single reflector using FRED optical software (Photon Engineering Tuscon USA). (b) Model of the distribution of flux at the focal plane [14].

The ray tracing modelling results shown in Figure 7(b) indicate a peak flux of just under 700 kW/m^2.

Fabrication

The schematic layout of the lamp/reflector array is shown in Figure 8(a). The frontal view of 6 modules symmetrically arranged around a central module and a cross-sectional top view is shown. The central module (no.4) is fix mounted and is only adjustable in the forward horizontal direction. Modules 3 and 5 are hinge mounted allowing them to rotate sideways on a vertical axis. Modules 1, 2, 6 and 7 are able to be rotated about their vertical and horizontal axis.

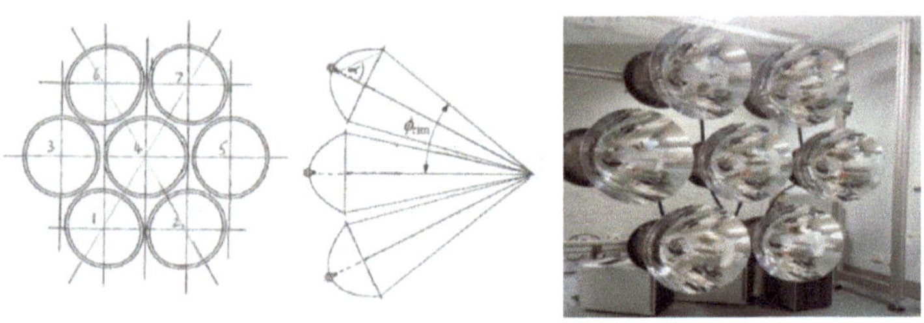

(a) (b)

Figure 8. (a) Schematic of the positioning of the 7 module array, front and side views. (b) Solar simulator consisting of seven metal halide lamps each coupled to precision ellipsoidal reflector.

Each lamp/reflector module had a combined weight of 25 kg's and required adequate support and strength to be safely fixed and held in position whilst maintaining their position during operation. Temperatures of the supporting structure were measured around the lamp and reflector support and after an hour of operation reached 140°C. The assembled array is shown in Figure 8(b). Other peripheral equipment includes the seven high voltage ignitors each wired to a ballast or power supply.

Each ignitor is required to be located in close proximity to the lamp socket and connected by heavily insulated high voltage cable rated for an ignition voltage of 40,000 volts during start-up. Each of the modules can be operated individually depending on the level of heat flux required. In addition each of the modules can be positioned to target a single common focal point or arranged so as to not overlap but to cover a larger area within a volume receiver. This would result in lower concentrated flux intensities and a more uniform flux distribution.

Flux Measurement

To avoid spectral errors in the measurement of the flux distribution [15, 16], flux measurement involved a circular foil heat transducer also known as a Gardon type water cooled (Model:Vatell TG1000-1) heat flux gauge. The thermal flux measured by the transducer is proportional to the temperature difference between the centre of the foil disc and the edge. The disk is bonded to a circular opening in a cylindrical heat sink. The foil is made of Constantan and the heat sink is Copper. Exposing the thermogauge transducer to thermal flux produces a proportional voltage. The thermogauge is coated with a high temperature colloidal graphite with an emissivity of 0.82 and calibration constant of 0.782 that compensates for the spectral absorptance of the coating [15, 16]. The gauge, mounted on an aluminium plate, provided continuous heat flux readings while incremental movements were made in the x, y and z directions. Care was taken to ensure that the surface coating was not damaged as well as ensuring that convective heat loss was minimized to maximise gauge accuracy. The flux measurements taken for each individual lamp are shown in Figure 9a. Each of the lamp/modules reached a peak flux ranging from 117 kW/m^2 to 148 kW/m^2. The total thermal flux with all lamps focused to a common point is shown in Figure 9b, reaching a peak of 927 kW/m^2 and with an aperture diameter of 175 mm the input power received is 12 kW.

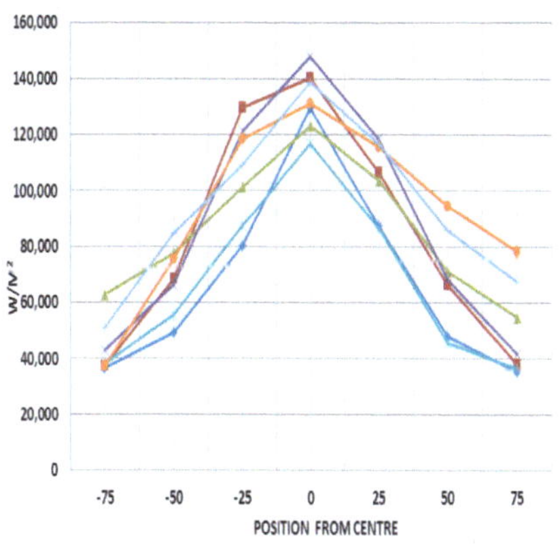

(a)

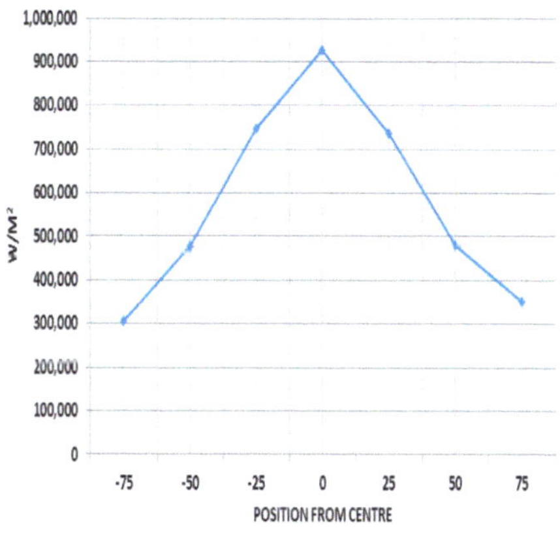

(b)

Figure 9. Distribution of the thermal flux (a) each of the seven lamps measured at the focal plane. (b) with all lamps seven combined. Units are in watts/m^2.

Performance

A theoretical calculation of the stagnation temperature based on a measured maximum heat flux value of 927 kW/m^2 is given by Equation 1. The stagnation temperature is the highest temperature that an ideal blackbody receiver would achieve when the energy is being absorbed as fast as it is re-radiated.

$$q = \sigma T^4 \qquad (1)$$

where q is the radiated flux and σ is the Stefan-Boltzmann constant.

Such high thermal fluxes would result in a corresponding stagnation temperature of 2011 K. By installing a CPC and assuming this would increase the flux concentration by a factor 2 [3], the resulting stagnation temperature is 2390 K and an input power of 24 kW. This demonstrates the effect of the T^4 where large increases of heat flux are required to produce relatively small increases in temperature. Individual lamp currents and voltages were not measured however according to the manufacturer's specification [10], the rated lamp voltage is 124 V and lamp current is 55 amps (6820 lamp wattage). Allowing 10% for fluctuations, power factor losses and ohmic losses at the electrodes (Osram specs), all the power can be accounted for. Experimental measurements and calculations have verified that on average each lamp delivered 1714 watts of power delivered through a 175mm diameter aperture, an average reflectivity of 89%, a non-scattered light factor of 63% and only 50% of the arc light entering a 175mm diameter aperture, correlates well to the rated power of 6046 watts per lamp. This results in a lamp to aperture (175 mm) transmission efficiency of 25%. An installed CPC would result in a transmission efficiency of 51%. Experiments are currently being conducted where these high thermal fluxes are directed into a hybrid designed receiver/reactor with varying configurations and the thermal effects are observed and measured.

The average annual solar DNI or direct normal irradiation is the quantity of direct solar radiation per unit area that is intercepted by a flat surface that is at all times pointed directly at the sun. On average the annual solar DNI for Melbourne are in range of 1600 kWh/m^2 to 1800 kWh/m^2 , local hourly measurement were of particular interest in this study. For purposes of comparison, random measurements of solar flux using a TES 1333 Solar Power Meter were taken on clear days during the months of August and September from 10am to 4pm directed at the sun and ranged from 900 W/m^2 to 1112 W/m^2. On cloudy days this was substantially less. From this it is fair to assume an amount of one solar constant or 1000 watts per square meter as the local Melbourne solar resource. Therefore to concentrate 24 kW of direct sunlight using parabolic mirrors, allowing 10% for reflective losses, would require a mirror collection area (or aperture size) of approximately 30 square meters.

Summary

Historically high flux solar simulators have used xenon lamps as their light source primarily for their short arc length and high power however MH lamps and their associated drives are significantly lower in cost than the equivalent power rated xenon lamp system. In addition MH lamps have a better spectral match to natural sunlight and are much safer to handle and operate than xenon lamps. Operational and design parameters that maximise the transfer efficiency include a truncation angle in the range of 50° to 70°, lamps with short arc length, increased truncation diameter, shorter focal lengths and a decreased tilt angle. With all lamps commonly focused, almost 1000 suns of concentration were achieved. A peak flux of 927 kW/m^2 was measured and a total of 12 kW of power was delivered through a 175 mm diameter aperture. Ray tracing results compare reasonable well with experimental results and lamp power transmission efficiency to a 175 mm aperture was 25%. Installation of a CPC would double the power delivered to a 175mm aperture increasing the transmission efficiency closer to 50%. Metal halide lamps coupled with precision ellipsoidal reflectors will produce a flatter, less intense flux profile that closely mimics both the spectral distribution of sunlight as well as generating a wider focal hot spot as do the long focal lengths of commercial sun focused optical systems.

References

1. Unvala B.A., and Maries A., Radiant Heating Using an Ellipsoidal Reflector 1974. J. Phys. E: Sci. Inst. 7.349

2. Jaworske D.A., et.al., Alignment and Initial Operation of an Advanced Solar Simulator J. of Spacecraft and Rockets Vol. 33, No. 6, November-December 1996.

3. Petrasch J. Coray P. Meier A. Brack M. Haberling P. Wuillemin D. Steinfeld A. A Novel 50 kw 11,000 suns High-Flux Solar Simlulator Based on an Array of Xenon Arc Lamps. 2007, J. of Solar Energy Eng.. Vol.129.

4. Hirsch D. Zedtwitz P.V.Osinga T. Kinamore J. Steinfeld A. A New 75kw High-Flux Solar Flux Solar Furnace and Xenon High Flux Solar Simulator. German Aeros. C, J.of Solar Energy Eng. Feb.2003 Vol. 125.
5. Kuhn P., Hunt A., A new solar simulator to study high temperature solid-state reactions with highly concentrated radiation. Solar Energy Materials, Vol.24.1991.
6. Krueger K.R., et.al. Design of a New 45 kWe High-Flux Solar Simulator for High-Temperature Solar Thermal and Thermochemical Research. J. of Solar Energy Eng., Vol.133.2011
7. www.forida energy.ufl.edu
8. www.dlr.de.
9. Lovegrove K. "Solar Thermal Energy Systems in Australia". The international Journal of environ studies 2006.
10. OSRAM -Metal Halide Lamps Tech.Sheet. DO 4404.
11. Alxneit I., Dibowski G., R12.5 Solar Simulator Evaluation Report. Project SFERA., August 2011.
12. Optiforms, Temecula California USA
13. PROMES Odeill France
14. FRED Optical Software Engineering, www.photonengr.com
15. Alxneit I. Error Analysis Of The Radiative Power Determined From Flux Distributions Measured With a Camera in a Xe Arc Lamp Based Solar Simulator. J. of Solar energy Eng. Nov. 2012.Vol. 134.
16. Ballestr!ına J., Ulmerb S., Moralesa A., Barnesc A., Langleyc L. C.,Rodr!ıgueza M. Systematic Error In The Measurement Of Very High Solar Irradiance. Solar Energy Mat. & Solar Cells 2003.

Energy Technology 2015
Carbon Dioxide Management and Other Technologies

Recycling and Sustainability Update

Energy Technology 2015
Carbon Dioxide Management and Other Technologies

RECYCLING AND
SUSTAINABILITY UPDATE

Waste

Session Chairs:

Jeffrey S. Spangenberger

Energy Technology 2015: Carbon Dioxide Management and Other Technologies
Edited by: Animesh Jha, Cong Wang, Neale R. Neelameggham, Donna P. Guillen, Li Li, Cynthia K. Belt, Randolph Kirchain, Jeffrey S. Spangenberger, Frank Johnson, Andrew J. Gomes, Amit Pandey, and Peter Hosemann
TMS (The Minerals, Metals & Materials Society), 2015

THE REMOVAL OF PHOSPHATE AND AMMONIA NITROGEN FROM WASTEWATER USING STEEL SLAG

Xu Ping, Yang Liyun*, Renaguli Aikebaier, Bai Hao

Department of Ecological Science and Engineering, School of Metallurgical and Ecological Engineering, University of Science and Technology Beijing, Beijing 100083, China.
*Corresponding author: wendy6393@163.com

Keywords: Steel slag, Phosphate, Ammonia nitrogen, Removal capacities, Removal mechanism.

Abstract

The steel slag was used to remove phosphate and ammonia nitrogen from wastewater in this paper. A series of batch experiments were designed to focus on investigating the removal capacities of these two kinds of pollutants, and establish the overview about pollutants removal mechanisms. The study shows that the removal rate of phosphate (99%) was significantly higher ammonia nitrogen (22.3%). At 20 ℃, the experimental data of two kinds of pollutants could be fitted by using both Langmuir and Freundlich models. The maximum removal capacities of phosphate and ammonia nitrogen with initial concentrations of 2.5-12.5 and 40-240 mg/l were 1.39 and 10.29 mg/g, respectively. The tendency of concentrations of phosphate, Ca^{2+} and pH in solution verified that the principal removal mechanism of phosphate is formation of calcium phosphate precipitation.

Introduction

As the main waste of iron and steel production, steel slag has been recycled in cement and concrete block production, road backfill and other fields [1]. But its efficiency was very low, which resulted in China's stock of steel slag reaching 1.12 billion tons in 2012 [2]. Steel slag has the effect of filtration, adsorption and precipitation on the pollutants in the water because of its loose structure, large pore, stable chemical composition, and it contains the free CaO, MgO and Fe [3-4]. Now, researchers have begun to explore the use of steel slag on removal of phosphate (PO_4^{3-}) and ammonia nitrogen (NH_4^+-N) in wastewater [5-8], but the corresponding removal mechanisms have not been elaborated. In addition, the influences of parameters did not include particle size of the steel slag. Aiming at these problems, this paper will carry out the following research: one is to investigate the removal mechanisms of phosphate and ammonia nitrogen in wastewater; second is to consider the influence of the steel slag particle size on pollutants removal.

Materials and methods

Slag collection and preparation

The samples of steel slag used in this study were supplied by Laiwu Steel, Shandong, China. It was first broken in ball mill and then through 100 mesh, 80 mesh, 60 mesh, 40 mesh and 20 mesh sieve respectively. Five steel slags with different particle size were put into plastic bags for subsequent experiments. X-ray fluorescence spectrum (XRF) was performed to determine the

* Corresponding author

chemical composition of the slag, and the microstructure of the sample was analyzed by scanning electron microscope- X-ray energy spectrometer (SEM-EDX).

Release kinetic experiments

The solubility property of steel slag in water was investigated in a series of batch experiments, which is to study the kinetic of Ca^{2+} and OH^- release from slag under time gradient. For each experiment, 4 g of slag with particle size 40 mesh was immersed in 100 ml of deionized water contained in a flask. The bottle was placed on a vibration thermostat and shaken at 200 rpm under 20 ℃. In order to establish the equilibrium time, previous solutions were agitated for 30, 60, 90, 120, 150, 180 and 210 min. The test solutions were centrifuged and filtered to separate the slag from solution, and analyzed the pH values and total Ca^{2+} concentrations immediately. The capacities of Ca^{2+} and OH^- release from slag were calculated by the equation (1), where Q_t and X^t are the capacities of Ca^{2+} and OH^- release (mg/g) and concentrations of solutions (mg/l) at time t, respectively, V and M are the volume of solution (l) and the mass of slag (g), respectively.

$$Q_t = \frac{X^t V}{M} \quad (1)$$

PO_4^{3-} and NH_4^+-N adsorption experiments

The effects and mechanisms of PO_4^{3-} and NH_4^+-N removal from solutions were studied in a series of batch experiments. These solutions were prepared with deionized water, NH_4Cl and KH_2PO_4. Take a series of slag with different particle size (20, 40, 60 80 and 100 mesh) and dosage (1.0, 2.0, 3.0, 4.0, 5.0 and 6.0 g) into 250 ml flasks each containing 100 ml solution with initial PO_4^{3-} and NH_4^+-N concentrations (10 and 100 mg/l, respectively). The pH of solutions was adjusted at 7.0 with 0.1 mol/l HCl and NaOH. The bottle was placed on a vibration thermostat and shaken at 200 rpm for 90 min under 20 ℃. In addition, PO_4^{3-} and NH_4^+-N adsorption experiments with variations of initial PO_4^{3-} concentrations 2.5, 5.0, 7.5, 10.0 and 12.5 mg/l, NH_4^+-N concentrations 40, 80, 100, 120, 160, 200 and 240 mg/l, pH 5, 6, 7, 8, 9, 10 and 11, reaction time 30, 60, 90, 120, 150, 180 and 210 min were performed on 4 g of slag with particle size 40 mesh. After the adsorption experiments, the samples were centrifuged and filtered. And then the separated solutions were chemically analyzed for residual PO_4^{3-}, NH_4^+-N, Ca^{2+} and pH under different initial concentrations.

The adsorption capacity of PO_4^{3-} and NH_4^+-N Q (mg/g) were calculated from equation (2), where V is the volume of the solution (l), M is the mass of slag (g), C_0 and C_e are the initial concentration and final concentration (mg/l), respectively.

$$Q = \frac{(C_0 - C_e)V}{M} \quad (2)$$

Description of models

The adsorption data can be interpreted using several relationships for expressing the isothermal adsorption effect of PO_4^{3-} and NH_4^+-N more clearly. Langmuir and Freundlich models were used to simulate the experimental data when metal ions uptake occurs by a monolayer sorption. The Langmuir model assumes that the surface of the sorbent to be homogeneous with no interaction between adsorbed ions. The Freundlich model assumes that the uptake of metal ions which can interact occurs on a heterogeneous surface. The corresponding equations with linear forms are as follows:

Langmuir:
$$\frac{1}{Q}=\frac{1}{K_A Q_m}\frac{1}{C_e}+\frac{1}{Q_m} \qquad (3)$$

where Q is the amount sorbed at equilibrium (mg/g), that is the ratio of the mass solute adsorbed and the mass adsorbent, Q_m is the monolayer saturation capacity (mg/g), C_e is the equilibrium concentration (mg/l), and K_A is the equilibrium constant related to the sorption capacity (l/mg).

Freundlich:
$$\log Q=\frac{1}{n}\log C_e+\log K_F \qquad (4)$$

where K_F and n are the Freundlich constants, and n is greater than 1.

Analytical methods

Phosphate analyses were performed according to the ammonium molybdate spectrometric method (GB 11893-89). Ammonia nitrogen was measured by a multi-parameter controller (Merck Company, Germany). Ca concentration in solution was determined by EDTA titration method (GB 7476-87). Solution pH was measured using a pH meter (pH 340i, WTW, Germany). All chemicals used were of analytical grade. All sorption experiments were conducted in triplicates and the data were expressed as the average value.

Results and discussion

Chemical characteristics of the steel slag

The main chemical compositions of the slag for this study measured by XRF were CaO, Fe_2O_3 and SiO_2 (Table I). The microstructure of steel slag using SEM-EDX analyses are shown in Figure 1. As shown, the porosity of slag is larger and the structure is loose.

Table I. Chemical composition of the steel slag

Components	SiO_2	Al_2O_3	MgO	CaO	P_2O_5	TiO_2	Fe_2O_3	MnO
Values (%)	14.446	2.034	5.831	53.044	0.994	0.782	19.209	2.369

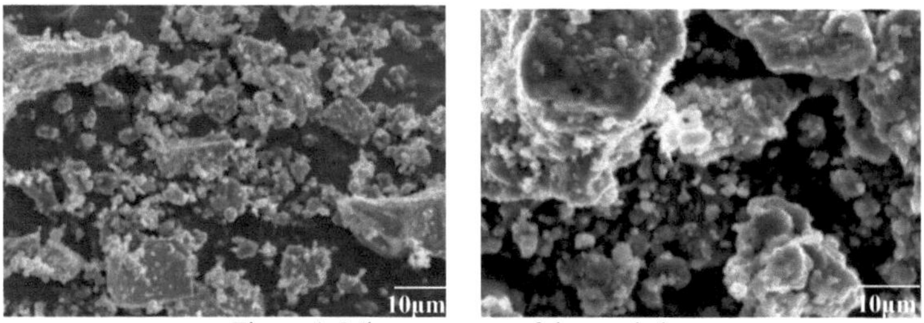

Figure 1. Microstructure of the steel slag

Kinetics of Ca^{2+} and OH^- release

The kinetics of $Ca^{2+}$and OH^- release from slag are shown in Figure 2. The total Ca^{2+} concentrations and pH values in the solutions have risen to a certain height when shaking time range between 30 to 60 min. And then, the change trend of release was not strong with the increase of time. The same trend of $Ca^{2+}$and OH^- release shows that CaO dissolved from the slag, which can be explained by equation (5).

$$CaO + H_2O \rightarrow Ca^{2+} + 2OH^- \quad (5)$$

According to equation (5), the highest CaO content in slag could contribute to the higher total Ca^{2+} concentrations and pH values. Moreover, the shaking time can be set up to 90 min in the removal experiments of PO_4^{3-} and NH_4^+-N according to the equilibrium time in Figure 2.

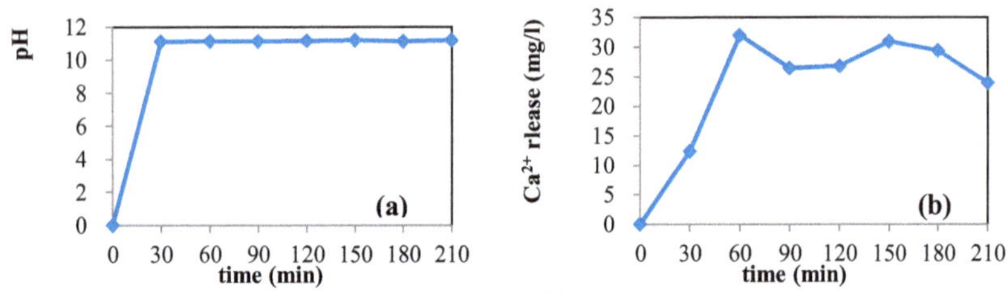

Figure 2. Kinetics of Ca^{2+} and OH^- release: total Ca^{2+}concentrations (a) and pH values (b) of the solutions.

PO_4^{3-} and NH_4^+-N removal from solutions

Effect of particle size of slag. When slag with 2.0 g size range between 20 to 100 mesh was added in the solutions with initial concentration of PO_4^{3-} and NH_4^+-N were 10 and 100 mg/l, respectively, the residual concentrations in solutions at equilibrium are shown in Table II. As can be seen, the effect of particle size of slag on PO_4^{3-} removál was lesser, and all removal rate of PO_4^{3-} were above 99%. However, the removal efficiency of NH_4^+-N was strongly influenced by particle size of slag. The maximum removal capacity of 31.22% occurs when the particle size was 40 mesh. And the removal capacity of PO_4^{3-} and NH_4^+-N ranged from 0.450 to 0.498 mg/g and 1.165 to 1.750 mg/g, respectively.

Table II. The effect of particle size on removal of PO_4^{3-} and NH_4^+-N

Particle size (mesh)	PO_4^{3-}			NH_4^+-N		
	Residual concentration (mg/l)	Adsorption capacity (mg/g)	Removal rate (%)	Residual concentration (mg/l)	Adsorption capacity (mg/g)	Removal rate (%)
20	0.024	0.499	99.76	70.5	1.475	25.40
40	0.023	0.499	99.77	65.0	1.750	31.22
60	0.010	0.450	99.90	69.8	1.510	26.14
80	0.021	0.499	99.79	75.7	1.215	19.89
100	0.032	0.498	99.68	76.7	1.165	17.80

Effect of dosage of slag. When slag with 40 mesh dosage range between 1.0 to 6.0 g was added in the solutions with initial concentration of PO_4^{3-} and NH_4^+-N were 10 and 100 mg/l, respectively, the residual concentrations in solutions at equilibrium are shown in Table III. The removal rate of PO_4^{3-} increases with the increase of slag dosage, and it reached above 99% when the dosage of slag was 4.0 g. By contrast, slag dosage has a large effect on removal of NH_4^+-N. The removal rate in peak appeared when the dosage was 4.0 g, and then it decreases with a further increase in the dosage. And the removal capacity of PO_4^{3-} and NH_4^+-N ranged from 0.166 to 0.861 mg/g and 0.217 to 1.117 mg/g, respectively.

Table III. The effect of dosage on removal of PO_4^{3-} and NH_4^+-N

Dosage (g)	PO_4^{3-}			NH_4^+-N		
	Residual concentration (mg/l)	Adsorption capacity (mg/g)	Removal rate (%)	Residual concentration (mg/l)	Adsorption capacity (mg/g)	Removal rate (%)
1.0	1.386	0.861	87.67	88.83	1.117	11.17
2.0	0.318	0.484	97.17	87.17	0.642	12.83
3.0	0.186	0.327	98.35	85.67	0.478	14.33
4.0	0.035	0.249	99.69	85.50	0.363	14.50
5.0	0.029	0.199	99.74	87.83	0.243	12.17
6.0	0.014	0.166	99.88	87.00	0.217	13.00

Effect of initial pH of solution. The solutions with initial concentration of PO_4^{3-} and NH_4^+-N were 10 and 100 mg/l, respectively, were adjusted into pH of 5 to 11 and treated by the slag with 4 g and 40 mesh, the residual concentrations in solutions at equilibrium are shown in Table IV. The initial pH of the solution has little effect on the removal of PO_4^{3-} because high removal rate (>98%) achieved at all the range of pH. On the contrary, the removal of NH_4^+-N was closely related to the initial pH of solution. The residual concentrations of NH_4^+-N decreased when the initial pH between 5 to 8, but increases after this value of initial pH. It was observed that the maximum NH_4^+-N removal capacity was 0.580 mg/g occurs on initial pH reached 8.

Table IV. The effect of initial pH on removal of PO_4^{3-} and NH_4^+-N

Initial pH	PO_4^{3-}			NH_4^+-N		
	Residual concentration (mg/l)	Adsorption capacity (mg/g)	Removal rate (%)	Residual concentration (mg/l)	Adsorption capacity (mg/g)	Removal rate (%)
5	0.19	0.245	98.1	85.5	0.363	14.50
6	0.073	0.248	99.3	79.5	0.513	20.50
7	0.036	0.249	99.7	80.0	0.500	20.00
8	0	0.250	100	76.8	0.580	23.20
9	0	0.250	100	78.2	0.545	21.80
10	0	0.250	100	80.0	0.500	20.00
11	0	0.250	100	81.2	0.470	19.80

Kinetics of PO_4^{3-} and NH_4^+-N adsorption. It was established in the above experiments that the best removal efficiency of PO_4^{3-} and NH_4^+-N can be reached on the particle size and dosage of slag was 40 mesh and 4.0 g, respectively, and the initial pH of solution was 8. Thus, a kinetic experiment of the simultaneous PO_4^{3-} and NH_4^+-N adsorption was carried out by the best ratio of steel slag and solution at various reaction times from 30 to 210 min. The experiment data are shown in Table V. The removal capacity of PO_4^{3-} was very high within the time range, especially since 150 min, it close to 100%. By contrast, the removal capacity of NH_4^+-N increased up to 90 min, but it fell slightly after this time. Thus, the maximum NH_4^+-N removal capacity was 0.583

mg/g when the reaction time reached 90 min. The removal capacity of NH_4^+-N ranged from 0.355 to 0.583 mg/g, significantly higher than that of PO_4^{3-}.

Table V. The effect of reaction time on removal of PO_4^{3-} and NH_4^+-N

Reaction time (min)	PO_4^{3-}			NH_4^+-N		
	Residual concentration (mg/l)	Adsorption capacity (mg/g)	Removal rate (%)	Residual concentration (mg/l)	Adsorption capacity (mg/g)	Removal rate (%)
30	0.146	0.246	98.00	84.8	0.380	15.2
60	0.064	0.248	99.40	83.2	0.420	16.8
90	0.077	0.248	99.30	76.7	0.583	23.3
120	0.030	0.249	99.70	78.5	0.538	21.5
150	0	0.250	100	85.5	0.363	14.5
180	0	0.250	100	85.8	0.355	14.2
210	0	0.250	100	83.5	0.413	16.5

Adsorption isotherms of PO_4^{3-} and NH_4^+-N. Before slag with 4 g, 40 mesh was added in the solutions, the initial concentration of PO_4^{3-} was adjusted to 2.5, 5.0, 7.5, 10.0 and 12.5 mg/l, while that of NH_4^+-N was adjusted to 40, 80, 100, 120, 160, 200 and 240 mg/l.

Table VI. The effect of initial PO_4^{3-} concentration on removal of PO_4^{3-}

Initial concentration (mg/l)	pH	Residual concentration (mg/l)	Adsorption capacity (mg/g)	Removal rate (%)	Ca^{2+} concentration (mg/l)
0	11.39	10	0	0	22.07
2.5	11.35	0.0055	0.062	99.8	19.25
5.0	11.29	0.0109	0.125	99.8	17.12
7.5	11.27	0.0218	0.187	99.7	16.19
10.0	11.28	0.0238	0.249	99.8	15.27
12.5	11.28	0.0291	0.312	99.8	13.67

Table VII. The effect of initial NH_4^+-N concentration on removal of NH_4^+-N

Initial concentration(mg/l)	Residual concentration (mg/l)	Adsorption capacity (mg/g)	Removal rate (%)
0	100	0	0
40	32.3	0.193	19.25
80	63.3	0.418	20.88
100	83.0	0.425	17.00
120	93.3	0.668	22.25
160	138.0	0.550	13.75
200	162.3	0.943	18.85
240	184.5	1.388	23.13

The isothermal adsorption data are shown in Tables VI and VII. The adsorption capacity of PO_4^{3-} increases with increasing initial concentration, and the peak was appeared on 12.5 mg/l. Moreover, the decrease in PO_4^{3-} concentrations was accompanied by a decrease in Ca^{2+} concentrations and pH values. Due to the precipitation of calcium phosphate depend on the pH values, Ca^{2+} and PO_4^{3-} concentrations [9-10], the same changes of these three data proved that the removal of PO_4^{3-} was mainly based on the formation of calcium phosphate precipitates. Equation (6) shows that the precipitation of calcium phosphate consumed Ca^{2+} and OH^-. The adsorption capacity of NH_4^+-N also shows the same change trend with PO_4^{3-}, which sharply increases with the increase of initial concentration. And the peak was appeared on 240 mg/l, which indicates that adsorption capacity of NH_4^+-N may have a further increase with the increase of initial concentration. The removal capacity of PO_4^{3-} and NH_4^+-N ranged from 0.062 to 0.312 mg/g and 0.193 to 1.388 mg/g, respectively.

$$5Ca^{2+} + 3PO_4^{3-} + OH^- \rightarrow Ca_5(PO_4)_3(OH)\downarrow \quad (6)$$

Table VIII. Langmuir and Freundlich parameters, correlation coefficient (R^2) for the sorption of PO_4^{3-} and NH_4^+-N

Adsorbing ion	Langmuir constants			Freundlich constants		
	Q_m (mg/g)	K_A (l/mg)*10^{-3}	R^2	K_F ($l^{1/n}g^{-1}mg^{1-1/n}$)	n	R^2
PO_4^{3-}	1.39	8617.79	0.9911	7.127	1.10	0.9753
NH_4^+-N	10.29	0.60	0.9509	0.006	1.02	0.8837

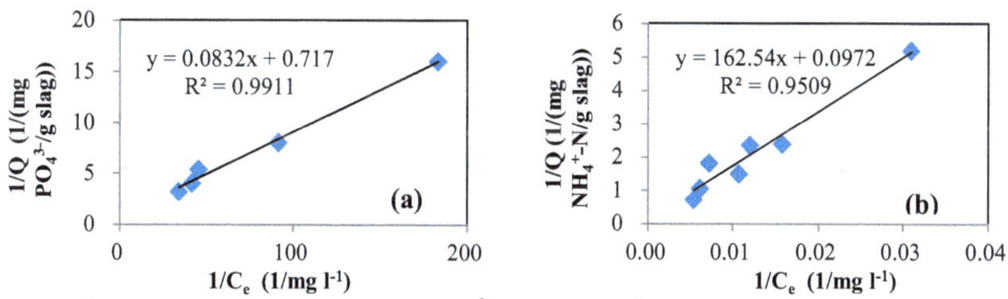

Figure 4. Langmuir plots for (a) PO_4^{3-} and (b) NH_4^+-N at different temperatures

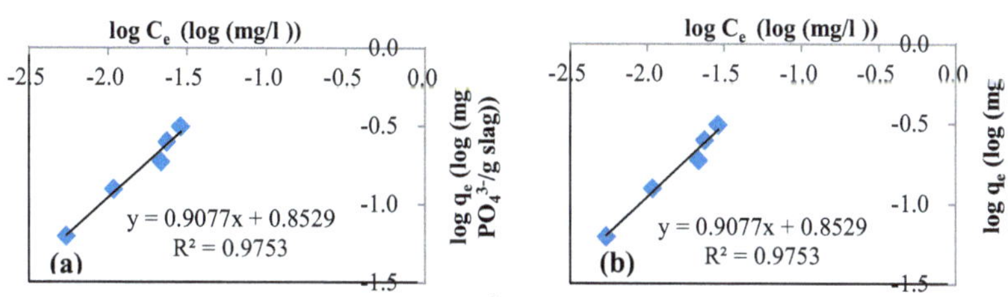

Figure 5. Freundlich plots for (a) PO_4^{3-} and (b) NH_4^+-N at different temperatures

The experimental data obtained from previous report fitted the linear relationship of both Langmuir (Figure 4(a) and (b)) and Freundlich (Figure 5(a) and (b)) well. The parameters calculated for Langmuir and Freundlich equations are listed in Table VIII. The fitting degree of two equations are both higher, but the equilibrium correlation coefficients of Langmuir model (R^2=0.9911 and 0.9509 for PO_4^{3-} and NH_4^+-N) were better than Freundlich model (R^2=0.9753 and 0.9509 for PO_4^{3-} and NH_4^+-N), which indicates that Langmuir model is more suitable for simulating the experimental data of this paper. Furthermore, the results also prove that the surface of slag is homogeneous and no interaction between adsorbed ions, PO_4^{3-} and NH_4^+-N was adsorbed by monolayer adsorption.

Conclusion

The removal efficiencies of PO_4^{3-} and NH_4^+-N in solutions treated by steel slag were investigated by a batch method. These experimental results shown that the removal efficiencies of PO_4^{3-} was obviously higher than that of NH_4^+-N, and the removal effects also existed differences with the same kind of pollutants when the parameters was different. The main parameters affecting the experiment results include the particle size and dosage of slag, initial pH of solutions, reaction time and initial concentration of pollutants. In general, the removal efficiencies of PO_4^{3-} and NH_4^+-N in 100 ml solutions were high treated by the steel slag with 40 mesh and 4.0 g when the reaction time was 90 min, the removal rate were 97.67 to 100% and 11.17 to 31.22%, respectively. Moreover, the previous experiments proved that the major PO_4^{3-} removal mechanism was precipitation of calcium phosphate complexes, and the main parameters influencing PO_4^{3-} removal were pH of solutions, Ca^{2+} and PO_4^{3-} concentrations. Thus, steel slag has a broad prospect in wastewater treatment which can be used as a cheap raw material for reducing general ion pollutants in water.

References

1. Y.G. Zhu et al., "The research progress of steel slag utilization in our country," *Iron and Steel Scrap of China*, (4) (2007), 25-29.

2. X. Zhao, "Impervious surface storm runoff pollution and control research in Suzhou high-tech zone," (Suzhou University of Science and Technology, 2012).

3. L. Huang, "Vertical undercurrent steel slag applied to artificial wetland to deal with high ammonia nitrogen sewage test research" (China University of Mining and Technology, 2009).

4. H. Motz, J. Geiseler, "Products of steel slags: an opportunity to save natural resources," *Waste Management*, 21 (3) (2001), 285-293.

5. Cristian Barca et al., "Yves Andres. Phosphate removal from synthetic and real wastewater using steel slags produced in Europe," *Water Research*, 46 (7) (2012), 2376-2384.

6. J. Yang et al., "Converter slag-coal cinder columns for the removal of phosphorous and other pollutants," *Journal of Hazardous Materials*, 168 (1) (2009), 331-337.

7. V.K. Jha et al., "Hazardous ions uptake behavior of thermally activated steel-making slag," *Journal of Hazardous Materials*, 114 (2004), 139-144.

8. V.K. Jha et al., "Utilization of steel-making slag for the uptake of ammonium and phosphate ions from aqueous solution," *Journal of Hazardous Materials*, 156 (2008), 156-162.

9. I. Joko, "Phosphorus removal from wastewater by the crystallization method," *Water Science and Technology*, 17 (2-3) (1984), 121-132.

10. L.I. Bowden et al., "Phosphorus removal from wastewaters using basic oxygen steel slag," *Environmental Science and Technology*, 43 (7) (2009), 2476-2481.

Energy Technology 2015: Carbon Dioxide Management and Other Technologies
Edited by: Animesh Jha, Cong Wang, Neale R. Neelameggham, Donna P. Guillen, Li Li, Cynthia K. Belt, Randolph Kirchain, Jeffrey S. Spangenberger, Frank Johnson, Andrew J. Gomes, Amit Pandey, and Peter Hosemann
TMS (The Minerals, Metals & Materials Society), 2015

SUSTAINABLE RECYCLING TECHNOLOGIES FOR BAUXITE RESIDUE (RED MUD) UTILIZATION

Ji Whan Ahn[1*] Thenepalli Thriveni[1], Seong Young Nam[2]

[1]Mineral Processing Division, Korea Institute of Geoscience and Mineral Resources (KIGAM), Gwahagno-124, Yuseongu-gu, Daejeon-305350.

[2]Department of Energy Resources Engineering, Inha University, Incheon.

Keywords: Bauxite residue (red mud), utilization, calcium sulfoaluminate (CSA) (Green cement)

Abstract

In alumina production or alumina refining, bauxite is converted to alumina (aluminium oxide) using the Bayer Process. Throughout the entire history of alumina production there has been a desire to utilize the bauxite residue created in the Bayer process with either by recovering additional products from it or using it. The red mud applications falls into several categories such as rare earths recovery, manufactured construction materials (cement, bricks, tiles, aggregate blocks and wood substitute) and bulk impermeable material for covering landfill. Considerable effort has been expended in finding applications for bauxite residue but a number of key factors affect the feasibility and economics of its adoption. The use of the carbonation can be an advantageous solution for overcoming problems associated with red mud storage and the emissions of several thousand tons of CO_2 from aluminium manufacturers each year. The key utilization of bauxite residue or red mud is to prepare "Calcium Sulfoaluminate (CSA) (Green cement)".

Introduction

In alumina production or alumina refining, bauxite is converted to alumina (aluminium oxide) using the Bayer process [1]. Aluminium is integral to our everyday lives; it's used in trains, planes, cars and buses, laptops, mobile phones, DVDs, building construction, outdoor furniture, screw cap bottle tops, beverage cans, aluminium foil and so much more. Growth in primary aluminium production continues to be driven by china and the Arabian Gulf. Fig.1 shows the global primary aluminium production in 2012 was a record 46 million tonnes [2]. Fig.2a&b shows the primary aluminium consumption by regions and end use. Bauxite residue, sometimes referred to as 'red mud', and is the by-product of the Bayer process. In the Bayer process, bauxite ore is dissolved using sodium hydroxide (caustic soda). Alumina is then extracted from the solution and the residual bauxite residue is sent to a storage area. The characteristic red colour of bauxite residue results from the high concentrations of naturally occurring iron compounds in the original bauxite ore.

In recent years for certain rare earth elements have re-awakened interest in their extraction from bauxite residue (red mud). Recycling is an integral part of any waste management system as it represents a key utilization alternative to reuse and energy recovery (waste-to-energy). Recently, Indonesia government will restrict and possibly ban export bauxite to China due to their new policies i.e increased to export tax to 20% by 2012. Production will decrease due to scattered resources, inefficient mining practices, competition from more profitable palm oil production and resources to support local refineries.

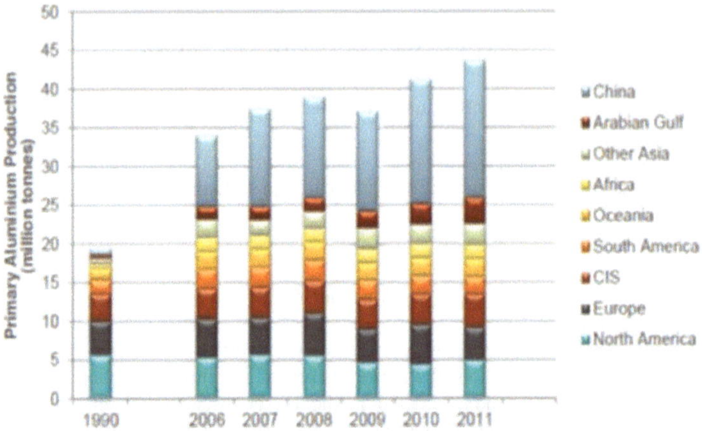

Figure.1. Global primary aluminium production, 1990 & 2006-2012 (SOURCE: IAI & CRU)

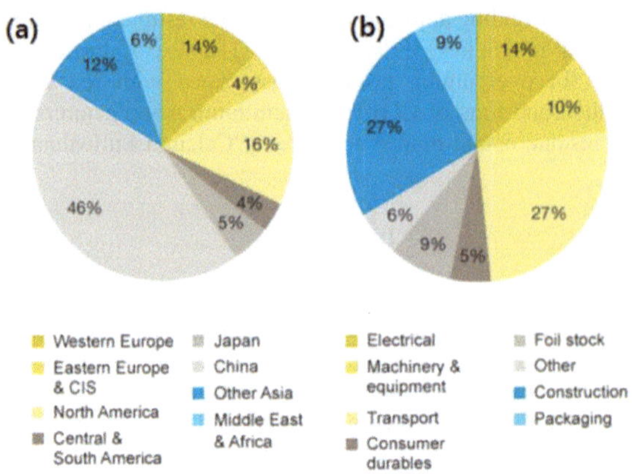

Figure.2 (a) primary aluminium consumption in 2013 by region (b) by end use.

Alumina Processing

Refining alumina (aluminium Oxide) is the second step in the production of aluminum. Alumina is refined from bauxite using a chemical process known as the Bayer process. This process proceeded by 4 ways [3].

i) digestion-bauxite is finely ground in mills and then mixed with a hot, caustic soda solution which dissolves the alumina contained in the bauxite

ii) Clarification-the solution of alumina and caustic soda passes into rows of thickener tanks, where the solid impurities sink to the bottom as a fine red mud is a by-product of the process. It is washed several times with water, neutralized, and then stored at the residue management area. The clear solution is sent to precipitation for the alumina to be recovered from it

iii) precipitation-alumina trihydrate is added to the alumina solution in a line of precipitation tanks to seed the formation of additional alumina trihydrate crystals

iv) calcination-the alumina trihydrate crystals are washed, filtered and then heated in gas-fired kilns at temperatures greater than 1,100^{0}C to remove water molecules, creating a fine white powder known as alumina.

Bauxite Residue (Red Mud)

Throughout the entire history of alumina production there has been a desire to utilize the bauxite residue created in the Bayer process with either by recovering additional products from it or using it. The red mud applications falls into several categories such as rare earths recovery, manufactured construction materials (cement, bricks, tiles, aggregate blocks and wood substitute) and bulk impermeable material for covering landfill. Considerable effort has been expended in finding applications for bauxite residue but a number of key factors affect the feasibility and economics of its adoption. The main application areas that have been evaluated are briefly summarized below:

Table I. The typical composition of bauxite residue [4]

Contents	Typical Composition
Al_2O_3	15-19%
Fe_2O_3	33-40%
SiO_2	10-15%
TiO_2	4-6%
CaO	3-9%
MgO	0.3-1.0
Na_2O	7-11%
V_2O_5	0.2-0.4%
P_2O_5	0.5-1.0%
CO_2	2-3%

As shown in Table I, bauxite residue contains a considerable amount of titania. Although this concentration is less to utilize but this titanium is a metal and alloying ingredient of strategic importance in defense and in the aerospace industries and also utilizes in producing medical implants and chemical catalysts etc.

Utilization

China, above 10% utilizes the bauxite residue for metal recovery or for brick production. In Australia ALCOA and Japan tried to utilize red mud in the building and construction materials produced from it and as an additive in cement production. The main application areas that have been evaluated are briefly summarized below: i) cement production ii) Road Construction iii) Brick Production iv) iron production v) acid mine drainage and heavy metals absorption vi) recovery of rare earth elements vii) green cement (Calcium Sulfo aluminate (CSA)) production from bauxite residue

Cement Production from Bauxite Residue (Red Mud) (KIGAM, Korea)

Here, we reported the production of calcium sulfoaluminate (CSA) based cement from red mud. The CSA based cement synthesized by sintering process (Fig. 3 & Fig.4). This process requiring less energy and time and reducing costs in terms of setting time and changes in mechanical strength. The available work in the area is immense and there are different strategies and cementations systems suggested[5-9].

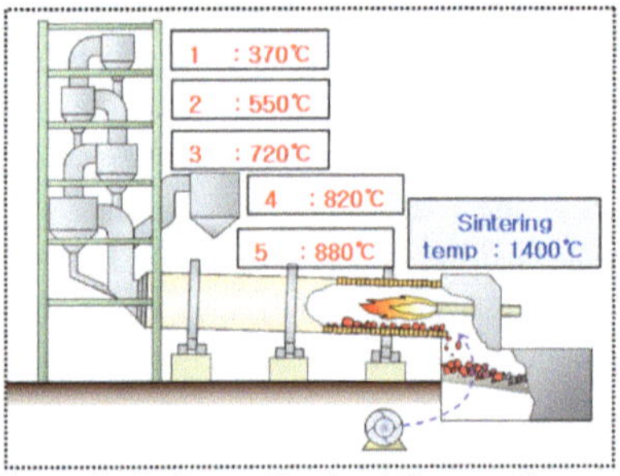

Figure.3. Ordinary Portland Cement Process

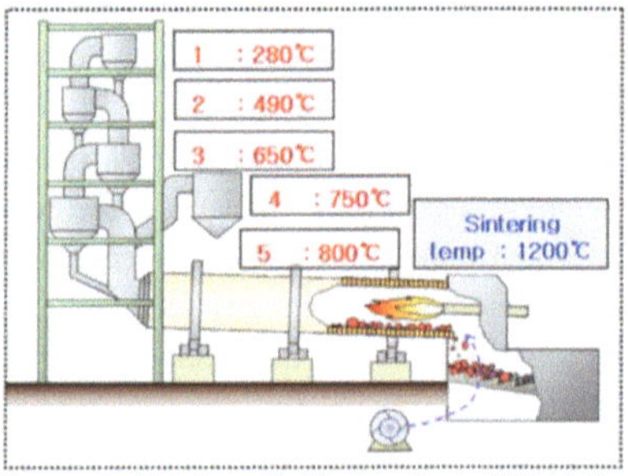

Figure.4. Calcium Sulfoaluminate(CSA) Process

The high functional cements have been focused on, with increase on the need for multistorey building, complex constructions and so on. The calcium sulfoaluminate is a representative cement with high functions. The aluminate resource like bauxite is used as main raw material to manufacture calcium sulfoaluminate(CSA) (Fig.5). CSA properties are hydraulic reaction with water take place instantaneously. Contribute expansion, high strength, and rapid hardening to materials. The applications of CSA is 1. Expansion (compensation for shrinkage, self-leaving, 2. Rapid hardening (ex: tunnel, water proof, reclamation by solidification, 3. High Strength (Bridge, High rise building, marine structure).

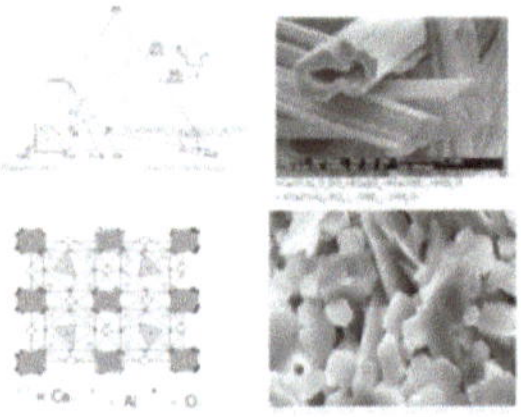
Figure.5. Calcium Sulfoaluminate (CSA) from Coal Ash & Bauxite residues

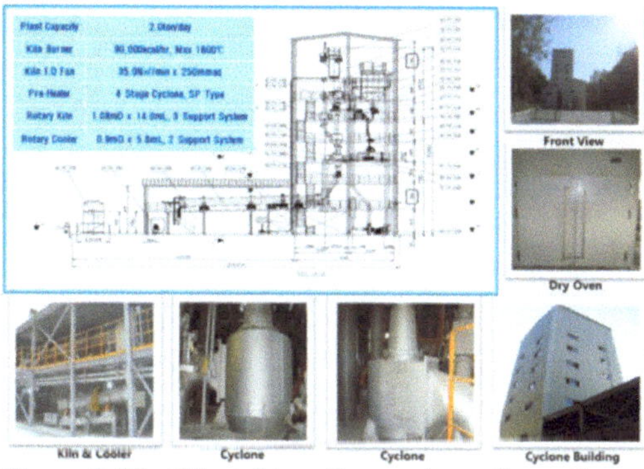

Figure.6. Pilot Plant, 2 tons/day produces Green Cement.

Fig.8 & 9 shows the difference between ordinary Portland cement and CSA cement process and the materials and conditions presented in Table II.

Table II. Materials and Conditions of OPC and CSA

Materials	OPC	CSA	
Sintering Temp(^{0}C)	1300-1400	1200[8]	1100[8]
Injection Rate	120	100	100
Total Energy(kcal/kg-cl)	700-720	620	500
Recycling Rate(%)	25-30	25	10
CSA Mineral Synthesis Rate(%)	-	14-16	14-16

Advantages of CSA from Red Mud

Significant benefits can be derived from the lower energy consumption during production (temperature between 700-500^{0}C) and the reduced CO_2 emissions per mass of cement produced. In addition, due to their low porosity and ability of ettringite (Fig.7&8) to bind heavy metals, CSA cements and their blends with Portland cement are of interest in the field of hazardous waste encapsulation. The major concern with CSA cement is associated with the main hydrating phase, ettringite, which is said to be prone carbonation. This work aimed at the synthesis of the calcium sulfoaluminate phase exhibits high strengths. More ettringite formation or expansion after drying it gives less cracking for buildings.

Figure.7 Hydration effect (After 1 hour) of ettringite

Figure.8 Hydration effect (After 12 hour) of ettringite

The Bauxite Residues (red mud) are a red slurry consisting of the un-dissolved portion of the bauxite ore, produced on a dry-basis at an almost 1:1 mass ratio with alumina, amassing to a total of 100 to 120 million tonnes per year globally. So this is one of the major biggest environmental problems. For this, the alternative and more economic method is recycling of bauxite residue.

This preliminary data confirms the present study have indicated that the red mud carbonation at ambient conditions (temperature of 25±1°C and atmospheric pressure) could present an effective option for the carbon dioxide sequestration. The carbonation efficiency of red mud is 2.18 g CO_2/kg. The results are indexed a promising parameter for CSA is strength and it would be an interesting extension. The use of this type of system could thus become a source of additional income for aluminum manufacturers and reduce environmental problems associated with red mud storage.

General CO_2 solidification is just eliminating the hydration but our developed carbonation technology is not only eliminating the hydration but also ecofriendly CO_2 capture with high efficiency. The application of our technology is mainly synthesis of CSA cement rom red mud. Now our investigation of the basic characterization studies of rare earth presence in bauxite residue and ore deposit origins is ongoing.

Acknowledgments

This research was support from the Energy Technology Development Project [2013T100100021] of the Korea Institute of Energy Technology Evaluation and Planning, financed by the Ministry of Trade, Industry and Energy.

References

[1] World Aluminium, (2013) Global life cycle inventory data for the primary aluminium industry 2010 data, published pp.1-53.

[2] International Aluminium Institute, (2012) "Results of the 2012 Anode Effect Survey". pp.1-24.

[3] Sustainable development report, (2012) Rio Tinto Alcan Yarwun, pp.1-66.

[4] Kozeky Laszlo., (2012) Utilization of the red mud, pp.1-14.

[5] Y.Pontikes,G.N., (2013) Angelopoulos, Bauxite residue in cement and cementitious applications: Current status and a possible way forward, Resources, Conservation and Recycling, 73, pp.53-63.

[6] G.Stavrinides, (2010) Alternative cements in concrete construction. University College London: Assessment, Prospects & Commercialisation Strategies [MSc.]; 2010

[7] Schneider, M., Romer, M., Tschudin, M., Bolio, H., (2011) Sustainable cement production present and future, Cement and Concrete Research, 41, pp. 642–50.

[8] Shi, C., Jiménez, A.F., Palomo, A., (2011) New cements for the 21st century: the pursuit of an alternative to Portland cement, Cement and Concrete Research, 41, pp.750- 63.

[9] Juenger, M.C.G., Winnefeld, F., Provis, J.L., Ideker, J.H., (2011) Advances in alternative cementitious binders, Cement and Concrete Research, 41, pp.1232-1243.

Energy Technology 2015
Carbon Dioxide Management and Other Technologies

RECYCLING AND
SUSTAINABILITY UPDATE

Recycling

Session Chairs:

Randolph Kirchain

Energy Technology 2015: Carbon Dioxide Management and Other Technologies
Edited by: Animesh Jha, Cong Wang, Neale R. Neelameggham, Donna P. Guillen, Li Li, Cynthia K. Belt, Randolph Kirchain, Jeffrey S. Spangenberger, Frank Johnson, Andrew J. Gomes, Amit Pandey, and Peter Hosemann
TMS (The Minerals, Metals & Materials Society), 2015

INVESTIGATION ON RECYCLING OF Ag FROM Pb-Cu-Ag ALLOY BY VACUUM DISTILLATION

Bingyi Song[1, 2, 3], Wenlong Jiang[1, 2, 3], Bin Yang[1, 2, 3, *], Baoqiang Xu[1, 2, 3], Qitong Yang[1, 2, 3], Shuai Xu[1, 2, 3], Dachun Liu[1, 2, 3]

[1]The National Engineering Laboratory for Vacuum Metallurgy, Kunming University of Science and Technology, Kunming 650093, PR China
[2]Key Laboratory for Nonferrous Vacuum Metallurgy of Yunnan Province, Kunming 650093, PR China
[3]State Key Laboratory of Complex Nonferrous Metal Resources Clear Utilization, Kunming 650093, PR China

Keywords: Crude silver, Pb-Cu-Ag alloy, Vacuum distillation, Separation

Abstract

The feasibility of recycling of Ag from Pb-Cu-Ag alloy by vacuum distillation was analyzed at six different temperatures (973, 1023, 1073, 1123, 1173, and 1223K) under the vacuum of 1-20Pa. The saturated vapor pressure, separation coefficient and gas-liquid phase equilibrium were calculated. The results show that this method is feasible. The influence of distillation temperature and time on the result of experiment was also investigated. The results show that Ag content in the residue and Pb and Cu content in the volatile increase with the increasing distillation temperature and time. The enriching times of Ag and the removal rate of Pb are larger than 32 and 99.9%, respectively at 1223K corresponding to 30-120min. The optimum distillation condition is the temperature of 1223K and time 60 min, and the high purity of Ag (93.3 wt. %) is reached.

Introduction

Anode slime, which is mainly composed of elements Pb, Cu, Ag, etc., is of great economic and social value due to enrichment of precious metals, rare metals and other valuable metals. The interest in extracting noble metals from anode slime has grown with the development in electrorefining, and the exhaustion of the native mineral resources. Ag, as noble metal, can be extracted from copper and lead anode slime. There are a great many extraction methods. However, traditional methods, used widely in the past for recycling of valuable metals from copper and lead anode slime such as pyrometallurgy-electrolysis process, fire metallurgy and hydrometallurgy process, the combined flotation hydrometallurgical technology, and hydrometallurgical method, etc., are not consummate because of the high metal content in tailing, low metal recovery, long flow sheet, high energy consumption and evident environment pollution, etc. [1]. Therefore, an effort must be made to find more advanced technology without a negative impact on the environment for recycling of Ag from copper and lead anode slime. Vacuum metallurgy is a clean technology for secondary metal production, which has many advantages, such as high metal recovery, short flow sheet and good environmental protection, etc. It can eliminate the disadvantages of traditional metallurgical processes. Moreover, it can produce new products to meet the needs of the development of high technology [2]. At present, the research on vacuum metallurgy is successfully studied in refining and recycling of various non-ferrous crude metals and alloys by Dai [3], Liu [4], Xu [5], Li [6], and Ali [7] etc.

The purpose of this study was to recycle Ag from Pb-Cu-Ag alloy by vacuum distillation. In this paper, we studied the evaporation behavior of elements in the alloy and enrichment factor of Ag during vacuum distillation. The effects of distillation temperature, distillation time were investigated.

Experimental principles

Saturated vapor pressure

The difference of the vapor pressure of Pb, Cu and Ag in pure state is the necessary prerequisite to determine whether Pb-Cu-Ag ternary alloy can be separated by vacuum distillation. The calculation of the saturated vapor pressure of pure substances has the following expression [8]

$$\lg p^{*} = AT^{-1} + B\lg T + CT + D \tag{1}$$

Where p^{*} is the saturated pressure of pure substances, Pa, A, B, C and D are evaporation constants. The evaporation constants of Pb, Cu and Ag can be obtained from the relevant literature [8], and substituting A, B, C and D into the Eq. (1):

$$\lg p^{*}_{Pb} = -10130T^{-1} - 0.985\lg T + 13.285 \tag{2}$$

$$\lg p^{*}_{Cu} = -17520T^{-1} - 1.21\lg T + 15.335 \tag{3}$$

$$\lg p^{*}_{Ag} = -14400T^{-1} - 0.85\lg T + 13.825 \tag{4}$$

Using the three equations, the saturated vapor pressure of components of Pb-Cu-Ag alloy can be calculated easily in the temperatures range of 973-1223 K, as shown in Fig.1.

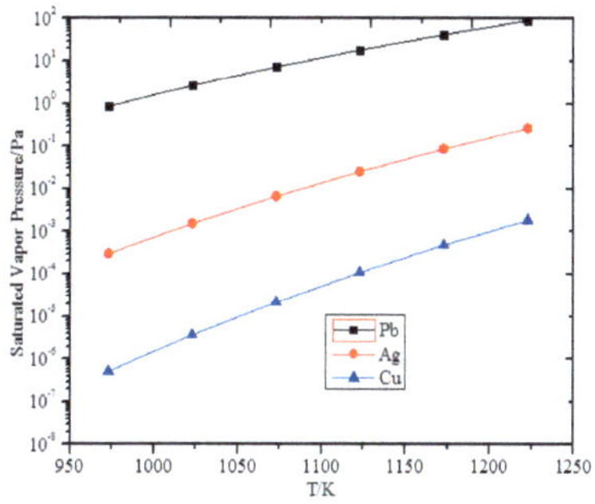

Fig.1 Saturated vapor pressure of pure substance Pb, Ag, and Cu

As seen from the figure 1, the saturated vapor pressure of Pb, Cu and Ag increased as the temperature increased, and the saturated vapor pressure of Pb and Ag was larger than that of Cu at the same temperature. Pb, Ag, therefore, evaporated more into the gas phase, and more Cu remained in the liquid phase during vacuum distillation for Cu-Pb alloy and Cu-Ag alloy.

Separation coefficient

In order to qualitatively determine the feasibility and separation degree of different elements in Pb-Cu-Ag alloy during vacuum distillation, Dai and Yang [8] proposed the separation coefficient β according to thermodynamic equilibrium theory. And the separation coefficient was expressed as follows:

$$\beta_A = \frac{\gamma_A}{\gamma_B} \cdot \frac{p_A^*}{p_B^*} \tag{5}$$

Where γ_A and γ_B are the activity coefficient of A and B components, respectively; p_A^* and p_B^* are the saturated vapor pressures of A and B in pure state, respectively. When β>1 or β<1, the separation of alloys could happen, but when β=1, it could not. Therefore, the separation coefficient β can be used to predict whether the components may be separated by vacuum distillation for all alloys [9].

Vapor-liquid equilibrium diagram

In order to further study the separation effect and element distribution of Pb-Cu-Ag alloy, the vapor-liquid phase equilibrium of the Pb-Cu-Ag alloy was investigated. Dai and Yang [8] presented vapor-liquid equilibrium diagram, its expression is as follow:

$$A_g = \left[1 + \left(\frac{B}{A}\right) \cdot \left(\frac{\gamma_B}{\gamma_A}\right) \cdot \left(\frac{p_B^*}{p_A^*}\right)\right]^{-1} \tag{6}$$

For Cu-i binary alloy

$$Cu_g = \left[1 + \left(\frac{i_l}{Cu_l}\right) \cdot \left(\frac{\gamma_i}{\gamma_{Cu}}\right) \cdot \left(\frac{p_i^*}{p_{Cu}^*}\right)\right]^{-1} \quad (7)$$

Where Cu_g and Cu_l are Cu content in vapor and liquid phase; i_l is the content of i in liquid, γ_i and γ_{Cu} are the activity coefficients of i and Cu; p_i^* and p_{Cu}^* are the saturated pressure of pure element i and Cu, respectively.

The activity coefficient required for Eqs. (5) and (7) can be calculated by molecular interaction volume model (MIVM) [10]. Thus, the separation coefficient of β_{Pb}, β_{Ag} and vapor-liquid phase equilibrium diagram can be calculated according the Eqs. (5) and (7), plotted in Fig.2 and Fig.3.

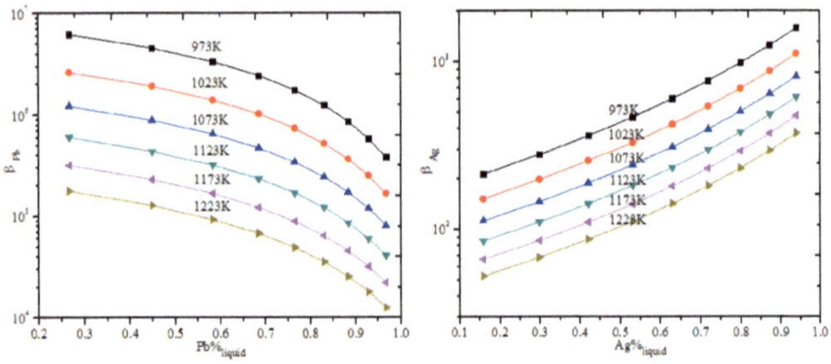

(a) (b)

Fig.2 Separation coefficient of Pb and Ag at different temperatures, (a) Cu-Pb alloy, (b) Cu-Ag alloy

Figure 2 shows the separation coefficient of Pb and Ag in Cu-Pb and Cu-Ag systems at different temperatures, respectively. It could be found that the separation coefficient of Pb and Ag in Cu-Pb alloy and Cu-Ag alloy reached up to 10^4 and 10^2, which were far bigger than 1, it was shown that Cu and Pb (Ag) can be concentrated and purified in liquid and gas phases respectively by vacuum distillation.

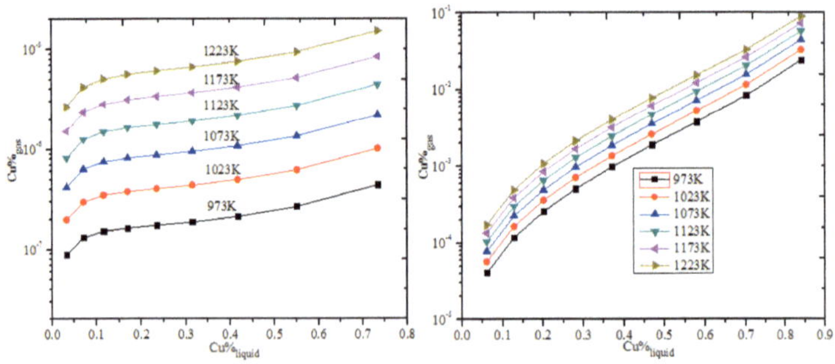

(a) (b)

Fig.3 Vapor-liquid equilibrium diagram of Cu at different temperatures, (a) Cu-Pb alloy, (b) Cu-Ag alloy

Similarly, vapor-liquid equilibrium diagram of Cu at different temperatures is shown in figure 3. It can be seen from Fig. 3 that the content of Cu in vapor phase increased with rising temperature. In Figure 3(a), you see that the content of Cu in vapor phase was 10^{-5} wt.%, while it was 73.40 wt.% in the liquid phase at 1223 K, and in Figure 3(b), it reached 0.0909 wt% in the vapor phase, while in the liquid phase, it was 84.13 wt.% at 1223 K. This behavior indicates that Pb and Ag will evaporate into vapor phase, while Cu remains in the liquid phase, Pb and Ag, therefore, can be separated from Cu by vacuum distillation. And the above systems will have good separation efficiency.

Experimental

Experimental Materials

The material was provided that the products were gained by vacuum distillation from anode slime, and the chemical composition of the material is listed in Table I.

Table I Chemical composition of material

Element	Pb	Cu	Ag
Content (%)	88.82	0.18	2.85

Equipments and Procedure

Equipment: For the vacuum distillation furnace used in this experiment, it is mainly composed of six parts: pit furnace, retort-flange assembly, evaporator-condenser set, circulating water control system, vacuum resistance furnace and vacuum system using rotary pumps. The schematic diagram can be seen from the literature [12].

Procedure: The separation of Pb-Cu-Ag ternary alloy was carried out in a clean room of vacuum furnace. In the process of distillation, nearly 50 (100) g of raw Pb-Cu-Ag ternary alloy was loaded into the bottom graphite crucible and placed inside the vacuum furnace, which was then evacuated to a vacuum of 15 Pa using vacuum pump. The temperature was increased to target temperature and then thermal insulation for a period of time. After completion of the distillation process, the furnace was cooled to room temperature to solidify respectively the purified residue Cu and the condensed Pb and Ag. Water cooling system was working over the process.

Finally, the samples were collected from the volatile and residue. Flame atomic absorption spectroscopy (FAAS) and chemical analysis methods were used to analyze the contents of Pb, Cu and Ag in both the volatile and residue.

Results and discussion

The effect of distillation temperature

The influence of distillation temperature on the result of experiment was investigated when distillation time was 30min. The quantitative analysis was carried out by FAAS and chemical analysis method, which were given in the Table II. At the same time it also calculates the enrichment factor of Ag, and the removal rate of Pb, the result were plotted in Fig. 4.

$$N_{Ag} = \frac{\omega_t}{\omega_0} \tag{8}$$

Where N_{Ag} is the enriching times of Ag, ω_0 and ω_t are the content of Ag in raw materials and residue, respectively.

$$\eta_{Pb} = \frac{W_0 - W_t}{W_0} \times 100\% \tag{9}$$

Where η_{Pb} is the removal rate of Pb, W_0 and W_t are the mass of Pb in raw material and residue, respectively.

Table II Quantitative analysis results of Pb-Cu-Ag alloy after vacuum distillation

Number	Feeding amount/ g	Time/ min	Temperature/ K	constituents after distillation /%							
				residue				volatile			
				mass/g	Pb	Cu	Ag	mass/g	Pb	Cu	Ag
1	50	30	973	41.52	91.27	0.17	4.53	2.35	86.63	0.0011	0.57
2	50	30	1023	39.63	90.27	0.18	6.95	2.23	87.41	0.0033	0.42
3	50	30	1073	21.62	89.74	0.24	8.06	9.47	88.91	0.0065	0.37
4	50	30	1123	4.68	70.39	0.82	25.32	18.89	91.51	0.012	0.32
5	50	30	1173	3.84	67.32	1.11	30.24	17.29	93.15	0.015	0.29
6	50	30	1223	1.04	0.33	6.56	93	32.31	99.75	0.032	0.19

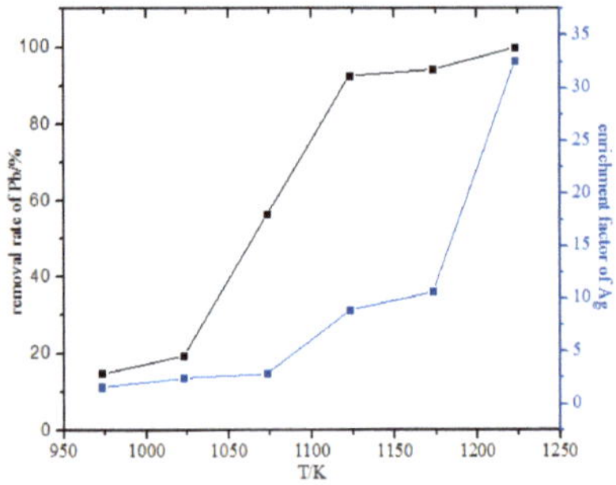

Fig. 4 Effect of temperature on enriching times of Ag and removal rate of Pb

Table II summarizes the experimental conditions and the content of components of Pb, Cu, and Ag in the volatile and residue phases, respectively. The content of Pb in vapor phase increases with increasing distillation temperature while the content of Pb in liquid phase decreases. Meanwhile, the content of Ag in vapor phase decreases while the content of Ag increases in liquid phase as distillation temperature increases. This is because the saturated vapor pressure of Pb is larger than that of Ag and Cu at the same temperature, Pb preferentially evaporates into the gas phase, and the saturated vapor pressure of Pb increases quickly with the increase of temperature, which shows that temperature rising promotes full and evaporation of Pb, while Ag and Cu are restrained in liquid phase. This behavior indicates that the higher the distillation temperature is, the better the separating effect has. When temperature rises from 1173K to 1223K, the content of Pb in vapor phase plummets from 67.32% to 0.33% while the content of Ag in vapor phase surges from 30.24% to 93.00%. A conclusion was thus reached that the optimal condition for the separation of Pb-Cu-Ag alloy is temperature of 1223K.

Fig. 4 shows that the relationships between the distillation temperature and enriching times of Ag and removal rate of Pb. Here you can see that volatilization rate, enriching times of Ag and removal rate of Pb increase as the temperature increases. When the temperature is between 1023K and 1123K, removal rate of Pb increases very quickly, Pb is removed a lot. When it is larger than 1173K, the enrichment factor of Ag dramatically increases, while the content of Ag rises to 93.00% from 30.24% at 1123K, although the amplitude is low for removal rate of Pb, so the more Ag can be obtained at 1223K, and this temperature is better than other else for economic reasons. This conclusion also provides in table II.

<u>The effect of distillation time</u>

In order to achieve the maximum enrichment of Ag, the effect of the holding time on volatilization of each component was carried out when the temperature was 1223K. The quantitative analysis was carried out by FAAS and chemical analysis method, which were given in the Table III. Similarly, the volatilization rate of alloy, the enrichment factor of Ag, and the removal rate of Pb were plotted in Fig. 5.

Table III Quantitative analysis results of Pb-Cu-Ag alloy after vacuum distillation

Number	Feeding amount (g)	Time (min)	Temperature (K)	constituents after distillation /%							
				residue				volatile			
				Mass (g)	Pb	Cu	Ag	Mass (g)	Pb	Cu	Ag
1	100	30	1223	2.39	2.53	6	91.41	70.13	99.1	0.0033	0.32
2	100	60	1223	2.34	0.59	6.06	93.3	70.41	99.35	0.0038	0.43
3	100	90	1223	2.18	0.22	6.16	93.6	69.65	99.43	0.0042	0.53
4	100	120	1223	2.17	0.077	6.21	93.69	71.18	99.57	0.0057	0.66

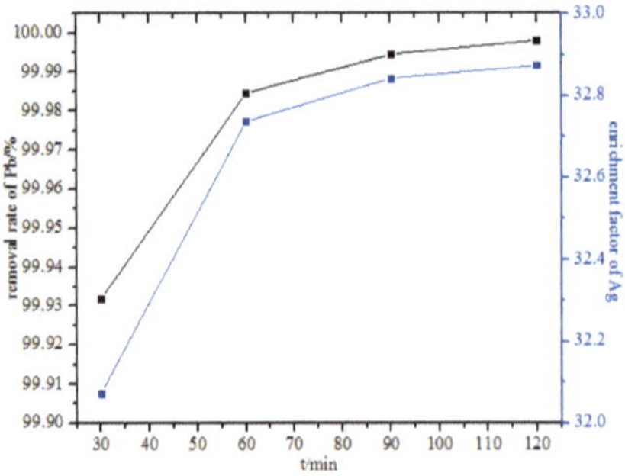

Fig.5 Effect of the holding time on enriching times of Ag and removal rate of Pb

Table III summarizes the experimental conditions and the content of components of Pb, Cu, and Ag in the volatile and residue phases, respectively. As can be seen from Table III, the content of Ag and Pb in the vapor phase increases as time goes on, and the content of Ag and Cu in the liquid phase increase while the content of Pb decreases, which indicates that Pb evaporates into vapor phase, and Ag and Cu remain in liquid phase. In addition, as you can see from figure 5, when the time is between 30 min and 60 min, the enrichment factor of Ag and the removal rate of Pb increase rapidly, and longer than 60 min, they increase with small gradient, so the better time of Ag enrichment is 60 min when the rate of volatilization is 0.43%, viz. Ag has been more enrichment and less volatilization. This is because the content of Ag in liquid phase is less than 94% after distillation, the purity is a little low although Ag has been greatly enriched compared with the raw material, so electrolytic refining is imperative to obtain pure Ag. For electrolytic refining, suitable conditions are Ag>75%, Au<20%, Cu0.5%～7%. It is on this premise that the time can be chosen in order to save energy.

Conclusions

1) It is feasible to obtain crude Ag by vacuum distillation. Experiments for separation of Pb-Cu-Ag alloy were executed in the temperature range of 973−1223K. The results show that the content of Pb in vapor phase increases with the increasing distillation temperature and content of Ag in liquid phase.

2) The effects of distillation, distillation time on the separation of alloy were carried out. And distillation temperature has very significant effects on the removal rate of Pb and the enrichment factor of Ag.

3) The optimum distillation condition for separation experiment is temperature of 1223K and time 60 min, and the purity of Ag (93.3 wt. %) is reached when the rate of volatilization is 0.43%.

Acknowledgements

This work was supported financially by the NSFC (Grant No. 51104079 & No. U1202271), the Fundamental Research of Yunnan province (Grant NO. 2013FZ033), the Cultivating Plan Program for the Leader in Science and Technology of Yunnan Province (Grant No. 2014HA003), the General Program of Natural Science

Foundation of China (Grant No. 51474116), the Program for Innovative Research Team in University of Ministry of Education of China (Grant No. IRT1250).

References

[1] H. Y. Yang, X. J. Li, L. L. Tong, G. B.Chen, "Process mineralogy of copper anode slime," The Chinese Journal of Nonferrous Metals, 24(1) (2014), 269–278.

[2] L. X. Kong, B. Yang, Y. F. Li, B. Q. Xu, "Application of molecular interaction volume model for phase equilibrium of Sn-based binary system in vacuum distillation," Metallurgical and Materials Transition A, 45(10) (2014), 4405–4410.

[3] Y. N. Dai, "The separation of each element in crude metal and alloy by vacuum distillation," Vacuum Metallurgy, (1) (1990), 14–28.

[4] D. C. Liu, B. Yang, Y. N. Dai, G. S. Du, W. Q. Tang, L. Han, "Extracting Indium Metal from Indium-Enriched Slag," Chinese Journal of Rare Metals, 29(4) (2005), 574–577.

[5] B. Q. Xu, B. Yang, D. C. Liu,Y. N. Dai, W. H. Ma, "The Study on Recovering Metal Zinc from the hot Galvanizing Slag by Vacuum Distillation," Non-ferrous mining and metallurgy, 23(4) (2007), 53–55.

[6] L. Li, D. C. Liu, H. J. Zhou, B. Yang, Q. Luo, Y. N. Dai, "Enrichment of Ag and Sb by Vacuum distillation of Ag-containing Pb-Sb Multi-Component Alloys," Journal of Vacuum Science and Technology, 32(7) (2012), 593–598.

[7] S. T. Ali, K. S. Rao, C. Laxman, N. R. Munirathnam, T. L. Prakash, "Preparation of high pure zinc for electronic applications using selective evaporation under vacuum," Separation and Purification Technology, 85 (2012),178–182.

[8] Y.N. Dai and B. Yang, *Vacuum Metallurgy of Non ferrous Metal Materials*, (Metallurgical Industry Press, Beijing, 2008) (in Chinese).

[9] L. X. Kong, B. Yang, Y. F. Li, B. Q. Xu, L. Han, D. C. Liu, Y. N. Dai, "Application of molecular interaction volume model in separation of Sn-Zn alloy by vacuum distillation," Journal of Central South University, 20(12) (2013), 3372–3378.

[10] D. P. Tao, "A new model of thermodynamics of liquid mixtures and its application to liquid alloys," Thermochim. Acta, 363 (2000), 105–113.

[11] R. Hultgren, et al, *Selected Values of the Thermodynamic Properties of Binary Alloys*, (ASM, Metals Park, OH, 1973).

[12] L. X. Kong, B. Yang, B. Q. Xu, Y. F. Li, L. Li, D. C. Liu, Y. N. Dai, "Application of molecular interaction volume model in separation of Pb-Sn-Sb ternary alloy by vacuum distillation," Trans. Nonferrous Met. Soc. China, 23 (2013), 2408–2415.

Energy Technology 2015: Carbon Dioxide Management and Other Technologies
Edited by: Animesh Jha, Cong Wang, Neale R. Neelameggham, Donna P. Guillen, Li Li, Cynthia K. Belt, Randolph Kirchain, Jeffrey S. Spangenberger, Frank Johnson, Andrew J. Gomes, Amit Pandey, and Peter Hosemann
TMS (The Minerals, Metals & Materials Society), 2015

RECYCLING OF SINTER PLANT OFFGAS CLEANING SYSTEM DUST BY PRE-AGGLOMERATION

Naiyang Ma[1]

[1]ArcelorMittal Global R&D - East Chicago Laboratories; 3001 E Columbus Dr.; East Chicago, IN 46312, USA

Keywords: Sinter plant, Dust, Recycling, Pre-agglomeration, Offgas

Abstract

Fine particle size and high level of alkalis are two barriers for recycling of ironmaking sinter plant offgas (OG) cleaning system dust back into the sintering process. The fine particles in sinter feed mix might cause permeability deterioration of sinter bed, and hence result in productivity decrease of sinter machines. Pre-agglomeration of fine sinter feed has been proven to be an effective technology to solve the problems of decrease of sinter bed permeability. Testing of pre-screening followed by pre-agglomeration of a sinter plant OG dust is reported in this paper.

Introduction

An ArcelorMittal sinter plant is equipped with an electrostatic precipitator (EP) in its offgas cleaning system. The electrostatic precipitator dedusts the sinter plant offgas which carries large amount of dust particles from the sintering process. The dust collected by the electrostatic precipitator is referred to as EP dust. The EP dust is dry, contains fairly small particles and often has a high level of alkali presence. The small particle size and high level of alkali presence are two major barriers for recycling of the EP dust back into the sintering process.

It is well known that high percentage of fines in sinter feed mix can cause permeability deterioration of sinter bed, and hence result in productivity decrease of sinter machines. Due to poor recyclability of the EP dust, the sinter plant under this study has stored a large quantity of the material on the ground. The EP dust is not only taking over land, but also places the plant at a risk of possible environmental liability in the future. Therefore, how to convert the EP dust into a recyclable material becomes an urgent subject.

To cope with problems of fines in sinter feed, there are three strategies in general. The first strategy is to optimize process parameters to mitigate negative effect of fines in sinter feed on sintering process. For example, add burnt lime to improve micro-pelletization [1], and use permeability bars to maintain high permeability of sinter bed [2]. The second strategy is to pelletize sinter feed with dedicated pelletizers and sinter the pelletized sinter feed. HPS (hybrid pelletized sinter) technology belongs to this category [3]. In recent years, the third strategy has been widely tested and applied. This strategy is to pre-agglomerate only fine materials with stronger binders. After the pre-agglomerated materials are cured, they are blended into regular sinter feed. This strategy is specifically well for dealing with some fine recycling materials which often have small particle size and small quantities. In this study, pre-agglomeration was chosen to solve the problem of the sinter plant EP dust and test results are reported.

METHOD

Sampling

A 50 gallon drum of the sinter plant EP dust sample was taken from the stored pile. Individual samples were taking from eight locations in the sinter plant EP dust storage yard as shown in Figure 1.

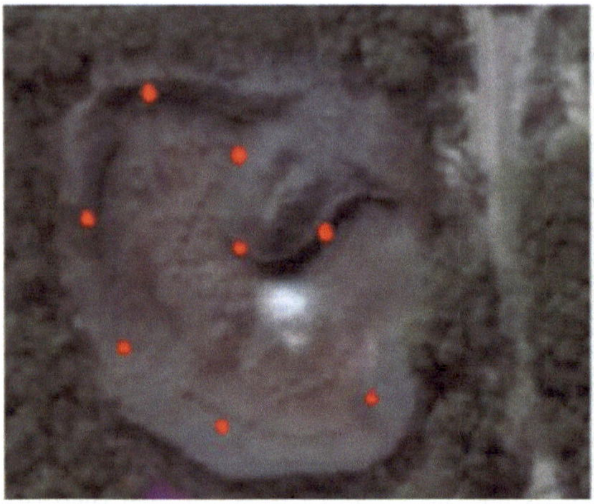

Figure 1 Bird view of the sinter plant EP dust storage yard and sampling locations as marked

A front loader was used to take samples from top, middle and bottom layers of each sampling location. The individual samples were mixed and divided, and a representative 50 gallon sample was obtained in the end.

Characterization

Representative sub-samples were first taken from the raw sample. One of the sub-samples was dried in an oven whose temperature was set at 105 °C. The sample was kept in the oven until no noticeable weight loss. A sub-sample was taken from the dried sample and was ground below 80 mesh (0.18 mm) with a BICO ring mill. The ground sample was analyzed for chemical composition.

Particle size distribution of the sinter plant EP dust was analyzed by wet screening with sieve apertures 6.3 mm, 1.18 mm, 0.6 mm, 0.25 mm, 0.18 mm, 0.075 mm and 0.045 mm. The screened samples were dried in an oven and size distribution was then recorded. Each sized sample was ground below 80 mesh and analyzed for chemical composition.

Small-scale agglomeration tests with a tire pelletizer

Small-scale agglomeration tests of the EP dust were conducted first on a small tire pelletizer. Around 1.4 kg as-received EP dust sample was used in each test, and 1 – 5 mm agglomerates were produced. Agglomerates larger than 5 mm were shredded and returned to the tire pelletizer.

Particles smaller than 1 mm were screened out and returned to the pelletizer for further agglomeration. The tire pelletizer is shown in Figure 2. It stands upright and is not tilted. Its speed is adjustable, but was 40 rpm when the agglomeration tests were conducted.

Figure 2 Tire Pelletizer

Bentonite, burnt lime and type I Portland cement were used as binders in the small-scale agglomeration tests.

Large-scale agglomeration tests with a disc pelletizer

Based on the results of small-scale agglomeration tests with the tire pelletizer, large-scale agglomeration tests with a disc pelletizer were then conducted with selected binder combinations. Around 9 kg EP dust sample was used in each large-scale test. 2 – 5 mm agglomerates were made. After screening, over-size agglomerates were shredded and returned into the disc pelletizer for further agglomeration along with under-size ones. The disc pelletizer has 90 cm diameter and 20 cm depth. It was tilted at 58° from horizontal and ran at 22 rpm when the agglomeration tests were conducted. The disc pelletizer is shown in Figure 3.

Property tests of cured agglomerates

Green agglomerates were cured for at least 24 hours. Cured agglomerates were first tested for fines generation by screening a 100 g sample of cured agglomerates with a 1 mm or 2 mm sieve. Percentage of under-size agglomerates was used as a measure of loss, which was the complement of yield.

Figure 3 Disc pelletizer

A 200 g sample of cured agglomerates was then dropped from 5 meter height onto a steel plate, and the dropped agglomerates were screened with a 1 mm or 2 mm sieve. Percentage of under size fines in the dropped agglomerates was used as a measure of resisting dropping impacts.

Lastly, 200 g water was sprayed onto a 200 g sample of cured agglomerates, the wet agglomerates were naturally dried for at least 48 hours, and the dry agglomerates were screened with a 1 mm or 2 mm sieve. Percentage of under size fines in the agglomerates was used as a measure of resisting rainfalls.

RESULTS AND DISCUSSION

Chemical composition of the EP dust

The sinter plant EP dust contains about 44% Fe, 7.7% C and 11% CaO, which are beneficial components for sintering and ironmaking. Undesired components with significant levels in the sinter plant EP dust include potassium (0.46%), sulfur (0.459%), lead (0.24%) and silica (7.02%).

Size distribution

Sieve analysis of as-received EP dust with wet screening is presented in Table I. It can be seen that about 12% EP dust is larger than 6.3 mm, and less than 73% is smaller than 0.6 mm.

Effects of particle size on compositions of the EP dust

Effects of particle size on contents of some components in the EP dust are graphed in Figures 4 – 6. From Figures 4 – 6, one can clearly see the strong dependences of Fe, C, SiO_2, and S contents on particle sizes. Large particles of the EP dust contain higher S and less Fe than small particles, while carbon and silica contents in the dust follow truncated Gaussian distributions, i.e. higher in middle-size particles and lower in both large and small particles.

Table I Size distribution of the sinter plant EP dust, %wt

>6.3 mm	11.9
6.3 mm - 1.18 mm	15.6
1.18 mm - 0.6 mm	5.7
0.6 mm - 0.25 mm	16.2
0.25 mm - 0.18 mm	6.2
0.18 mm - 0.075 mm	20.8
0.075 mm - 0.045 mm	10.1
< 0.045 mm	13.5

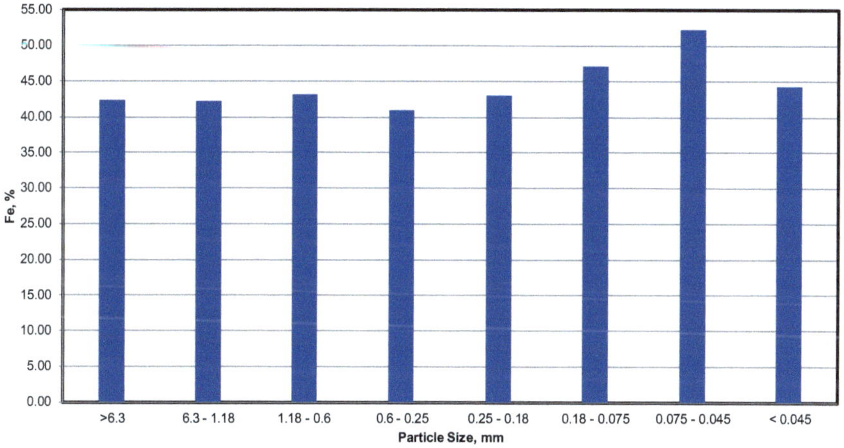

Figure 4 Effect of particle size on Fe content in the sinter plant EP dust

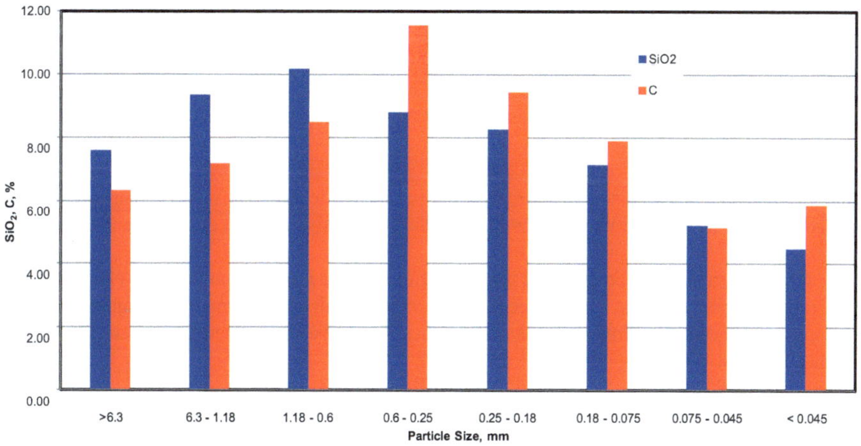

Figure 5 Effects of particle size on silica and carbon contents in the sinter plant EP dust

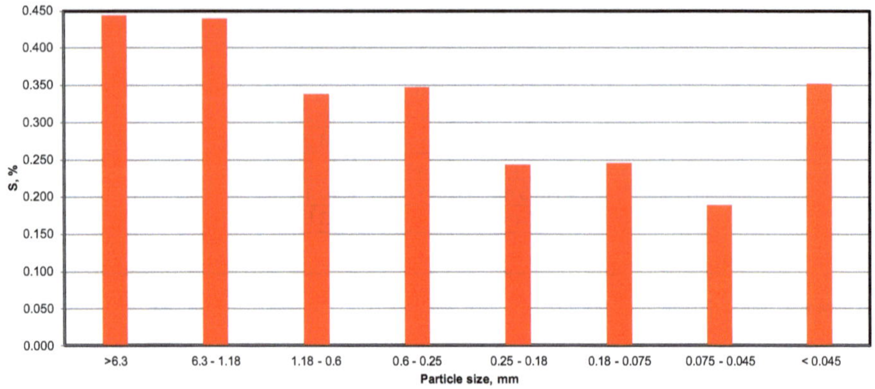

Figure 6 Effect of particle size on S content in the sinter plant EP dust

It seems that K_2O also changes with particle size, i.e. higher K_2O content in large particles and lower K_2O content in small particles. As a matter of fact, this is due to the effect of wet screening. Potassium in sinter plant offgas cleaning system solid wastes exists, in general, as potassium chloride which rather easily dissolves into water [4]. Since smaller particles have higher specific surface area, potassium chloride in smaller particles can dissolve into water much faster than in large particles. This is the reason why in the screened EP dust samples, large particles contain more potassium than smaller particles.

A dry screening test of the raw EP dust was then conducted with a 6.3 mm sieve. Both +6.3 mm and -6.3 mm samples were analyzed. The analytical results showed that K_2O contents in both samples were almost identical, which confirmed that K_2O content in the EP dust is not size dependent.

Above analysis suggests that alkali in the EP dust can be mitigated in two ways: (1) screening out +6.3 mm large particles so that about 12% K_2O can be removed, and (2) washing the -6.3 mm fine EP dust with water so that most of the potassium in the EP dust can be removed. Washing test on sinter plant offgas cleaning system dust showed that about 90% alkali in the dust dissolved into water within about 15 minutes [4].

Though alkali in the EP dust can be easily washed out, the technology is not recommended since subsequent dewatering and wastewater treatment are expensive and not environment-friendly.

The +6.3 mm product can be recycled into blast furnaces for iron, flux and fuel. When using the +6.3 mm material as a blast furnace feed, blast furnace alkali loading must be ensured under limit and appropriate central chimney size in the furnace should be maintained for effective removal of alkali.

A dry screening test of the as-received EP dust was conducted and the results give 11.98% +6.3 mm over-size product which is very close to the result of wet screening 11.9% +6.3 mm. Therefore, the sinter plant EP dust can be efficiently screened with a 6.3 mm aperture screen even though the EP dust contains about 15% moisture.

Results of agglomeration tests with the small tire pelletizer

Agglomeration tests with the small tire pelletizer went very well. Even without adding any binders, the sinter plant EP dust can be easily agglomerated.

Properties of cured agglomerates are presented in Table II. In the table, for all cases small numbers of – 1 mm or 2 mm fines suggest good properties of the agglomerates. It can be seen that (1) more binders added result in better quality of agglomerates; (2) 5% Portland cement gives the best quality of agglomerates; (3) burnt lime can help hold dust particles together very well, but it does not significantly help the agglomerates to resist breakage from dropping and soaking.

Chemistry of the binders used in the tests is presented in Table III. Clearly, bentonite contains too much silica and potassium oxide. Silica, potassium oxide and sulfur are all in quite high levels in Portland cement. In order to minimize the input of unfavorable components from the binders, but still maintain excellent resistance of agglomerates to dropping and soaking, a combination of burnt lime with Portland cement could be used in agglomeration of the EP dust.

Table II Experimental results of EP dust agglomerates

Experiment No.	Bentonite, %	Burnt Lime, %	Portland Cement, %	Moisture in Green Agglomerates, %	-1mm % after 24 h Curing	-1 mm % after 5 m Drop Test of Cured Agglomerates	-1mm % after Soaked in Water
1	0	0	0	9.6	33.1	58.3	40.3
2	0.5	0	0	16.6	24.6	43.5	25.6
3	1	0	0	16.8	16.3	38.4	20.5
4	1	2	0	18	9.3	38.4	11.5
5	0	0	2	18.9	9.6	34.4	13.4
6	0	0	5	17	7.5	12.8	7.5
7	0	2	0	18.7	8.6	35.7	18
8	0	5	0	19.3	6.2	31	15.6

Table III Chemical analysis of the binders

Binder	Fe	SiO_2	Al_2O_3	TiO_2	CaO	MgO	MnO	Na_2O	K_2O	P	Sr	Cr_2O_3	ZrO_2	LOI	C	ε
Bentonite	2.73	68.16	14.76	0.34	4.18	2.32	0.05	1.50	2.05	0.03	0.02	< 0.01	< 0.01	11.05	0.54	0.072
Burnt Lime	1.73	1.23	0.46	0.05	92.55	1.87	0.06	< 0.1	< 0.01	0.02	0.02	< 0.01	< 0.01	24.69	3.68	0.087
Portland Cement	2.04	19.25	4.86	0.27	63.68	4.01	0.17	< 0.1	0.60	0.05	< 0.01	< 0.01	< 0.01	4.65	0.76	1.122

Results of agglomeration tests with the disc pelletizer

Results of agglomeration tests with the disc pelletizer are presented in Table IV. Clearly, it is feasible to make quality agglomerates with 2% burnt lime plus 1 - 3% Portland cement.

Table IV Results of agglomeration tests with the disc pelletizer

Experiment No.	Burnt Lime, %	Portland Cement, %	Moisture in Green Agglomerates, %	-1mm/2mm % after 24 h Curing	-1mm/2mm % after 5 m Drop Test of Cured Agglomerates	-1mm/2mm % after Soaked in Water
9	2	1	15.4	4.8/13.8	24.8/43	6.9/16.4
10	2	2	15.3	7.3/26.6	17.5/42	8.8/30.3
11	2	3	14.3	3.1/13.4	9.6/23.6	4/16

CONCLUSIONS

1. The sinter plant EP dust contains about 44% Fe, 7.7% C and 11% CaO, which are beneficial components for sintering and blast furnace ironmaking. In addition, the dust also contains considerable undesired components, including 0.46% potassium oxide, 0.459% sulfur, 0.24% lead and 7% silica.
2. There are about 12% particles in the EP dust greater than 6.3 mm, and 73% smaller than 0.6 mm.
3. There are strong dependences of compositions of the EP dust on particle sizes. Large particles contain more sulfur and less iron than small particles. Carbon and silica contents in the EP dust follow truncated Gaussian distribution, i.e. carbon and silica concentrations are higher in middle-size particles and lower in both large and small particles.
4. Screening the sinter plant EP dust with a 6.3 mm screen can remove about 12% potassium oxide and 16% sulfur. Washing the -6.3 mm fine EP dust can further remove most of the potassium.
5. The sinter plant EP dust can be easily agglomerated. More binders will result in stronger agglomerates. 5% Portland cement shows the best results under the tested conditions. However, for minimizing input of silica, potassium and sulfur to sinter feed mix, combination of burnt lime with Portland cement is preferred.

REFERENCES

[1] K. Higuchi et al., "High-productivity operation of commercial sintering machine by stand-support sintering," *Nippon Steel Technical Report*, 2006, No. 94: 31-35.

[2] E. Marliere et al., "Improved permeability of sinter mix by combined optimization of feeder system, permeability bars and charging monitoring," *Technical Steel Research*, 2007.

[3] W. Borges, et al., "Application of HPS (Hybrid Pelletized Sinter) process at Monlevade Works," *Revue de Metallurgie*, 2004, No. 3: 189-194.

[4] N.-Y. Ma, "Production of high-grade potassium chloride from a sinter plant baghouse dust," *TMS EPD Congress 2009*, 927-929.

Energy Technology 2015: Carbon Dioxide Management and Other Technologies
Edited by: Animesh Jha, Cong Wang, Neale R. Neelameggham, Donna P. Guillen, Li Li, Cynthia K. Belt, Randolph Kirchain, Jeffrey S. Spangenberger, Frank Johnson, Andrew J. Gomes, Amit Pandey, and Peter Hosemann
TMS (The Minerals, Metals & Materials Society), 2015

RECOVERY OF METALS FROM WASTE PRINTED CIRCUIT BOARDS BY LEACHING WITH 1-ETHYL-3-METHYL-IMIDAZOLIUM HYDROGEN SULFATE IONIC LIQUID

Tuğba Selcen Atalay[1], Ayfer Kılıçarslan[1], Muhlis Nezihi Sarıdede[1]

[1]Yildiz Technical University, Department of Metallurgical and Materials Engineering, Davutpasa Campus, Esenler, Istanbul, 34210, Turkey

Keywords: Printed circuit board, Ionic liquid, Leaching, Recovery, Copper, Iron

Abstract

The interest for the ionic liquids, which are environmentally friendly, is increasing day by day because of their high performance in recovery of metals. In this work, 1-ethyl-3methyl-imidazolium hydrogen sulfate ([Emim]HSO_4) ionic liquid and its aqueous solutions were used as leaching media for dissolution of waste printed circuit boards (WPCBs) to recover of copper in the presence of hydrogen peroxide (H_2O_2) as oxidant. The effects of ionic liquid concentration, leaching time and temperature on the copper recovery rate were investigated. By the way dissolution behavior of iron was also determined. Leaching temperatures were selected as 25°C and 50°C. The results showed that recovery rates of copper and iron decreased with increasing ionic liquid concentration. Leaching time and temperature had positive effect on copper and iron dissolution. Higher dissolution rates were obtained with heat treated samples than not heat treated samples.

Introduction

Waste electrical and electronic equipment which is increasing day by day causes environmental problems due to their hazardous content and very attractive for their valuable metal contents. Therefore, recycling of electronic wastes has great importance in recent years. Waste printed circuit boards (WPCBs) which are the most of the electronic equipment, comprise various important metals, especially copper and precious metals [1].

There are a lot of techniques of recycling WPCBs. Pyrometallurgical processing, hydrometallurgical method, bio-technology, supercritical fluid technology and physical-mechanical processing are the methods for recycling PCBs [2-3]. Physical-mechanical processing was employed to recover valuable metals from WPCBs by mechanical separation process consisting of milling, size classification, gravity separation, and two-step magnetic separation [4]. Hydrometallurgical methods generally contain the leaching with acid or alkaline solutions [5]. For instance, H_2SO_4-$CuSO_4$-NaCl solutions was used to extract metals from WPCBs and thus Cu, Ni, Ag, Pd recovered more than 90% [6].

Ionic liquids (ILs), which have been accepted as a new green chemical revolution, are aprotic and polar solvent [7]. Ionic liquids have many important properties such as thermal stability, negligible vapor pressure, high conductivity and nonflammability [8]. Ionic liquids composed of organic cations and inorganic/organic anions can be used for recycling of WPCBs [9]. Huang et al. have studied the leaching of copper from WPCBs using 1-butyl-3-methyl-imidazolium hydrogen sulfate ([Bmim]HSO_4). The results showed that copper leaching rate has been affected by particle size but ionic liquid concentration has no important effect on it. Also the leaching rate increased with increasing time and temperature [10].

In this work, 1-ethyl-3methyl-imidazolium hydrogen sulfate ([Emim]HSO_4) has been used to leach metals from WPCBs and the effect of leaching temperature, time and ionic liquid concentration on copper, iron and zinc recoveries from two different samples of WPCBs has been investigated.

Experimental

Waste Preparation

Experiments were carried out with two different WPCB samples. First group was ground in a ball mill and applied heat treatment in 250°C (sample HT), while the second one was crushed using different crusher, hammer mill (sample NHT). After screening, four different sample fractions were obtained: $F_1 < 0.075$ mm; 0.1 mm $< F_2 <$ 0.25 mm; 0.25 mm $< F_3 <$ 0.5 mm; and $F_4 >$ 0.5 mm.

Aqua regia digestion which is less hazardous, appears best for general purpose analysis, was applied during 5 hours for three elements in PCB [11]. Table I shows changes of metal content of samples according to particle size. It can be seen from Table I that the metal contents of WPCBs changes with particle size substantially for both HT and NHT samples. As two samples of WPCBs particle size increased from 0.075 mm to 0.5 mm, the content of copper and iron increased. However, the contents of copper, zinc, and iron decreased when particle size increased from 0.5 mm.

Table I. Major metal contents of WPBCs

	F_1	F_2	F_3	F_4
Sample HT				
Cu %	4.4	23.81	23.16	4.6
Fe %	1.89	1.008	0.66	0.36
Zn %	0.67	0.45	0.4	0.68
Sample NHT				
Cu %	0.67	5.16	22.35	15.83
Fe %	1.33	1.1	1.18	0.48
Zn %	0.27	0.09	0.15	0.04

* $F_1 <$ 0.075 mm, 0.1 mm $< F_2 <$ 0.25 mm, 0.25 mm $< F_3 <$ 0.5 mm and $F_4 >$ 0.5 mm

Leaching Test

Leaching process was carried out in shaking water bath with 100 ml solutions at 25°C and 50°C. The sample fraction has the highest copper content (F_3) was used in the leaching experiments (Table I). Concentration (v/v) of ionic liquid - EmimHSO_4 was changed between 20% and 80% with 20°C intervals. Hydrogen peroxide (10%) was used as oxidant for all tests. Solution samples were taken at 15, 30, 60, 90 and 120 minutes during dissolution. Metal content passed into the leach solution were analyzed in the atomic absorption spectrometer.

Results and Discussion

Effect of Ionic Liquid Concentration

Fig. 1 shows the effect of ionic liquid concentration on copper recovery ratio at 50°C with 100 g/L pulp density for sample HT and Fig. 2 for sample NHT under the same conditions. It can be seen from the figures that ionic liquid concentration highly influences copper leaching recovery from WPCBs. Fig 1 and Fig 2 presented that the copper leaching rate decreased significantly as ionic liquid concentration increased from 20% to 80%. This descent is noticeable for sample HT. Also increasing leaching time mainly increased copper recovery rate.

During solving of WPCBs, iron in significant amount went into the solution. Changing of iron recovery rate depending on time and ionic liquid concentration during 2 hours is given in Fig 3 for sample HT and in Fig 4 for sample NHT. The figures indicated that increasing ionic liquid concentration had a negative effect on iron dissolution, since iron extraction decreased from 69.43% to 2% when ionic liquid concentration decreased from 80% to 20% for 120 min. Higher dissolution for iron took place for sample HT than for sample NHT.

The reason in decreasing recovery rate of metals is viscosity of ionic liquid. IL concentration increases along with its viscosity and it causes decrement of metal extraction [8].

Besides all, the experiments have indicated that zinc showed the highest dissolution rate (> 99%) for both samples.

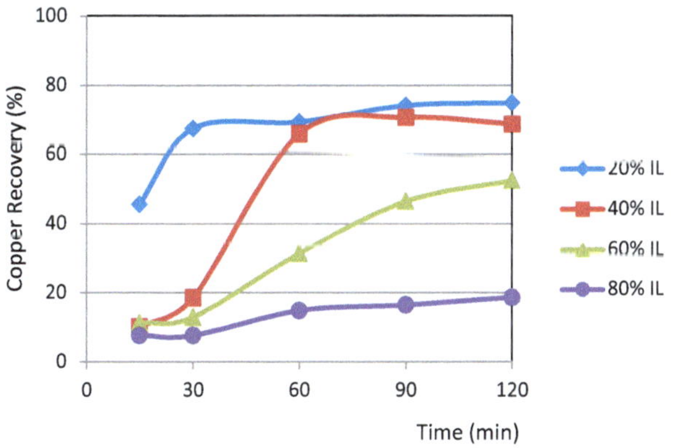

Figure 1. The effect of ionic liquid concentration on copper recovery rate for sample HT at 50°C

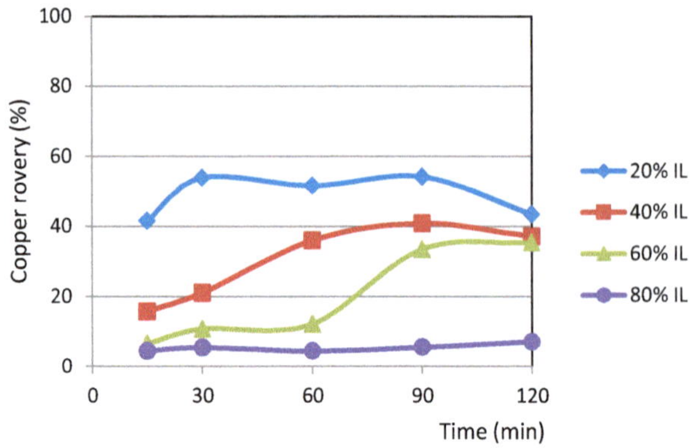

Figure 2. The effect of ionic liquid concentration on copper recovery rate for sample NHT at 50°C

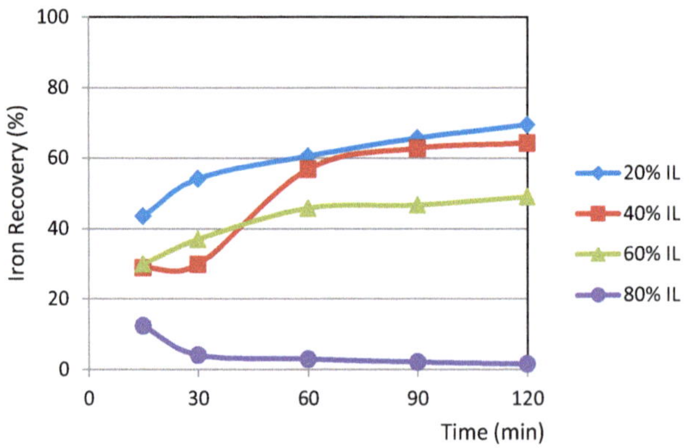

Figure 3. The effect of ionic liquid concentration on iron recovery rate for sample HT at 50°C

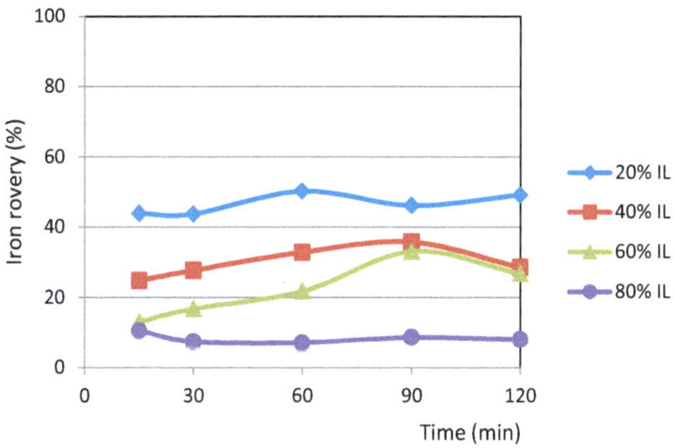

Figure 4. The effect of ionic liquid concentration on iron recovery rate for sample NHT at 50°C

Effect of Temperature

The leaching behavior of copper and iron as a function of leaching temperature (25°C and 50°C) using 20% ionic liquid concentration and 100 g/L pulp density. The effect of leaching temperature on copper recovery rate was shown in Fig. 5. It can be seen that copper recovery rate increased when temperature was raised from 25°C to 50°C. For sample HT, copper recovery rate was 66.85% for 25°C and 74.94% for 50°C after leaching time of 2h. Value of copper recovery ratio of sample NHT is lower than sample HT but the effect of temperature for sample NHT is more remarkable than sample HT.

Same situation is also current for dissolution of iron. Changing of iron recovery rate with temperature and time was given in Fig. 6. It demonstrates that iron extraction much increased compared with copper when temperature was raised from 25°C to 50°C for two samples. Iron recovery rate was 28.33% for 25°C but 69.43% for 50°C when sample HT was dissolved at the end of the tests. Also it is obtained as 49.06% for 50°C but 19.09% for 25°C for sample NHT after 120 min.

The following factors provide increasing copper as temperature is raised: the solubility of copper ions increases in ionic liquid, the solubility of oxygen increases in ionic liquid, the increasing acidity and ionic strength [12].

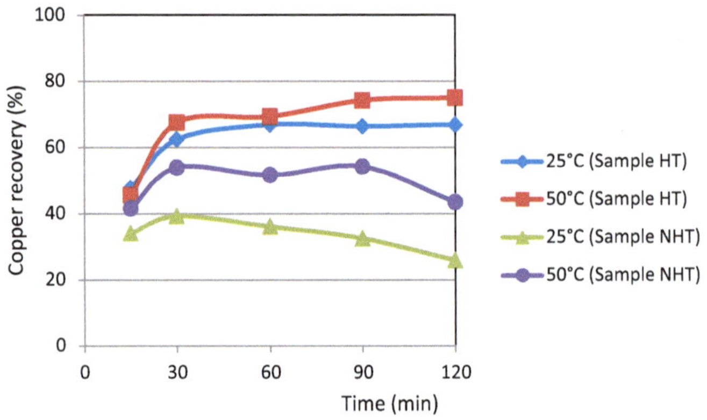

Figure 5. The effect of temperature on copper recovery rate (IL const: 20%)

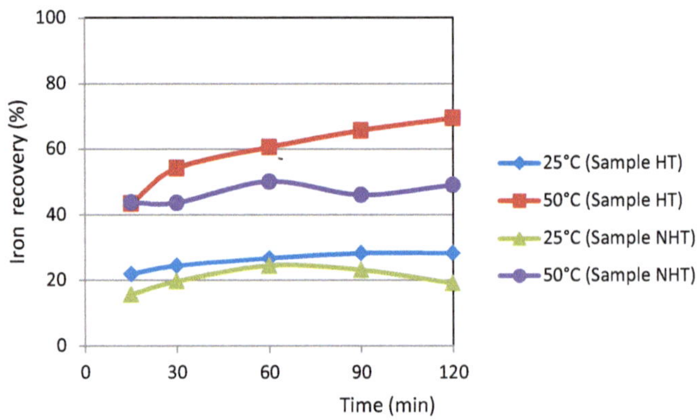

Figure 6. The effect of temperature on iron recovery rate (IL const: 20%)

Conclusions

The following conclusions can be drawn from this study which has investigated leaching of copper and iron in waste printed circuit boards with solution of ionic liquid, 1-ethyl-3methyl-imidazolium hydrogen sulfate - $EmimHSO_4$.

1. Increasing ionic liquid concentration has negative impact on copper and iron recovery rate. Ionic liquid concentration of 20% is the best rate for recovery of both metals.
2. When leaching time increases recovery rates of copper and iron also increase for all concentrations except 80%.
3. Leaching temperature showed significant positive influence on copper recovery and especially on iron recovery for both samples.
4. Heat treated samples showed more copper and iron solubility as compared to samples not heat treated.

References

1. I. Birloaga et al., "Study on The Influence of Various Factors in the Hydrometallurgical Processing of Waste Printed Circuit Boards for Copper and Gold Recovery" *Waste Management*, 33 (2013), 935-941.

2. Y. Zhang et al., "Current Status on Leaching Precious Metals from Waste Printed Circuit Boards" *Procedia Environmental Sciences*, 16 (2012), 560-568.

3. J. Cui, L. Zhang, "Metallurgical Recovery of Metals from Electronic Waste: A Review" *Journal of Hazardous Materials*, 158 (2008), 228-256.

4. J-M. Yoo et al., "Enrichment of the Metallic Components from Waste Printed Circuit Boards by a Mechanical Separation Process Using a Stamp Mill" *Waste Management*, 29 (2009), 1132-1137.

5. P.M.H. Petter, H.M. Veit, A.M. Bernardes, "Evaluation of Gold and Silver Leaching from Printed Circuit Board of Cellphones" *Waste Management*, 34 (2014), 475-482.

6. E.Y. Yazici, H. Deveci, "Extraction of Metals from Waste Printed Circuit Boards (WPCBs) in H_2SO_4- $CuSO_4$- NaCl Solutions" *Hydrometallurgy*, 139 (2013), 30-38.

7. P. Zhu et al., "Treatment of Waste Printed Circuit Board by Green Solvent Using Ionic Liquid" *Waste Management*, 32 (2012), 1914-1918.

8. T. Dong et al., "Leaching of Chalcopyrite with Brønsted Acidic Ionic Liquid" *Hydrometallurgy*, 99 (2009), 33-38.

9. T. Guo-Cai, L. Jian, H. Yi-Xin, "Application of Ionic Liquids in Hydrometallurgy of Nonferrous Metals" *Transactions of Nonferrous Metals Society of China*, 20 (2010), 513-520.

10. J. Huang et al., "Leaching Behavior of Copper from Waste Printed Circuit Boards with Brønsted Acidic Ionic Liquid" *Waste Management*, 34 (2014), 483-488.

11. I.O. Ogunniyi, M.K.G. Vermaak, D.R. Groot, "Chemical Composition and Liberation Characterization of Printed Circuit Board Comminution Fines for Beneficiation Investigations" *Waste Management*, 29 (2009), 2140-2146.

12. J.A. Whitehead et al., "Application of 1-alkyl-3-methyl-imidazolium Ionic Liquids in the Oxidative Leaching of Sulphidic Copper, Gold and Silver Ores" *Hydrometallurgy*, 88 (2007), 109-120.

Energy Technology 2015: Carbon Dioxide Management and Other Technologies
Edited by: Animesh Jha, Cong Wang, Neale R. Neelameggham, Donna P. Guillen, Li Li, Cynthia K. Belt, Randolph Kirchain, Jeffrey S. Spangenberger, Frank Johnson, Andrew J. Gomes, Amit Pandey, and Peter Hosemann
TMS (The Minerals, Metals & Materials Society), 2015

APPLICATION OF 1-METHYLIMIDAZOLIUM HYDROGEN SULFATE IONIC LIQUID TO THE OXIDATIVE LEACHING OF COPPER AND ZINC IN INDUSTRIAL BRASS DROSS

Ayfer Kılıçarslan[1], Muhlis Nezihi Sarıdede[1]

[1]Department of Metallurgical and Materials Engineering, Yildiz Technical University, Davutpasa Campus, Esenler, 34210 Istanbul, Turkey

Keywords: 1-methylimidazolium hydrogen sulfate, Oxidative leaching, Copper recovery, Zinc recovery.

Abstract

In recent years, usability of ionic liquids in the extraction processes of ores and secondary raw materials has been attracted due to their promising potential as green solvents. In this study, 1-methylimidazolium hydrogen sulfate (HmimHSO$_4$) ionic liquid and its aqueous solutions have been used as leaching media for the leaching of brass dross to recovery of copper and zinc in the presence of hydrogen peroxide (H_2O_2) as oxidant. The leaching experiments were carried out in the temperature range from 40 to 80°C at ambiance pressure with 100 g/L pulp density. The leaching behavior of copper and zinc has been discussed and the influences of ionic liquid concentration, leaching time and temperature on the metal recovery rate have been investigated. The results indicated that high copper and complete zinc extraction has been achieved under the optimum conditions.

Introduction

Today, brass drosses are common residues generated in the pyrometallurgical process of brass production in the world. These residues contain mostly high amount of metal oxides such as copper oxide (CuO), zinc oxide (ZnO) and also other oxides and trace metals [1]. Therefore, due to the increasing demand, recovering zinc and copper from these types of residues is increasing requirement in the brass industry. Although many researches have been conducted on conventional hydrometallurgical recovery of zinc and copper from these types of residues, acidic or basic leaching may not be environmentally friendly due to the high consumption of solvent and pollution of water [2, 3].

Recently, ionic liquids (ILs) have a growing interest as novel solvents for the sustainable dissolution, extraction and separation of metal ions [4-6]. They are organic salts generally with a low melting point (<100°C) [7]. A remarkable variation is possible in both anionic and cationic part of the IL, thus it is possible to create an IL with desired properties to meet the requirements of a particular process [8]. They have excellent properties including negligible vapor pressure, non-flammability, wide liquid range and thermal stability to solubilize a wide range of solutes including metal salts and complexes. Especially, their ability to dissolve a variety of non-ferrous oxides and hydroxides at low temperatures makes the ILs attractive and green alternatives to the traditional aqueous acid/alkaline solutions used in hydrometallurgical processes which are associated with large volumes of effluent production and require elevated temperatures and pressures [9]. So far, ionic liquids have been studied as solvents in different leaching systems for the extraction of copper, gold and silver from their concentrates or secondary resources [10-12].

The aim of this study is to developed a new and green leaching method to recover copper and zinc from industrial brass dross using an ionic liquid namely 1-methylimidazolium hydrogen sulfate ($HmimHSO_4$) in the presence of hydrogen peroxide, considering effective parameters such as concentration of ionic liquid, leaching time and temperature.

Experimental

Brass dross used in the leaching experiments was provided from a Brass Manufacturing Company in Turkey. The chemical composition of the dross was determined by ICP-OES (Inductively Coupled Plasma Optical Emission Spectroscopy) and the results were reported in Table I.

Table I. Metal content of brass dross (wt.%).

Zn	Cu	Si	Al	Fe	Ca	Na
53.92	22.36	12.17	3.84	0.16	5.55	0.57

Mineralogical analysis of the dross was determined by X-ray diffraction spectrum as in Figure 1. XRD analysis of the dross indicated that it contains two main constituents; namely zinc oxide (ZnO) and copper oxide (CuO) with high amounts of quartz (SiO_2) and aluminum oxide (Al_2O_3). However, compounds of iron, calcium and other elements were present in much less amounts.

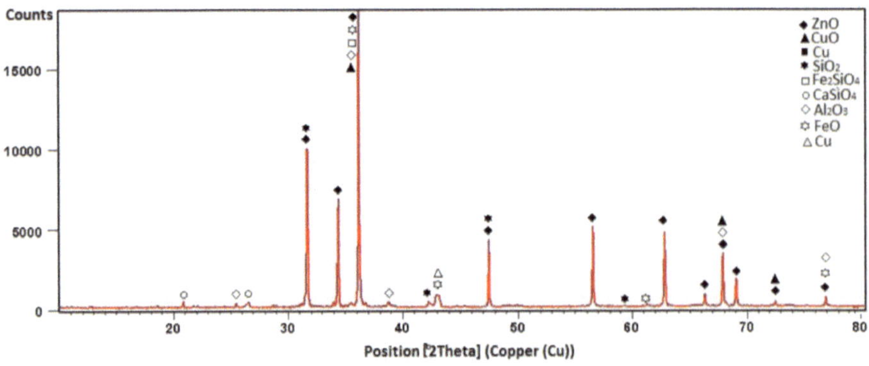

Figure 1. X-ray diffraction spectra of industrial brass dross

In order to leach copper and zinc from dross, 1-methylimidazolium hydrogen sulfate ($HmimHSO_4$) ionic liquid and its aqueous solution together with 20% (v/v) of hydrogen peroxide (H_2O_2) were placed in 250 ml glass beaker and mixing was provided by using a magnetic stirrer at constant rotation speed of 360 rpm for 2 h. All leaching experiments were carried out at solid/liquid (S/L) ratio of 100 g/L. During leaching test, sample solutions were taken out periodically to investigate the effect of the time on the copper and zinc dissolution. The solid/liquid separation was carried out by filtration. The leachate was analyzed for Cu and Zn using the atomic absorption spectrophotometer (AAS) to determine the leaching efficiency of metals. The effects of temperature, ionic liquid concentration and leaching time on copper and zinc recovery rate were investigated.

Results and Discussion

Copper Recovery

Brass dross was leached at temperatures varying from 40 to 80°C using various concentrations of ionic liquid to determine the effect of temperature and ionic liquid concentration. Figure 2 shows the effect of $HmimHSO_4$ ionic liquid concentration on copper recovery rates at 40°C as a function of leaching time. The copper recovery rate increases dramatically with decreasing ionic liquid concentration from 80 to 20% (v/v). When ionic liquid concentration is 80% (v/v), copper recovery rate is only about 17.98% after 120 minutes. Besides, a rapid increase was observed with decreasing ionic liquid concentration between 60% and 20% (v/v). Copper leaching recovery increased from 17.98% to 76.16% with decreasing the ionic liquid concentration from 80% to 60% (v/v) at 120 min. A smaller increase in copper leaching recovery from 70.69% to 76.16% after 120 minutes was observed when the ionic liquid concentration decreased from 60% to 40%. However, copper leaching recovery significantly increased from 76.16% to 95% with decreasing ionic liquid concentration from 40% to 20% (v/v).

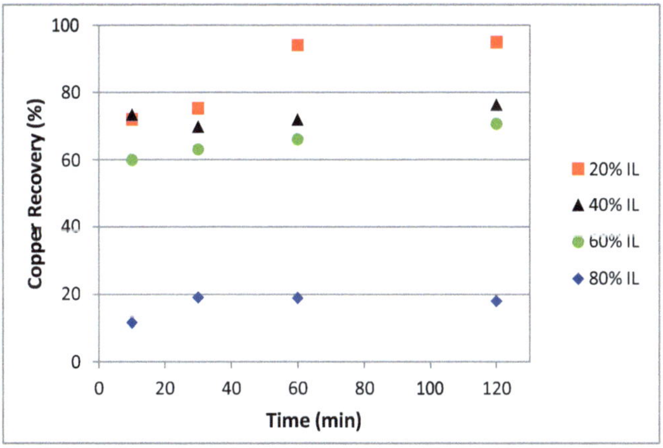

Figure 2. Effect of $HmimHSO_4$ concentration on copper recovery at 40°C

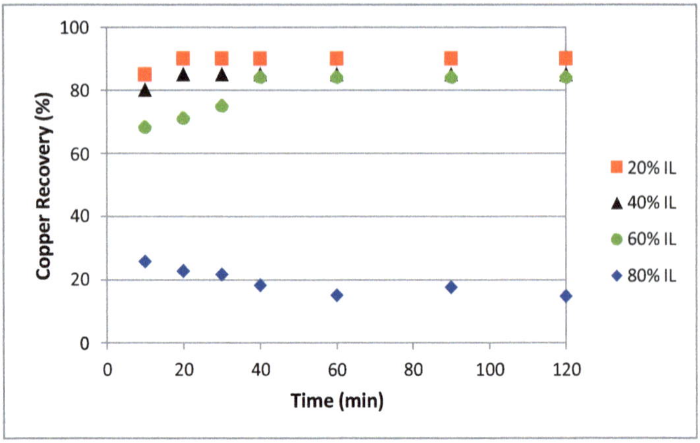

Figure 3. Effect of $HmimHSO_4$ concentration on copper recovery at 60°C

The effect of ionic liquid concentration on the copper recovery at 60°C and 80°C is given in Figure 3 and 4 respectively. The leaching recovery of copper at the temperature of 60°C and 80°C showed a similar trend as Figure 1 that copper recovery rate increased remarkably as the $HmimHSO_4$ concentration decreased from 80 to 20% for both leaching temperatures. The effect of increasing $HmimHSO_4$ concentration in water was to decrease the amount of copper recovered.

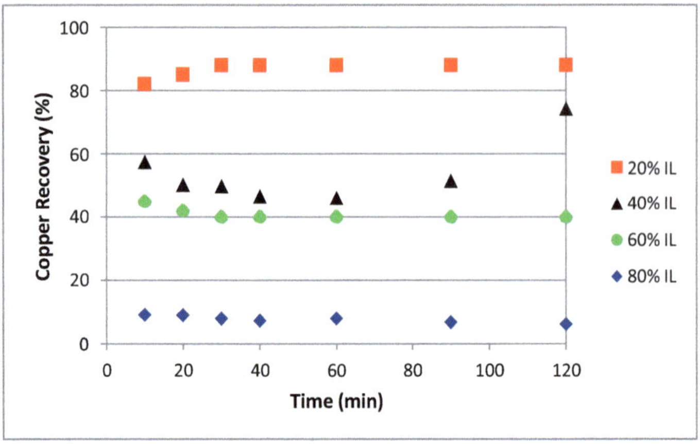

Figure 4. Effect of $HmimHSO_4$ concentration on copper recovery at 80°C

The decrement of copper recovery rate with increasing ionic liquid concentration is related to the increasing acidity of solution. When $HmimHSO_4$ is present in aqueous solution (20%, 40%, 60% (v/v)) with the mixture of 20% H_2O_2 (oxidant), the hydrogen sulfate anion can dissociate releasing H^+ ions and hence the $HmimHSO_4$ acts as an acid. On the other hand, in 80% $HmimHSO_4$ and 20% H_2O_2 mixture, the ability of the hydrogen sulfate anion to dissociate and act as an acid is reduced since proton hydration cannot occur in neat ionic liquid with oxidant in the absence water.

Increase of the leaching temperature from 40°C to 80°C has negative influence on copper recovery. The decrease of dissolved oxygen concentration as temperature increased affects the copper recovery. The oxygen solubility in the ionic liquids decreases as the temperature increases, which is consistent with oxygen solubility in aqueous systems [11].

Zinc Recovery

The effect of the $HmimHSO_4$ ionic liquid concentration on zinc leaching recovery is shown in Figure. 5, 6 and 7 for temperature of 40, 60 and 80°C, respectively. It is presented in Figure 1 that almost all zinc was recovered for all ionic liquid concentrations after 30 minutes at 40°C. It is evident that the ionic liquid concentration did not influence the zinc recovery rate significantly at moderately low temperatures.

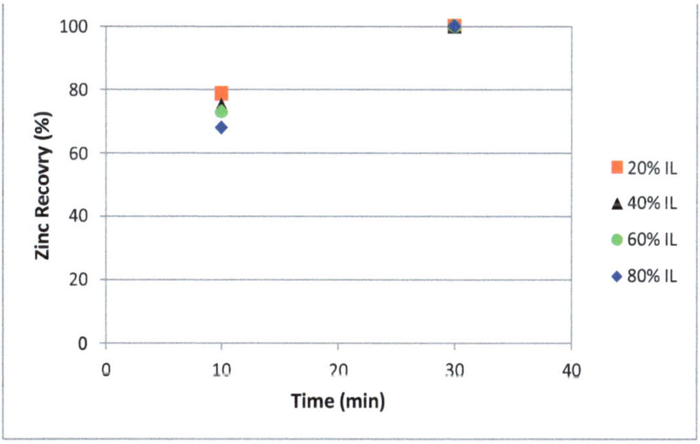

Figure 5. Effect of $HmimHSO_4$ concentration on zinc recovery at 40°C

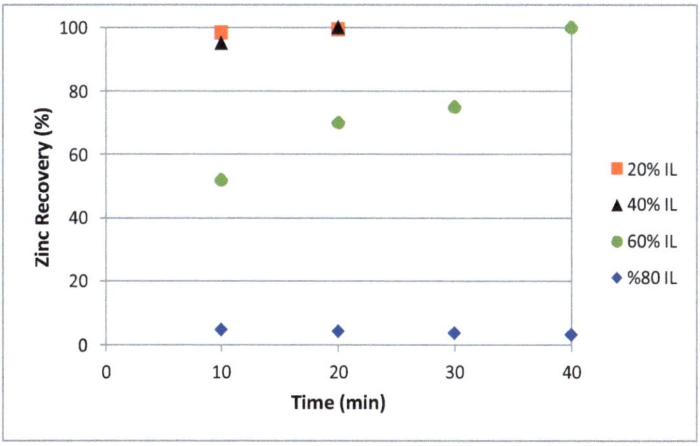

Figure 6. Effect of $HmimHSO_4$ concentration on zinc recovery at 60°C

It can be seen from Figure 6 that zinc recovery rate increased significantly from 3.20% to 100% when the ionic liquid concentration decreased from 80% to 60, 40 and 20% (v/v) after 40 minutes at 60°C. It is reasonable that zinc recovery rate increased as decreasing of ionic liquid concentration in solution since water content of initial leach solution reduced as $HmimHSO_4$ concentration increased.

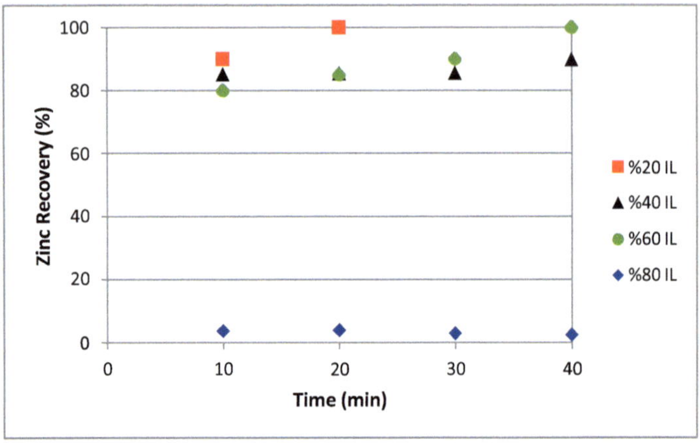

Figure 7. Effect of $HmimHSO_4$ concentration on zinc recovery at 80°C

The result shown in Figure 7 reveals that the ionic liquid concentration has a strong effect on zinc recovery for the concentration of 80% (v/v) $HmimHSO_4$, but there is no significant difference in zinc recovery rate for 20%, 40% and 60% (v/v) concentrations at 80°C. The high zinc recovery for these ionic liquid concentrations is similar to that achieved with at 60°C. The use of 80% $HmimHSO_4$ (v/v) and 20% H_2O_2 (v/v) as an initial leach solution in the absence of water limited the solubility of zinc. However, aqueous solutions consist of 20 and 60% (v/v) $HmimHSO_4$ and 20% (v/v) H_2O_2 is provided 100% zinc recovery rate after 40 minutes at 80°C.

Conclusion

Ionic liquid, namely 1-methylimidazolium hydrogen sulfate ($HmimHSO_4$) as a solvent was investigated to recover metal present in brass dross and following results obtained from the study.

- The ionic liquid $HmimHSO_4$ acts as an acid in leaching of dross for the recovery of copper and zinc. H_2O_2 is the suitable oxidant for the ionic liquid aqueous solutions for direct leaching of brass dross.
- Ionic liquid concentration showed a significant influence on copper and zinc recovery. Compared with concentrate $HmimHSO_4$ ionic liquid solution, its aqueous solutions are more effective for the recovery of copper and zinc.
- Copper recovery rate was up to 95%, under the conditions: 20% ionic liquid, 40°C and 2h.

- Almost 100% zinc recovery was achieved at 40°C for all used ionic liquid concentrations after 30 minutes. However when temperature at 60 and 80°C, the leach solution of 80% ionic liquid concentration has very limited zinc recovery due to the lack of water in the solution.

Acknowledgement

The authors are grateful for the financial support of the Scientific and Technological Research Council of Turkey (TUBITAK) (Project No. 113M241).

References

1. S. Timur et al., "Beneficiation of brass ashes, "Developments in Mineral Processing, 13(12) (2000), C12a-27-33.
2. O. Kahvecioglu et al., "Carbothermal Recovery of Zinc from Brass Ash," Miner. Process. Extr. Metall., 112 (2) (2003, 95-101.
3. I. M. Ahmed and J.A. Daoud, "Leaching and recovery of zinc and copper from brass slag by sulfuric acid", Journal of Saudi Chemical Society, (2012), (accepted).
4. D. Han and K. H. Row, "Recent Applications of Ionic Liquids in Separation Technology," Molecules, 15 (2010), 2405-2426;
5. A. McCluskey, G.A. Lawrance, S.K Leitch, M.P. Owen and I.C. Hamilton, "Ionic Liquids Industrial Applications for Green Chemistry", American Chemical Society, 818, 199-212, (2002).
6. D. Han and K. H. Row, "Recent Applications of Ionic Liquids in Separation Technology," Molecules, 15 (2010), 2405-2426.
7. Welton T, "Room-Temperature Ionic Liquids. Solvents for Synthesis and Catalysis," Chem Rev, 99 (1999), 2071-2084.
8. J. Martyn et al., "Ionic liquids. Green solvents for the future," Pure Appl. Chem., 72 (7) (2000) 1391-1398.
9. G. Tian, J. Li, Y. Hua, "Application of ionic liquids in hydrometallurgy of nonferrous metals," Trans. Nonferrous Met. China 20 (2010) 513-520.
10. J. A. Whitehead, G. A. Lawrence and A. McCluskey, "Green leaching: recyclable and selective leaching of gold-bearing ore in an ionic liquid," J. Green Chem, 6 (2004) 313-315.
11. J.A. Whitehead, J. Zhang, N. Pereira, A. McCluskey and G.A. Lawrance, "Application of 1-alkyl-3-methyl-imidazolium ionic liquids in the oxidative leaching of sulphidic copper, gold and silver ores," Hydrometallurgy, 88 (2007) 109-120.
12. T. Dong, Y. Hua, O. Zhang, and D. Zhou, "Leaching of chalcopyrite with Brønsted acidic ionic liquid," Hydrometallurgy, 99 (2009) 33-38.

Energy Technology 2015

Carbon Dioxide Management and Other Technologies

RECYCLING AND
SUSTAINABILITY UPDATE

Poster Session

Energy Technology 2015: Carbon Dioxide Management and Other Technologies
Edited by: Animesh Jha, Cong Wang, Neale R. Neelameggham, Donna P. Guillen, Li Li, Cynthia K. Belt, Randolph Kirchain, Jeffrey S. Spangenberger, Frank Johnson, Andrew J. Gomes, Amit Pandey, and Peter Hosemann
TMS (The Minerals, Metals & Materials Society), 2015

EXPERIMENTAL STUDY ON REDUCTION IN LOW GRADE LATERITIC NICKEL ORE MIXED WITH PICKLING SLUDGE

Feng Ya-hui[1], Sun ying[1], Zhang jing[1], Li Qiu-ju[1]

[1]Shanghai Key Laboratory of Modern Metallurgy & Materials processing, Shanghai University, shanghai 200072, China

Keywords: Laterite Nickel ore, Stainless steel pickling sludge, Reduction

Abstract

Steel pickling sludge disposal lead to the pollution of soil and groundwater. The method of sludge mixed with low-grade lateritic nickel ore was proposed to either recover the valuable metals such as iron, chromium, and nickel, and the tail was disposal environmental friendly. The technique of XRF, ICP, XRD and EDS were used to characterize the chemical composition, mineral phases and physical chemical properties. The carbon was used to reduce the compound of low-grade lateritic nickel ore and sludge. The temperature, ratio of ore/carbon and the capacity of sludge were investigated for improving to recovery metals. The results showed that the products were separated into 3 types. The metals were enrichment in Fe-Cr-Ni alloy particle and concentrate. The mass content of Cr and Ni in tailing was low than 0.1% which could be safe disposal.

Introduction

During the processes of surface treatment of stainless steel cold rolling products, large mixed acid pickling was generated. During the pickling process, the pickling acids are recovered through ion exchange and metal ions such as Fe^{3+}, Cr^{3+} and Ni^{2+} get into the acids forming the pickling waste water. Then, the pickling waste water is precipitated with lime, forming pickling hydroxide sludge. The sludge has high contents of calcium fluoride and water but low contents of iron, chromium and nickel. However, it would also pose a potential environmental problem. At the same time, with the development of the industry, nickel is widely used and will not be replaced in a long time in the future. On the other hand, nickel sulfide ore which can be easily used is depletion, so the use for laterite nickel ore is paid more and more attention.

In this paper, a method of sludge mixed with low-grade lateritic nickel ore was proposed to either recover the valuable metals such as iron, chromium, and nickel or settle the disposal and environmental problems. The main process of this technology was follows: drying, roasting prereduction. The minerals need to be roasted in the drying kiln to remove attached water, and then roasted in the rotary kiln at high temperatures which aims to reduce and recovery metals. At last, the nickelferrite should be obtained [1-4].

Experimental

Materials and Characterization

The samples of nickel laterite ore and pickling sludge for this experiment are all from Jiangsu province, China. The pure graphite powder (Analytical reagent, Carbon content is more than

99.85%) was used as reducing agent. The calcium oxide, silica and alumina were used to simulated coal ash.

The character methods used in this experiment include: chemical analysis (XRF), X-ray diffraction (XRD), scanning electron microscopy (SEM) coupled with energy dispersive spectroscopy (EDS), thermogravimetric and laser particle analyzer.

Table I and Table II show the chemical compositions of laterite nickel ore and pickling sludge by using XRF method. Fig 1 and Fig 2 show the XRD patterns of nickel laterite ore and pickling sludge, respectively. The chemical composition of laterite nickel ore was main Fe (21%) with the oxide of Fe_2O_3 and 14.6% Si as showed in table I.

Table I Composition of laterite nickel ore (wt %)

Elements	O	Fe	Si	Mg	Al	Ni	Cr	Mn	Co	others
Content	44.37	21.78	14.60	13.86	2.02	1.68	0.75	0.38	0.08	0.55

The chemical composition of the stainless steel pickling sludge is listed in Table II. The compositions in the form of metallic oxide by calculating contained approximately 13.13 mass% Fe, 2.02 mass% Cr, 0.9 mass% Ni, 25 mass% Ca.

Table II Composition of stainless steel pickling sludge (wt %)

Elements	O	Ca	Fe	S	Cr	Si	Ni	Cl	F	others
Content	42.10	25.69	13.13	11.81	2.02	1.16	0.90	0.67	0.65	1.87

The XRD pattern of nickel laterite ore shows that the main phases of it are $(Mg, Al)_3(Si, Fe)_2O_5(OH)_4$, $(Mg, Fe)_3(Si, Fe)_2O_5(OH)_4$, $Mg_3Si_2O_5(OH)_4$, $Al_2Si_2O_5(OH)_4$, $(Mg, Fe)_6(Si, Al)_4O_{10}(OH)_8$, and FeOOH et al. The fundamental structure of the ore is kaolinite, Magnesium silicate asbestos, montmorillonite and mica, the nickel, chromium and cobalt replaced other anode ions in the phyllosilicate.

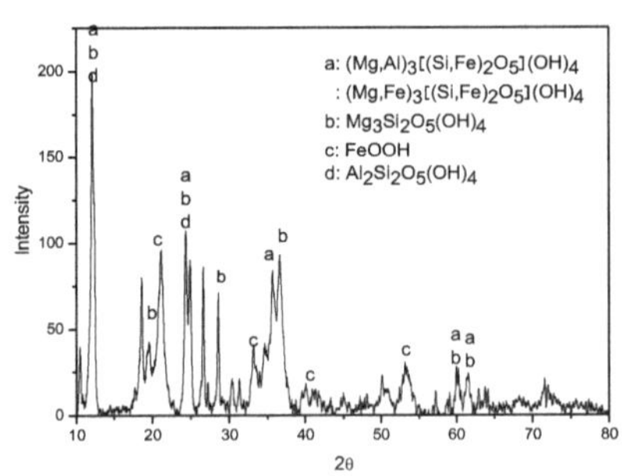

Fig 1 XRD pattern of nickel laterite ore

It is clear that in the XRD pattern of pickling sludge only the diffraction peaks of $CaSO_4 \cdot 2H_2O$ can be seen. It is because the metal elements are existing in the amorphous of hydroxide hydrate, so it is difficult to find in the diffraction peak diffraction peaks of the metals.

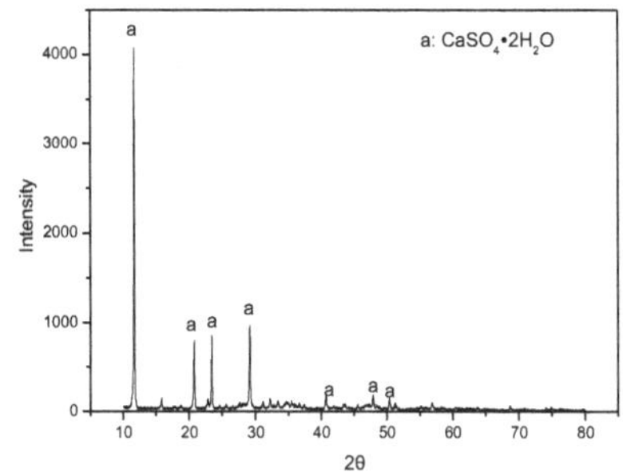

Fig 2 XRD pattern of stainless steel pickling sludge

Experimental processes

The drying and sieving of the ore is the first step. Then the ore after processeing was mixed with pickling sludge and reducing agent (graphite powder) in the certain proportion. The mixed samples were pressed into pellets. In the third step, the pressed pellets were roasted in the certain temperatures for reduction reactions. At last, the reacted samples were crushed and separated by magnetic separation.

This experiment studied the reduction temperature, ratio of ore/carbon, different sludge intake and reduction time on the effect of nickelferrite yield and grade. All of the following magnetic strength is 100 (mt) in this experiment. Fig.3 shows the process of the experiment equipment.

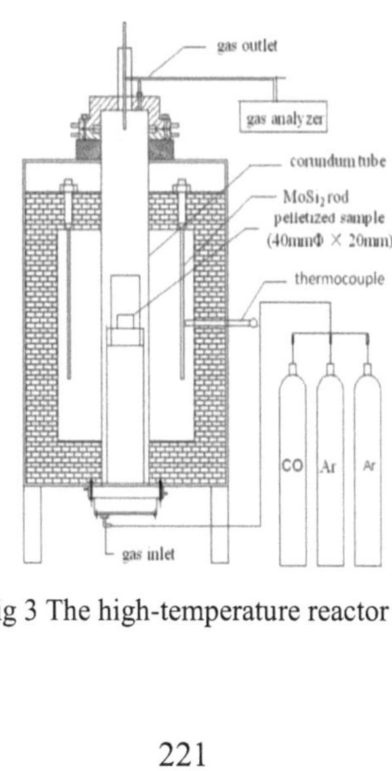

Fig 3 The high-temperature reactor

Results and Discussion

Effect of reduction temperature

The reaction temperatures were selected as: 1250℃, 1300℃, 1350℃, 1400℃, respectively. The other experimental conditions in the experiment were: carbon ratio (1.0), sludge intake (25%), reduction time (300 min), respectively.

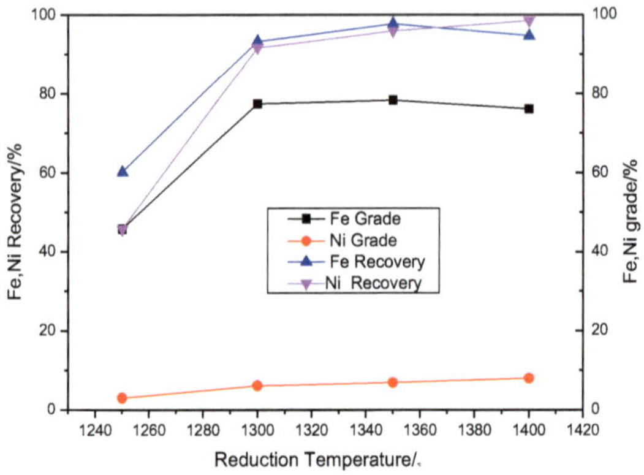

Fig 4 Effect of reduction temperature to the recovery and grade of Ni and Fe

Fig 4 shows that the reduction temperature had a significant impact on nickelferrite recovery and grade. The recovery and grade of the nickelferrite can be improved by raising the reaction temperature. In addition, the nickelferrite particles appeared in the condition of 1300℃, 1350℃, 1400℃ and the particle-size increased by the raising of temperature. The nickelferrite particle does not appear under the temperature of 1250℃. As the nickelferrite reduction temperature is far below 1250℃ on the thermodynamics, nickelferrite should be created at low temperatures in the reduction reaction. But the particles were too tiny for forming clumps to growing up. Therefore, the right temperature is a necessary condition for nickelferrite particle to growing up. The formation of nickelferrite particle needs to through the processes of crystallization nucleation, crystal nucleus grew up and merger.

Thus the main influences of the reduction temperature are reduced slag viscosity, increased the fluidity of slag generated during the reaction and promoted the formation of nickelferrite particles. The separation process of valuable metals and slag in the low-grade laterite nickel ore reduction reaction mainly depends on the size, shape, and the existing condition of nickelferrite. The large particle-size is conducive to the separation of valuable metals and slag. The small particle-size causes mosaic which leads to difficult separation.

Effect of carbon/ore ratio

The ratios of carbon/ore (%mass) were selected as: 0.8, 1.0, 1.2, 1.4, 1.6, respectively. The other experimental conditions in the experiment were: reduction temperature (1350℃), sludge intake (25%), and reduction time (300 min), respectively.

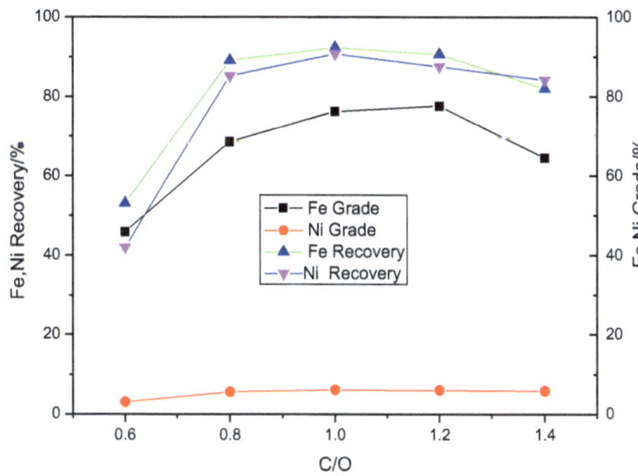

Fig 5 Effect of carbon/ore to the recovery and grade of Ni and Fe

Fig 5 shows that the recovery and grade of the nickelferrite increased when the carbon/ore increased at first. But when the carbon/ore continued to be raised, the recovery and grade showed the tendency of decline. It is similar to the situation of the laterite ore reduced singly.

The lack of carbon could limit the reduction reaction and leaded to low nickelferrite recovery and grade. But the excessive carbon also lead to a decline in the nickelferrite recovery and grade. This is due to excessive carbon against the process of metal crystal nucleus gathered and grew up.

Effect of different sludge intake

The percent of sludge intake were selected as: 8.3%, 16.7%, 5%, 33.3%, 40%, 50%, and 60%, respectively. The other experimental conditions in the experiment were: reduction temperature (1350℃), carbon/ore (1.0), and sludge intake (25%), respectively.

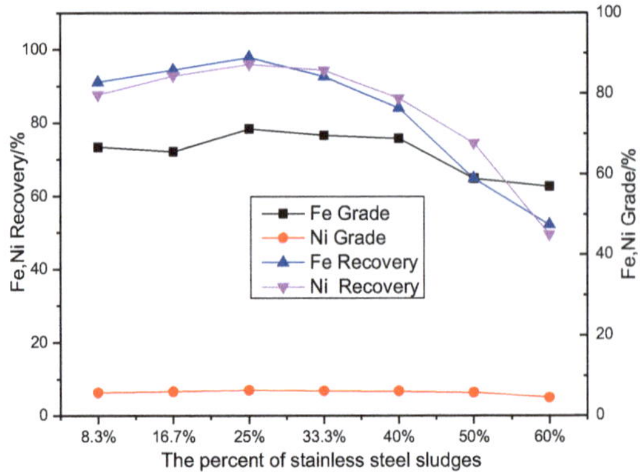

Fig 6 Effect of the percent of sludge to the recovery and grade of Ni and Fe

Fig 6 shows that the recovery and grade was on the rise when the percent of sludge was low and a downward trend appears with the increase of the percent. This is because the effect of auxiliary fusion was poor when the percent of sludge was low which lead to serious sticking of the metal and slag. But when the sludge was too much, at the high temperature the calcium silicate will be broken down into calcium oxide (CaO). Too much CaO also could cause serious sticking. The serious sticking of the metals and slag could inhibit the nucleation and growth of nickelferrite particles. So the appropriate percent of the sludge is very important for high recovery and grade.

Effect of reduction time

The reduction times were selected as: 120 min, 180 min, 240 min, 300 min, and 360 min, respectively. The other experimental conditions in the experiment were: reduction temperature (1350℃), sludge intake (25%), and reduction time (300 min), respectively.

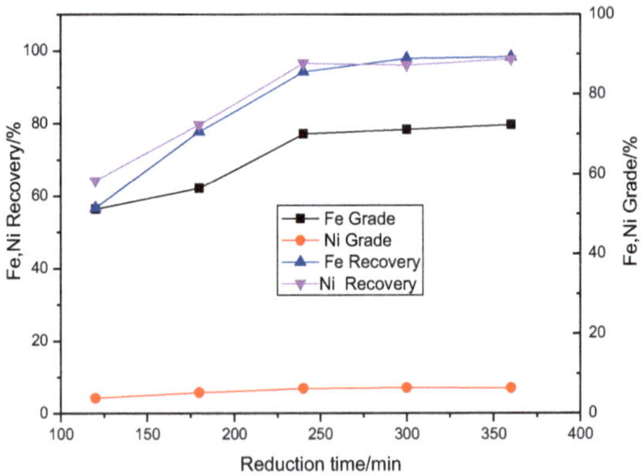

Fig 7 Effect of reduction time to the recovery and grade of Ni and Fe

Fig 7 shows that reaction time had an obvious impact on recovery and grade of nickelferrite, the extension of reaction time can improve the recovery and grade. In addition, when the reaction time reached 300 min and even more, there was no significant change of the recovery and grade. Experiments proved that nickelferrite particles need certain time to grow up, and it was nonsense to prolong the reaction time once the grow up processes haven been finished. The particles that have finished the grow up processes can be separated through magnetic separation, whereas the small particles were difficult to be separated due to their weak magnetic.

Nickelferrite products

There were two nickelferrite products which were nickelferrite particles and nickelferrite concentrate. The nickelferrite particle was a circular shape. The particle-size range was 1 mm to 10 mm, the large particle-size corresponds to the high reduction temperature and small corresponds to low reduction temperature. Fig 8 shows the morphology and XRD patterns of nickelferrite particles. Combined with XRD and XRF analysis results, the purity of the nickelferrite with little impurities was very high which can be used as raw materials of stainless steel directly.

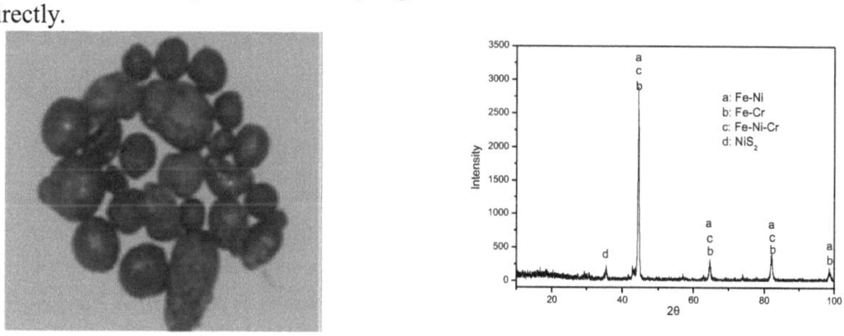

Fig 8 The image and phases of Ni-Fe alloy particle

Table III shows the chemical compositions of nickelferrite particles by using XRF method.

Table III Composition of nickelferrite particles (wt %)

Elements	Fe	Ni	S	O	Cr	Co	Al	Si	others
Content	80.91	7.85	4.26	3.71	2.50	0.33	0.21	0.13	0.10

In addition to nickelferrite particles, also got some nickelferrite concentrate. Table IV shows the chemical compositions of nickelferrite concentrate by using XRF method. The analysis obtained by SEM is shown in Fig 9. The results showed that the main ingredient still is Fe and Ni, but also contains a lot of gangue. In the SEM photograph one can see a lot of nickelferrite particles with gangue inlay.

Table IV Composition of nickelferrite concentrate (wt %)

Elements	Fe	O	Ni	Si	Cr	Mg	Al	S	others
Concent	67.15	17.81	7.61	1.88	1.48	1.11	0.91	0.65	1.40

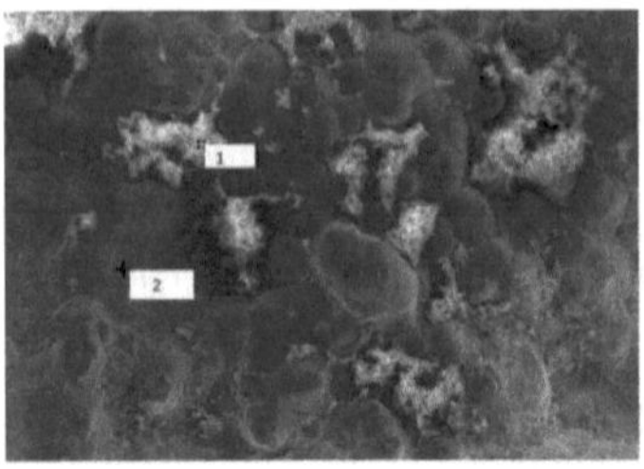

Spots	1	2
Fe	83.30	84.95
Ni	3.98	8.77
Cr	4.93	4.77
Si	7.78	1.03
Ca	0	0.48

Fig 9 SEM photograph of the nickelferrite concentrate

Conclusion

On the basis of analysis of raw material composition, content and structure, nickelferrite products were obtained by combining low-grade laterite nickel ore and stainless steel pickling sludge. This experiment studied the reduction temperature, ratio of ore/carbon, different sludge intake and reduction time on the effect of nickelferrite recovery. The optimum reaction conditions in the laboratory is: reduction temperature: 1350℃, carbon/ore: 1.0, reaction time: 300 min, and the percent of the sludge: 25%. Under this condition, the recovery rate of ferrite and nickel reached 96.07% and 97.86%, respectively.

The characterization of the products showed that the main products were as the form of the nickelferrite particles and nickelferrite concentrate. In the tailings, the main ingredient is gangue and the content of valuable metallic elements is very few. The tailings can direct emissions. The experiments proved that the use of stainless steel pickling sludge to treat low-grade laterite nickel ore is effective.

References

[1] WANG Hong, DENG Hai-Bo, LU Xiu-feng. Important Laterite Nickel Ore Resources in the World Present Situation and Exploitation [J]. GANSU METALLURGY, 2009, 2 (31): 20-24.

[2] Xiao Zhenmin. Status of exploitation of laterite type nickel ore and application of high pressure acid leaching technology in the world [J]. Chinese mining, 2002, 11 (1): 56-59.

[3] YANG Wei-jiao, MA Bao-zhong. Process on high pressure acid leaching of nickel laterites [J]. MINING & METALLURGY, 2011, 9 (20): 61-75.

[4] JIANG Ji-bo, WANG Ji-kun. Review on Progresses for Hydrometallurgy of Laterite-nickel Ore [J]. Hydrometallurgy of China, 2009, 3 (28): 3-11.

Energy Technology 2015: Carbon Dioxide Management and Other Technologies
Edited by: Animesh Jha, Cong Wang, Neale R. Neelameggham, Donna P. Guillen, Li Li, Cynthia K. Belt, Randolph Kirchain, Jeffrey S. Spangenberger, Frank Johnson, Andrew J. Gomes, Amit Pandey, and Peter Hosemann
TMS (The Minerals, Metals & Materials Society), 2015

RECYCLING WASTES IN THE ALUMINA AND THE CEMENT INDUSTRY

Nickolay Ilyoukha[1], Valentina Timofeeva[1]

[1]Academic Ceramic Center, 8, Frunse Str., Kharkov, 61002, Ukraine

Keywords: Composites Materials, Properties, Wastes, Recycling, Temperature.

Abstract

Main wastes of alumina and aluminum industry are not used at present and are being stored after partial treatment in special places. The paper will discuss the technology of treatment wastes as an additive for the cement industry. The solution of this complex problem will be of great importance for the purification of aerial, earth and water region in whole. In addition, as a result of waste treatment there will be special cements: high strength, extensible, gas-water proof, self-stressing needed in runway construction, in building tunnels and pits, and in the burial of nuclear wastes.

Introduction

Red mud is a waste material which results from alumina production during the Bayer process. Approx. 35-40% of the bauxite ore processed goes into waste as red mud. This is a cheap source of raw material for the manufacture of low-cost ceramic products such as building blocks, floor and wall tiles, sanitary ware and as an additive for cement. The chemical composition of red mud is as follows (Table I). Red mud is a very complex material. It is a mixture of several oxides and minerals such as hematite, sodium aluminum silicates, and rutile. These make the mud a potential raw material for the production of additives for cements.

Table I. Chemical composition of the calcined red mud

Oxidic compounds	Content (mass-%)
Fe_2O_3	38.1
Al_2O_3	27.01
SiO_2	14.15
TiO_2	5.01
Na_2O	6.03
K_2O	0.36
CaO	2.26
MgO	1.01
CO_2	4.02
SO_3	2.05

The particle size analysis of the calcined red mud showed that 85% of the particles are <10 μm (Figure 1). The calcined red mud ranged in size from 35 to 5 nm average particle diameter. The particle size analysis of the calcined red mud is given in Figure 2.

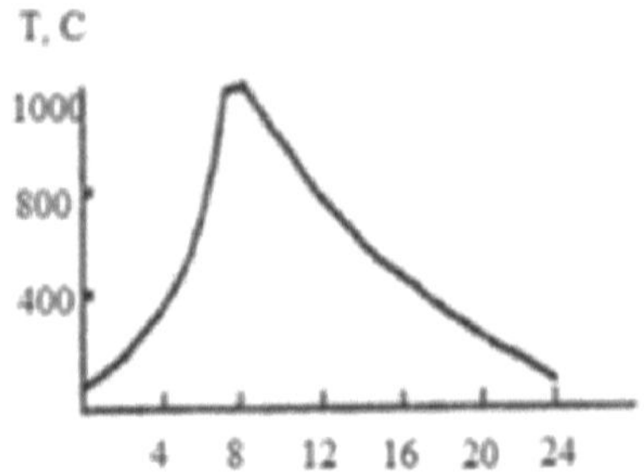

Figure 1. The firing treatment of red mud. The treatment of red mud is dependent on temperature and time

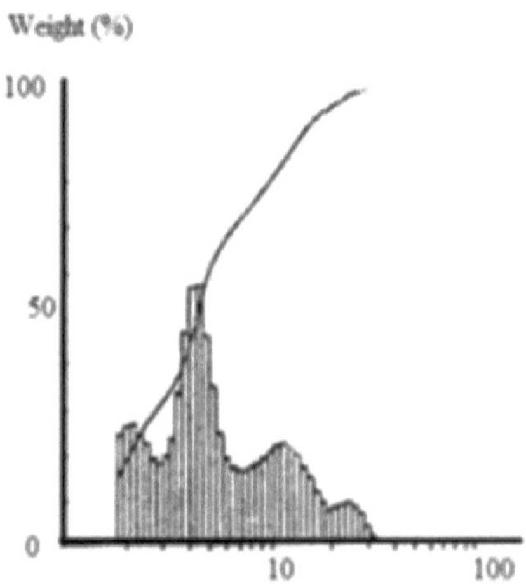

Figure 2. The particle size analysis of calcined red mud. X- Ray diffraction results of the calcined red mud

Clays Nonconditional Caolins

Ukraine has a great number of clay deposits. These clays can be used for traditional ceramics production and for production as an additive to cement. The physicoceramic date revealed that, in general, the clays sensitive to drying from this area, have significant plasticity (between 30 and 58%) and relatively high total shrinkage (between 6 and 21%) (Figure 3), have good to very good capability, average water adsorption (between 9 ang 16%) good to very good mechanical resistance (compressive strength between 325 and 355 daN/cm^2) (Figure 4), and bending strength between 90 and 460 daN/cm^2.

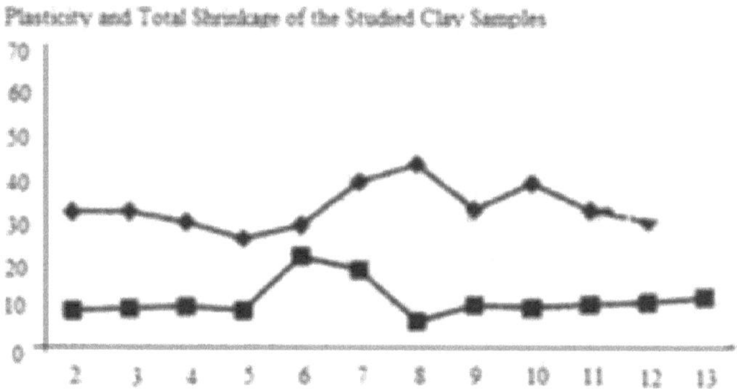

Figure 3. The Augustinic diagram of use of the clay deposits

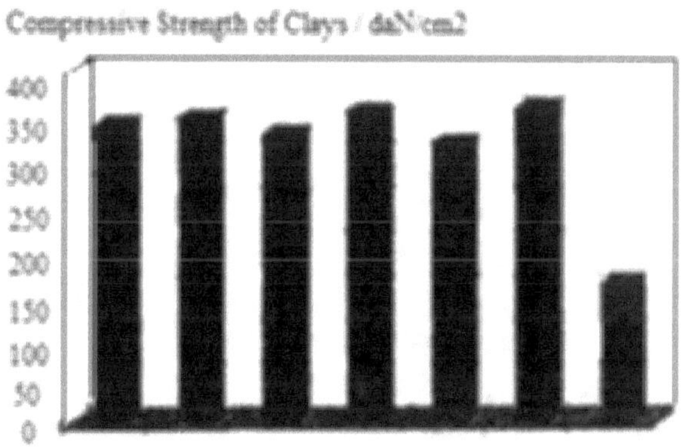

Figure 4. Comparative diagrams of physic – mechanical characteristics of clay deposits

Experimental procedure

The red mud sample was supplied by Nicolayew Plant (Ukraine), Kaolin wastes (region of Donetsk), Nepheline wastes (Atschinsk, Russia). The chemical analysis of the sample was accomplished by using an atomic absorption spectrophotometer or digital photometer and wet chemical methods (Table II). X-Ray diffraction studies on the sample were undertaken using a Philips X-Ray diffraction unit. The particle size distribution of the sample was measured by using a Sedigraph (5000D Micrometrics). The photographs of the samples were taken with an optical microscope (Olympus BH2-TMA).

Table II. Oxidic composition of wastes (mass - %)

No.	SiO_2	Al_2O_3	Fe_2O_3	MgO	Na_2O	CaO	K_2O
1	68.11	14.85	5.6	0.91	0.14	0.96	0.57
2	69	15	4.6	0.6	3.75	1.68	0.75
3	52	31	8	0	2.1	0	2.3
4	54.6	23	2.1	1.6	0.3	0.6	2.6

5	58.38	19	6.6	1.85	0.95	2.03	
6	64	14.4	6.7	1.1	1.13	2.14	1.27
7	57.03	18.57	7.37	2.22	1.04	2.16	2.88
8	64.88	12.51	5.13	2.29	1.63	5.99	2.1
9	65.07	16.48	6.08	1.18	0.74	1.18	1.66
10	51.62	27.97	3.92	1.09	1.59	0.76	1.59
11	14.76	26.94	37.6	1.01	5.83	2.06	0.34
12	57.21	20.02	1.44	0.94	0.55	1.96	3.2
13	53.43	32.07	5.82	2.06	0.05	2.14	0.19
14	59.1	25.64	7.1	1.28	0.06	2.27	0.85
15	50.12	29.78	5.04	1.00	0.04	0.45	0.7
16	43.02	27.68	3.09	0.6	11.62	2.04	9.34
17	6.86	35.9	0.9	4.39	0.15	48.9	0.06

The obtaining of additives from alumina wastes is based on a theoretical study of the system CaO - Al_2O_3 – SiO_2 - Fe_2O_3 [Figures 5].

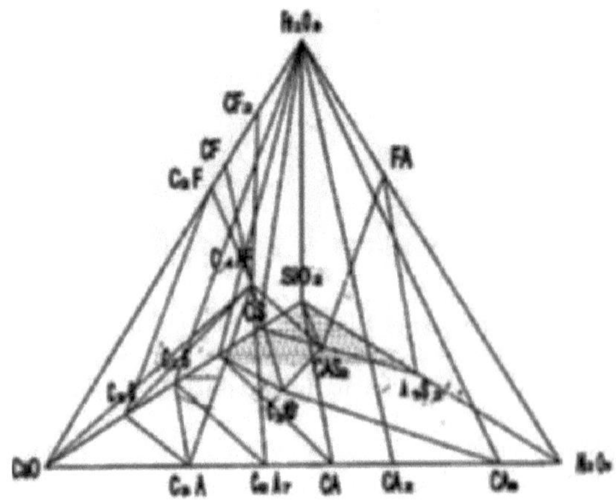

Figure 5. Optimum region of obtaining of special cement based on alumina wastes

It was found that pseudocuts containing double oxides and binder compounds (C_2S, C_3S, CA) have greater binder activity. The initial stage of hydration begins with leaching the surface atoms and active chemical adsorption of water molecules due to presence at active centers followed by their relation through forming OH hydroxide groups and a surfactant layer having specific surface area (to 2*10^6 sq.m/kg) and consisting of aluminum hydroxide. At the next stage the surfactant layer of double oxides adsorbs ions Ca+2. Chemical adsorption of the cations decelerates nucleation and promotes formation of epitaxy contacts on the surface of the double oxides as well as oriented growth of hydrate formations of hydroaluminates hydrosilicates of calcium. As a result of chemical adsorption process and oriented crystallization of hydrates around the double oxides grains there is formed a contact zone that is significantly higher than that of pure cement stone. This determines the high strength of the double oxides containing cement.

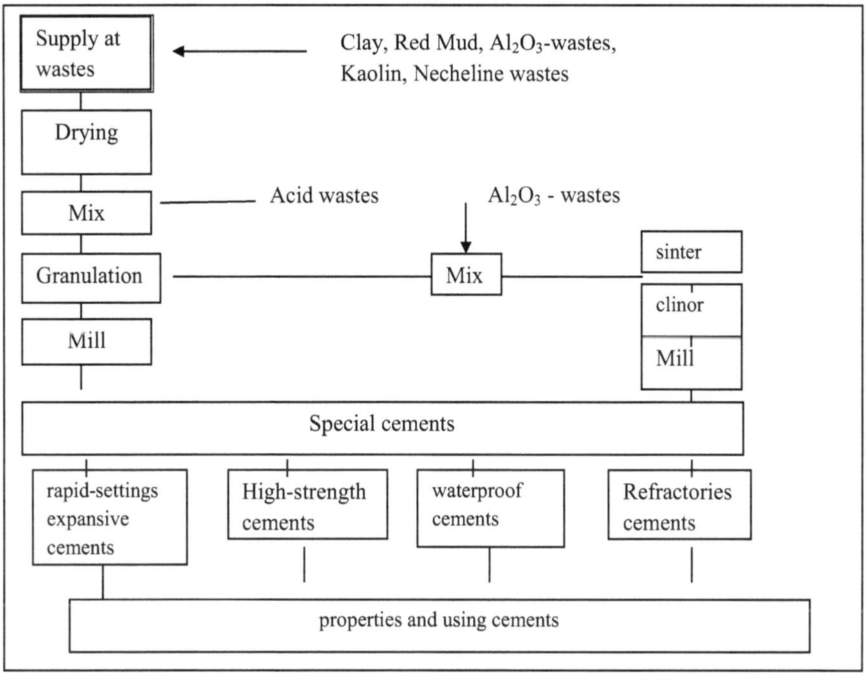

Figure 6. The scheme of the technology of recycling of wastes in alumina industry. Represents the technology of additives for cement industry

It is also reported that the results of the special cements hardening development is dependent on curing time duration, tempering water, amount, and hardening conditions. Compressive strength, mass and the volume variations, open porosity, and hydrates compositions were detected. The Blain's fineness test of powder was 3500 - 4000 cm^2/g, water-cement ratio of the normal thickness was 0.21, initial set was 15 min and final set was 25 min. The time dependence study of the cement hardening in the period from 1 to 28 days at relative humidity (w) 55+-20 % and at 21C was realized. It showed that after first day exposition the strength reached more than 50 MPa and after 28 days 80-90 MPa.

Hydration degree testing showed that at the final set, it was hydrated 25-30% of cement mass and after 4 hours 45-50%. After that during the period of 1 - 86 days the hydration degree was 50 - 55% and practically didn't change.

Refractories cements have compression strength from 40 MPa to 120 MPa, refractoriness, K, from 1873 to 2270. They are used for coating high temperature heat treatment items, in coating quartz glass tanks, in coating of furnaces for making special steels, H_2 - Furnaces, Petrochemistry reactors, and for making of crucibles used for melting of pure metals and alloys of rare-earth elements. Rapid-setting expansive cement distinguishes from similar materials by short setting time, stable expansion by the third day, high adhesion to rocks, improved resistance to aggressive media, water resistance during early hardening period, resistance to ground water, and handy application.

It is used in the construction of underground railways and mines, in geological prospecting and in the manufacture of special reinforced concrete structures.

Rapid-setting plugging cements features: short setting time at normal and high water/cement ratios: beginning within 5-12 minutes. Good mobility of cement paste during first minutes after mixing: spread 100-190mm. high strength characteristics during early hardening period: ultimate compression strength of specimens at the age of 4.24 hours and 28 days is 2-4 MPa; 9-12 MPa and 23-30 MPa, respectively, at w/c = 0.5.

High-strength cements are meant for production high-strength concretes of mark 600 and higher 700, 800, 1000, 1200, have high strength, water impermeability, and cold resistance.

Summary

It is possible to use the treatment wastes of industries as an additives for cement. Based on additives are developed special materials: high-strength, gas-waterproof, refractories, and rapid-setting expansive cements. Complex solution of the problem will be of great importance for purification of aerial, earth and water region in whole.

References

1. Braniski A.: Grundlegendle unterschiede zwischen calzium, stroncium - und Bariumzementem, - In: Zement- Kalk-Gips. 10(1969) - s.513-519

2. Ilyoukha N. Melnik M.: Refractory cements. - Buch(1984), Kijew, Verlag: Hochschule. S. 108-122.

3. Ilyoukha N., Timofeeva W.: Refractory cements. - in Proceeding of UNITECR'95 Congress, Japan, 1995.

4. Ilyoukha N., Timofeeva W.: Development of Zirconia Hydraulic Cements and their Application for Production refractory items. - in Proceeding of UNITECR'97 USA(1997 — P. 1089-1099.

5. Timofeeva W. Energiesaving technologie of cements based on additives of new generation. - in Proceeging of Academic. Kiev, 49, p. 3 00-3 22. – 2003

6. Timofeeva W., Ilyoukha N. Effect of alumosilicate synthetic additives on properties of heat resistant concrete. - in proceeding at International scientific and technological conference at Institute of refectories. - 2001. - p.24 – 28

7. Timofeeva W. Physicochemical grounds of strength structure formation of cements based on special additives. - in Noukoviy Visnyk Budivnytstva, Kharkiv. - 2003. - p.52 - 56

Energy Technology 2015: Carbon Dioxide Management and Other Technologies
Edited by: Animesh Jha, Cong Wang, Neale R. Neelameggham, Donna P. Guillen, Li Li, Cynthia K. Belt, Randolph Kirchain, Jeffrey S. Spangenberger, Frank Johnson, Andrew J. Gomes, Amit Pandey, and Peter Hosemann
TMS (The Minerals, Metals & Materials Society), 2015

EVALUATION OF THE SILVER RECOVERY FROM RADIOGRAPHIC FILMS IN A FILTER PRESS ELECTROCHEMICAL REACTOR

Pedro Alberto Ramirez Ortega[1], Victor Esteban Reyes Cruz[2], Laura Garcia Hernández[1], Diana Arenas Islas[1], Mizrraim Uriel Flores Guerrero[1] Luis Garcia Lechuga[1]

[1]Área Electromecánica Industrial, Universidad Tecnológica de Tulancingo
Camino a Ahuehuetitla No. 301 Colonia Las Presas, Tulancingo, Hidalgo, C.P. 43642, México
[2]Área Academica de Ciencias de la Tierra y Materiales, Universidad Autónoma del Estado de Hidalgo, Carretera Pachuca –Tulancingo Km. 4.5, Mineral de la Reforma, Hidalgo C.P. 42184, México

Keywords: Silver recovery, Radiographic films Industrial wastes, electrochemical reactor, A304 SS, DSA

Abstract

In this work the evaluation of a filter press type electrochemical reactor (REOV-01) with stainless steel (A304) electrode was realized, applied in the silver recovery from radiographic films. Macroelectrólisis studies were performed at different flow velocities with empirical and dimensionless correlations, and parameters such as: current efficiency, energy consumption Es and space-time yield. The results of the variation of the concentration of Ag at different flow linear velocities allowed the determination of the mass transfer coefficients of silver (obtained about the order of 10exp-5) for A304 SS electrode, which are consistent with the literature. The current efficiency values above 100% indicated that the electrochemical process is coupled to an electroless process. The results showed low energy consumption Es and high space-time yield values indicating that the reactor REOV-01 presents an excellent performance in the recovery of silver from radiographic films.

Introduction

In recent years, the industrial sector has shown an increase in the use of silver, due to its new applications in different areas. Nevertheless in sectors such as photography the consumption of silver has declined as a result of the rise of digital photography. However, the use of silver in radiographic applications has not shown greater variation (47.16 Moz), as the image quality is essential for diagnosis in various fields such as medicine. All the silver used in photographic applications is discarded completely (solid or in effluent). At the present, several countries have taken the task of recovering silver from industrial wastes, due to the economic value and environmental impact they represent. Their main sources of silver recovering are the wastes generated by the photographic sector. Various technologies have attempted to recover silver contained in these wastes; however, they have not fully satisfied with the requirements of maximum recovery, neither the environmental level (less than 5 ppm) of silver in effluents [1].

Nowadays, electrochemical reactors are used for various processes, including the recovery of precious metals. Among the most popular are the Electrocell AB, FM01-LC and FM21, used for studies in laboratory, pilot plant scale and industrial scale, respectively [2,3]. The reactor FM01-LC is a smaller version of the reactor FM21-SP (2100 cm2 electrode area) developed by ICI TM [4]. This type of reactor has shown its versatility, allowing the use of two or three-dimensional electrodes, membranes as well as baffles and turbulence promoters that allow a favored mass transfer [5,6]. This type of reactor is the best for the most of the electrochemical processes due to its simplicity of cell construction and the application of a variety of materials for its manufacture. Furthermore, the distributions of potential and current are reasonably uniform. Therefore, in this work the use of filter press type Electrochemical reactor (called REOV-01) was considered as a unit process for the recovery of silver present in radiographic films from the photographic industry. Previous studies [7] of silver recovery contained in the radiographic films, at current range from -40 to -125 mA, on the A304 SS and titanium electrodes, showed that both electrodes reach a current efficiency close to 100%. In these studies the values of current efficiency and space-time yield were better than those obtained in a previous work, indicating that this reactor provides an excellent performance in the silver recovery on A304 SS electrode.

Materials and Methods

The studies, here reported, used a filter press type electrochemical reactor (REOV-01), with a system of three electrodes (working, counter and reference). The capacity of the reactor is 280 mL. The working electrode was A304 SS (geometric area of 64.3 cm2), a saturated calomel electrode (SCE) as reference and a DSA mesh (Ti/RuO2) as counter electrode.
The hydraulic system was especially designed to work safely the nitrate solution (Figure 1) used to dissolve the silver of the radiographic films (SRF). A centrifugal Marathon Electric pump (1 HP of power) was used to maintain constant recirculation of the solution SRF. In this work the flow rate of the solution within the reactor REOV-01 was varied between 5 and 15 Lmin-1.

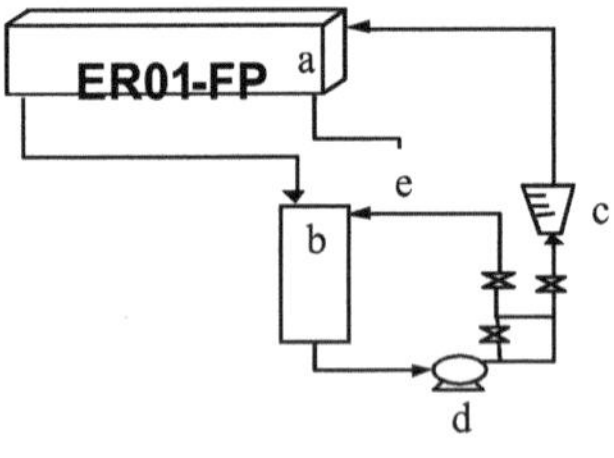

Figure 1. Flowchart of the electrorecovery process: (a) electrochemical reactor REOV-01; (b) electrolytic solution conteiners of 3 dm3; (c) flow meters for 0.5 – 4 dm3min-1; (d) centrifugal pumps of 1HP and (e) simple port.

The solution used for the studies was prepared in the laboratory with nitric acid solution 5%v/v and 250 g of radiographic films, giving an initial concentration of 2100 ppm Ag+. The macroelectrolysis studies were carried out to constant current during 120 minutes of electrolysis time, imposing a current of -40 mA on the A304 SS electrode. These current were selected from previous experiments (not showed in this work), with the purpose of avoiding hydrogen evolution. The electrochemical studies were carried out using a potentiostat-galvanostat PAR 263A connected to a power source KEPCO with capacity of 10 A. The used techniques were handled through the PowerSuit software provided by the same company.

Results and Materials and Methods

Figure 2 shows in logarithmic scale the decay of the normalized concentration of Ag+ of the solution SRF to different linear velocities, in function of electrolysis time for A304 SS electrode. The experiments used a current imposition of -40 mA on A304 SS electrode. It is important to mention that the experimental arrangement of the reactor and the electrolyte container differs from batch reactors in the presence of a recirculation device. The experimental setup considered in this work could be identified as a continuous reactor with recirculation, where the mathematical equation that relates the concentration variation with the flow, is very particular. However, considering that the solution remains in circulation and it changes with time, in addition that the volume of the cell is way smaller compared with the volume of the container makes the reactor behave like a batch reactor.

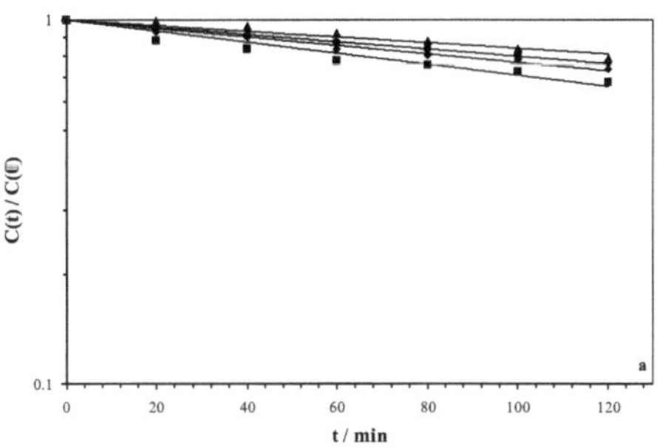

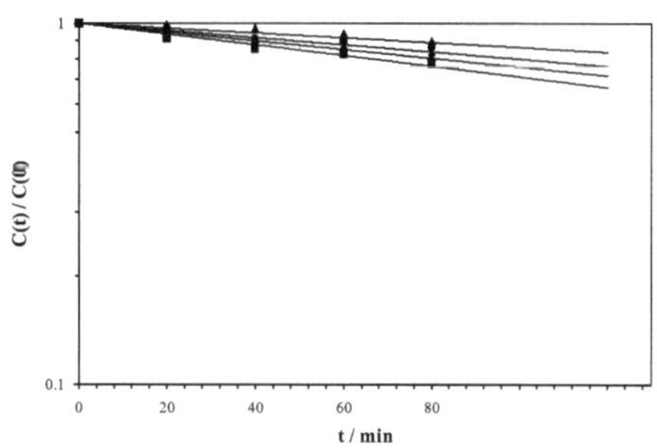

Figure 2. Log of the normalized concentration of Ag^+ of the solution SRF in function of electrolysis time for A304 SS electrode to current of -40 mA and flow linear velocity range of 5 a 15 $Lmin^1$

Figure 2 shows a linear behavior with the logarithm of the normalized concentration of Ag+, respect to the electrolysis time for the four flow velocities of the solution SRF on the A304 SS electrode. This behavior is consistent with the mathematical model of batch reactors and indicates that the reduction of silver is controlled by mass transport.

Under these conditions, the batch reactor equation to determine the mass transport coefficient is:

$$\ln\frac{C(t)}{C(0)} = -\frac{K_m A t}{V_T} \qquad (Ec.\,1)$$

Where C (t) is the concentration change respect to time, C (0) the initial concentration, Km is the mass transfer coefficient, A is the active area of electrode and V_T is the electrolyte volume.

From the value of the slope of each line (figure 2) and the equation Ec.1, the Km values for each electrode were calculated. Table I shows the Km values for the last three flow velocities imposed on the A304 SS electrode, considering that at these velocities the system is under an established flow regime. Table I indicates that the mass transfer coefficient (Km) for silver increases with flow linear velocities in A304 SS electrodes. Therefore, the results show the mass transfer coefficient to different flow linear velocity on electrode (2.55x10-5 ms-1 for the A304 SS).

Table I. Mass transfer coefficient (Km) for electroactive especies of Ag+ for A304 SS and Ti electrodes. The evaluation of Km is made considering an area of 37.4016 cm^2.

u (m/s)	**Km** (ms^{-1})
0.43	$1.68x10^{-5}$
0.54	$1.97x10^{-5}$
0.65	$2.55x10^{-5}$

The magnitude of the mass transfer coefficients Km (10-5 ms-1) obtained in this work is of the same order of magnitude to the ones obtained for the ferrocyanide in a reactor FM01 on Ni flat electrodes [8, 13, 14], indicating that this type of electrodes in filter press type electrochemical reactors (FM01), allows the adequate obtention of mass transfer coefficients of the present electroactive species between the electrode and solution. Generally is found that for a given electrode geometry, a certain electrolyte and under certain flow conditions, the Km is obtained through the following equation:

$$K_m = \alpha\, \upsilon^{\beta} \qquad (Ec.\ 2)$$

Where υ is the characteristic velocity (the electrolyte linear velocity or, the rotation velocity of a rotating electrode); the exponent β and the value of α are adjustment factors for the experimental data. For laminar flows the value of β is between 0.3 and 0.5 and higher β values are turbulent flows.

Figure 3 shows the behavior of log Km and log υ for electrode A304 SS on the SRF solution. In logarithmic scale, the mass transfer coefficient of Ag^+ has a linear increase with increasing flow linear velocity; also, with the last three flow rates, the mass transfer coefficients present the same order of magnitude. This indicates that the process is under an established flow regime. Experimental correlation expression is determined using the slopes in figure 3, which relates the mass transfer coefficient Km with the flow linear velocity υ and the equation Ec.2, where the β value indicates the type of flow regime in the reactor. The experimental correlations obtained for the A304 SS electrode is $Km = 3x10^{-5}\upsilon^{0.55}$:

The β values obtained indicate that the electrochemical reactor REOV-01 operates under a fully developed turbulent flow for A304 SS electrode.

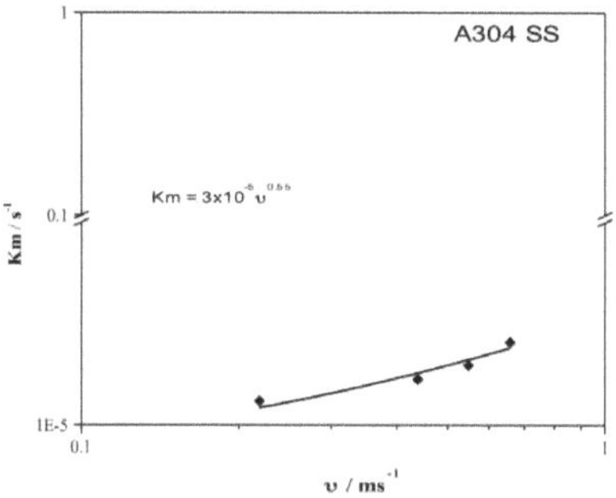

Figure 3. Variation of log Km Vs log υ to imposed a current of (a) -40 mA on A304 SS electrode, during an electrolysis time of 120 minutes.

Experimental correlations with dimensionless numbers Re-Sc-Sh for the silver recovery process in the reactor REOV-01.

Km values obtained from different flow rates (table I) are used to calculate the dimensionless numbers: Sherwood (Sh), Reynolds (Re) and Schmidt (Sc). The proposed correlation with dimensionless numbers is:

$$Sh = aRe^{b}Sc^{c} \qquad (Ec.\ 3)$$

Where **a,** is a constant of proportionality, **b** characterizes the flow regime and is assumed that **c** has a value of 1/3. The values of the constants **a,** **b** and **c** depend on both the geometry of the electrode and the flow conditions [8]. Table II shows the values of the physical properties of the solution used to obtain the corresponding correlations, Schmidt numbers (Sc) and the equivalent (hydraulic) diameter d_e. The diffusion coefficients values are from the literature [15] and the equivalent (hydraulic) diameter d_e of the cell was calculated using the following equation:

$$d_e = 2\,B\,S\,/\,(B + S) \qquad (Ec.\ 4)$$

Where S is the thickness of the electrode and B is the height of the electrode.

Table II. Physical properties of the SRF solution used to obtain the dimensionless numbers.

Physical propieties	A304 SS	Reference
Kinematic Viscosity	$0.01 cm^2s^{-1}$	(10,11)
Diffusion Coefficient of Ag (D_{Ag})	$2.26x10^{-5}cm^2s^{-1}$	(10)
Characteristic length (d_e)	1.58cm	This work

Figure 4 shows the behavior of the logarithm of the product ShSc-0.33 and the logarithm of Re on the A304 SS (curve a), Ti (curve b) an Ag (curve c) electrodes. It shows the relationship $ShSc^{-0.33}$ for both the electrodes (A304 and Ti) presenting a linear increase with increasing value of Re (curve b) in the last flow linear velocity. Using the slope of the curves of figure 4 and equation Ec.3 can be known the experimental correlations of dimensionless numbers for the A304 SS and Ti electrodes, these correlations are:

$$\text{A304} \qquad Sh = 0.0018Re^{1.024}Sc^{0.33} \qquad (Ec.\ 5)$$

$$\text{Ti} \qquad Sh = 0.0012Re^{1.163}Sc^{0.33} \qquad (Ec.\ 6)$$

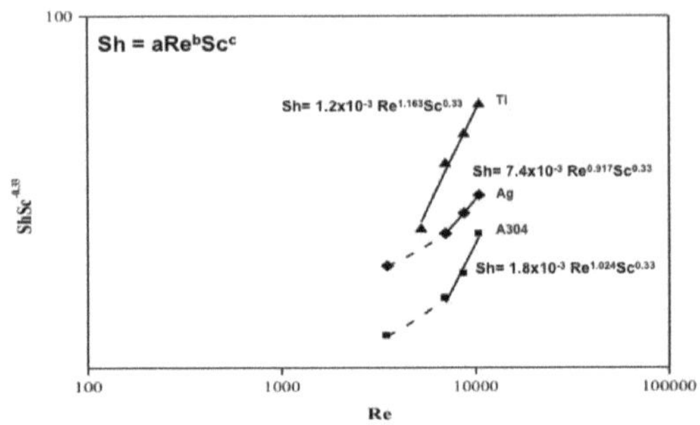

Figure 4. Variation of the log $ShSc^{-0.33}$ Vs log Re number for the silver recovery in an electrochemical reactor ER01-FP when imposing a current of -40 mA on A304 SS electrode and -65 mA on titanium electrode, during an electrolysis time of 120 minutes.

The values of the constant **a,** in both electrodes, are of the same order of magnitude. The same behavior occurs with the values of the constant b for the A304 (1024), which confirmed that the reactor ER01-FP operates under a fully developed turbulent flow.

Current efficiency ϕ^e

The values of current efficiency, ϕ^e, from the results of the macroelectrolysis studies with different flow linear velocity υ and maintaining a controlled current are shown in table III and were calculated using the equation Ec.7.

$$\phi^e = Wj \, / \, W_{total} \qquad (Ec.\ 7)$$

where Wj is the weight of metal j deposited and W_{total} is the weight of the deposit that should be obtained by imposing a determined current during certain time.
Table III shows the weight of metal j deposited increasing respect the flow linear velocity in A304 stainless steel electrode. On the other hand, for the A340 SS electrode, the current efficiencies obtained in the range of flow linear velocity studied are greater than 100%. This behavior in the current efficiency is attributed to the influence of the growth of silver nuclei on the surfaces of A304 SS electrode. The high current efficiency values indicated that the electrochemical process is coupled to an electroless process. This is confirmed according with the theoretical mass (W_{total}) that should be transformed by electrochemical process, which is less than the real mass (W_j) obtained from the surface during 120 minutes of electrolysis time.

Table III. Current efficiency ϕ^e respect the weight at different flow linear velocities imposed on A304 SS and titanium electrodes.

A304 SS			
υ (ms^{-1})	$W_{theoretical}$ (g)	W_{real} (g)	ϕ^e (%)
0.219	0.322	0.426	132.2
0.438	0.322	0.444	137.9
0.548	0.322	0.487	151.3
0.657	0.322	0.698	216.8

It's important to mention that the real mass has not the contribution of deposits from the previous macroelectrolysis, due to after each experiment the surfaces of the electrode were brushed and submitted to ultrasonic vibration during 15 minutes in order to maintain surfaces clean and free of traces. EDS revealed a purity of 99% of silver on the deposits obtained on the surface of A304 SS electrode. These results indicated that the electroless process variables affect the silver recovery. Although, the electroless process due to changing linear flow provides important information on the silver recovery, it is not the objective in this research.

Power consumption (Es) and space-time yield (ρ_{ST})

Using equations Ec.8 and Ec.9, the energy (Es) and space-time yield (ρ_{ST}) for silver recovery are obtained and they are presented in Table IV.

$$E_s = -(n\,F\,E_{cel} / \phi^e) \qquad (Ec.\ 8)$$

$$\rho_{ST} = \phi^e\, i\, M / n\, F\, V_R \qquad (Ec.\ 9)$$

Where n is the number of electrons, F is the Faraday's constant, E_{cell} is the cell potential, ϕ^e is the current efficiency, M is the atomic mass and V_R is the volume of the reactor.

Table IV. Energy consumption and space-time yield ρ_{ST} for silver obtained in the reactor REOV-01 when the flow linear velocity varies in the range of 0.219 to 0.657 ms^{-1} for A304 SS electrode and 0.328 to 0.657 ms^{-1} for the titanium electrode, electrolysis time of 120 minutes.

A304 SS		
i (mA)	E_s (kWh/Kg)	ρ_{ST} (Kg/sm^3)
-40	0.206	1.795×10^{-5}
-50	0.219	2.193×10^{-5}
-65	0.228	2.818×10^{-5}
-80	0.240	3.466×10^{-5}
-95	0.242	4.138×10^{-5}
-125	0.234	5.993×10^{-5}

Table IV shows the results of energy consumption and space-time yield ρ_{ST} for silver obtained in the reactor REOV-01 when the flow linear velocity varies in the range of 0.219 to 0.657 ms^{-1} for A304 SS electrode. It shows that energy consumption decreases with increasing of the flow linear velocity for electrode, while the space-time yield increases ρ_{ST}. The decrease in the consumption is due to decreased cell potential and increased current efficiency. On the other hand, the increase in space-time yield ρ_{ST} is due to the increase in current efficiency. This behavior is due to the surface changes with the silver deposit diminishing the cell potential and increasing the deposit rate.

The results show that the electrochemical reactor REOV-01 presents a good performance for the silver recovery from radiographic films on A304 stainless steel electrode, due to it complies with operational parameters of an electrochemical reactor, presents low energy consumption (about 10^{-2}), high space-time yields (about 10^{-5}) and high current efficiencies (over 100%).

Conclusions

Concentration decrease respect to the time obtained in this study indicated that the reactor REOV-01 behaves as a batch reactor. Furthermore, the Km values are of the order of 10^{-5} and agree with the literature [4], indicating that this type of reactor configuration with flat electrodes is a convenient way to obtain the Km. Through empirical and dimensionless equations was determined that the REOV-01 reactor operates under a fully developed turbulent flow. The similarity of the correlation coefficient **b** for both correlations (dimensionless and empirical) supports this fact. The correlations of dimensionless numbers obtained in this study can be used for future industrial scaling of the process of selective silver recovery from nitric solutions.

The current efficiency values were above 100% suggest that the electrochemical process is coupled to an electroless process. The magnitude of energy consumption, current efficiency and space-time yield meet the requirements necessary for the proper functioning of an electrochemical reactor, indicating the viability of using the reactor REOV-01 in the silver recovery from radiographic films.

Acknowledgments

The author Pedro A. Ramírez O, would like to express their gratitude to PROMEP for their economic support (136567).

References

1. Ramírez O.P. A, *Estudio electroquímico preliminar para depositar Ag proveniente de los desechos sólidos de la industria fotográfica y radiográfica.* Tesis de Licenciatura, UAEH, Hidalgo, México (2006).
2. K. Guenter, Silver Recycling from photographic solutions: review of the methods, Chemical Labor Betr. 32 (1981), p40-48.
3. M. I. Ismail, Electrochemical reactors, their science and technology, part A: fundamentals, electrolysers, batteries and fuel cells; Elseiver Science Publishers B. V., Ámsterdam, Holanda (1989).
4. A. A. Wragg y A. A. *Leontaritis*. Chem. Eng. J. 66 (1997)
5. D. Pletcher y F. C. Walsh. Industrial Electrochemistry, 2nd Edn., Chapman & Hall, London (1990).
6. D. A. McInnes, The Principles of electrochemistry. New York, USA (1990).
7. D. Pletcher. A First Course in Electrode Processes. RSC Publishing, UK (1991).
8. C. J. Brown, D. Pletcher, F. C. Walsh, J. K. Hammond y D. Robinson. Studies of Space-Averaged Mass Transport in the FM01-LC Laboratory Electrolyser. J. of Appl. Electrochem. 22 (1992) 631.
9. G. M. Zarkadas, A. Stergiou y G. Papanastasiou, Influence of citric acid on the silver electrodeposition from aqueous AgNO3 solutions, J. Electrochimica Acta 50 (2005) 5022-2031.

10. N. Nakiboglu, D. Toscali y I. Yasa, *Silver recovery from waste photographic films by enzematic method*, Turk J. Chem. 25 (2001) 349-353.

11. P. C. Holloway, K. P. Merriam y T. H. Etsell, *Nitric acid leaching of silver sulphide* precipitates, J. Hidrometallurgy 74 (2004) 213-220.

Energy Technology 2015: Carbon Dioxide Management and Other Technologies
Edited by: Animesh Jha, Cong Wang, Neale R. Neelameggham, Donna P. Guillen, Li Li, Cynthia K. Belt, Randolph Kirchain, Jeffrey S. Spangenberger, Frank Johnson, Andrew J. Gomes, Amit Pandey, and Peter Hosemann
TMS (The Minerals, Metals & Materials Society), 2015

LEACHING OF GOLD FROM PRINTED CIRCUIT BOARDS SCRAP OF MOBILE PHONES

Angela C. Kasper[1], Hugo M. Veit[1]

[1] LACOR–PPGE3M, Universidade Federal do Rio Grande do Sul, – Av. Bento Gonçalves, 9500, Setor 4, Prédio 74, 91501-970 Porto Alegre, Rio Grande do Sul, Brazil.
E-mail:angelakasper@globo.com

Keywords: Gold Leaching; Thiosulfate; Mobile Phone; Printed Circuit Boards

Abstract

Every year the natural gold reserves have declined worldwide. In parallel, has increased the demand for this metal, especially for use in the electrical and electronics industry. One way to minimize this problem would be the recovery of this metal contained in printed circuit boards (PCB) of mobiles phones that are discarded after use. In this work a commercial gold stripping (cyanide-based) and, alternatively, thiosulfate in ammoniacal medium, under several conditions of concentration, time and temperatures was tested to leaching gold. In leaching tests whole PCBs were used and the solutions obtained were analyzed by atomic absorption spectroscopy. The results show that, in the terms of efficiency, the commercial stripping is still the best alternative because with this product the extraction rate reached 88% of the gold present on the PCBs, while the extraction rates obtained with thiosulfate reached only 50%. However, the thiosulfate solutions have great environmental advantages.

Introduction

Data released by Gartner (Gartner, Inc.) shows that the volume of worldwide sales of mobile phones has exceeded 1.8 billion units in 2013, which represents an increase of 3.5% compared to 2012 [1]. Considering that the average time of use of a mobile phone is smaller than two years [2], it is verified that the amount of discarded handsets each year is very large.

These electronic handsets that reach the end-of-life may become an important source of secondary raw material for gold and other metals of high economic value. According Hagelüken & Corti (2010) the amount of gold on printed circuit boards (PCBs) of mobile phones can reach 300-350 g/ton. This means that the residues of these devices are significantly richer than gold ores which has, on average, 5 g/ton [3], which makes the recycling of these handsets very attractive from an economic standpoint. Besides, the economic benefits generated by recycling of these devices should be added to the environmental benefits, since each device may have more than 12 elements considered highly harmful to health and the environment, including cadmium, arsenic and lead [4, 5].

The extraction of gold ore (primary source) and recovery of this metal from WEEE (secondary source), by hydrometallurgical route, uses the cyanidation process. However, cyanide is a highly toxic component and therefore should be handled and disposed into the environment with care and after proper treatment.

Thus, this paper studied an alternative leaching to cyanide in the hydrometallurgical extraction of gold present in PCBs (printed circuit boards) from mobile phones. One of the leaching agents that have higher opportunities for industrial application is the thiosulfate solution in ammoniacal medium.

Compared to the cyanidation process, the thiosulfate process often has greater efficiency and versatility, combined with significantly less environmental impact [6]. Another advantage of thiosulfate leaching is the reduction of interference from other cations, such as copper, arsenic, antimony, zinc and nickel that in the conventional process, can consume cyanide, increasing the consumption of the reagents [7]. The disadvantages are the complexity of the process, the thiosulfate degradation, which increases the cost of reagents, and harmful effects of these degradation products on the leaching and recovery process [8].

The dissolution of gold by thiosulfate takes place by the formation of a complex aurothiosulfate [Au $(S_2O_3)_2^{3-}$]. In addition to thiosulfate, in this process, copper (II) sulfate and ammonium are used. Copper (II) sulfate acts as an oxidizing agent, catalyzing the reaction. The ammonium has the role of stabilizing copper II, through the formation of a cupric tetraamine complex [$Cu(NH_3)_4^{2+}$]. Furthermore, during this leaching process, a decomposition of thiosulfate can result in sulfur generation, that causes a passivation by gold sulfide formation. Therefore, the ammonium is also added to the leaching solution to prevent this formation [7,9, 10, 11, 12, 13].

The need to control several variables simultaneously (concentrations of ammonia, copper, thiosulfate, dissolved oxygen, besides temperature and pH) in the systems that utilize thiosulfate leaching make the process to be considered complex. However, the increased concern about the environmental problems caused by the use of cyanide in mining and the gold recovery from secondary sources made that which, from the 1990s, thiosulfate pass to be considered as a promising alternative for replacing the cyanide process [8].

Materials and methods

The printed circuit boards (PCB) of mobile phones used in this study (Figure 1) were obtained from defective or obsolete mobile phones collected in technical assistance stores.

Figure 1: Printed circuit boards of mobile phones

Initially, PCBs samples were leached with aqua regia to determine the gold content present in the boards. After, a commercial stripper (cyanide-based) was tested and the alternative leaching (the thiosulfate) were used.

In the aqua regia test a solid/liquid ratio of 1:20 was used, at a temperature of 70 (±2)°C for a leaching time of 2 hours. To commercial stripper (cyanide-base) a solid/liquid ratio 1:25, temperature of 30(± 2)°C and leaching time of 2 hours was used. In this work the results obtained with aqua regia were considered as a reference (ie 100% leaching of the gold).

In experiments conducted with thiosulfate, variations in the leaching time (24 hours and 4 hours) and the thiosulfate and ammonium concentrations were tested. In the 24 hour tests, 2M, 1M and 0.5M of thiosulfate with 2.5M and 1.5M of ammonium were studied. In the 4 hour tests, 0.2M, 0.15M and 0.1M of thiosulfate with 0.3M and 0.2M of ammonium were studied. The concentration of the copper sulfate was maintained at 10 mM in all experiments.

Furthermore, two different thiosulfate salts were tested, the sodium thiosulfate and the ammonium thiosulfate.

In the experiments of thiosulfate leaching the 1:25 solid/liquid ratio, temperature of 30(± 2)°C and pH ranging from 8.5 to 11.5 (as pH of the solution) was used.

After leaching experiments the samples were analyzed by Atomic Absorption Spectroscopy using a spectrometer SpectrAA 110/Varian.

Results and discussion

The results obtained with a commercial stripper (cyanide-based) show that the rate of leaching reached about 88% of gold, compared with aqua regia. This demonstrates the good performance of the commercial stripper, given that this rate of leaching was obtained in just 2 hours (tests with longer periods of leaching were performed but no significant increases were observed in the rate of leaching).

The results obtained in experiments using thiosulfate, as leaching agent, of gold present in PCBs of mobile phones will be shown and discussed below.

Sodium thiosulfate

In the figures 2 and 3 are shown the results of PCBs leaching with sodium thiosulfate. It can be observe that the results obtained in these experiments are consistent with the behavior described by Zhang (2004). According to Zhang the concentration of thiosulfate must be controlled in order to maintain an appropriate ratio of the ammonia concentration in the solution, so that copper can play its catalytic role [14].

The figure 2 shows the results obtained in the test with sodium thiosulfate from 24 hours of leaching. In these tests, leaching rates reached 50% of the gold of PCB for testing which used 1M and 2M thiosulfate with 2.5M ammonium and tests which used 0.5M thiosulfate with 1.5M ammonium.

The Figure 3 shows the results obtained in the test with sodium thiosulfate from 4 hours of leaching. In these experiments the best results were obtained using 0.1M thiosulfate and 0.3M ammonium, where was obtained leaching rates around 50% of the gold of PCBs.

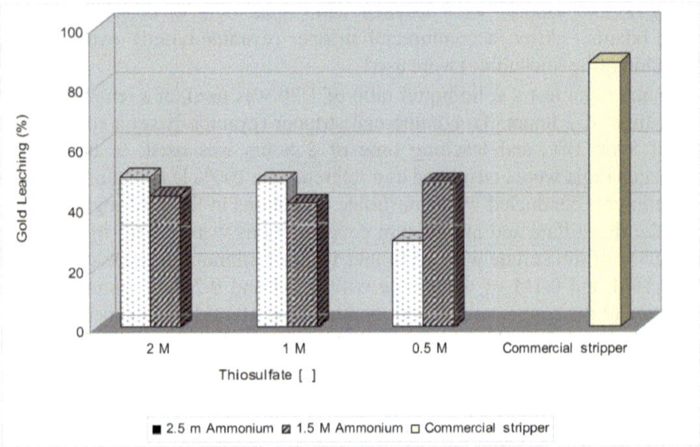

Figure 2: Results of PCBs leaching using sodium thiosulfate by 24 hours

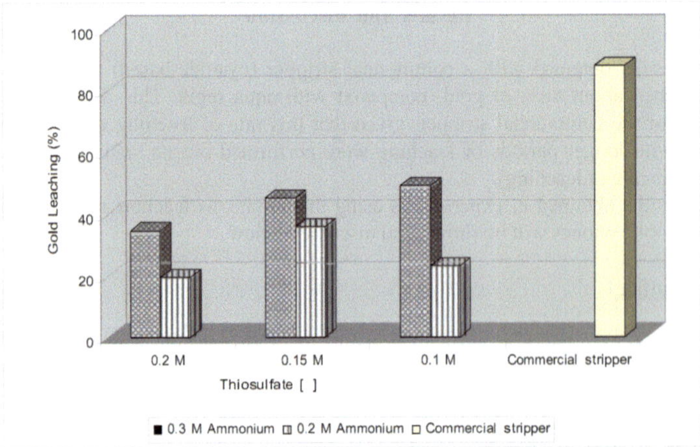

Figure 3: Results of PCBs leaching using sodium thiosulfate by 4 hours

In experiments with sodium thiosulfate as leaching agent of PCB it was observed that in both experiments (with longer time of leaching and higher concentrations of reagents and another with shorter time leaching and lower concentrations of reagents) leaching rates obtained were around 50% of the gold. These results indicate that the concentration of reactants would not be the determining factor for obtaining higher rates of gold leaching of PCB. A suitable ratio between the thiosulfate and ammonium concentrations is more important. However, the leaching rates obtained may still be considered low compared with those obtained with the commercial stripper. New experiments will be conducted in the future to prove this thesis. In these experiments may be tested, for example, longer periods of leaching using lower concentrations of reactants.

Ammonium thiosulfate

The figure 4 shows the results obtained in the test with ammonium thiosulfate from 24 hours of leaching. In this case the best results (a rate of 51% of gold leaching) were obtained when higher concentrations thiosulfate (2M) and of ammonium (2.5M) were used.

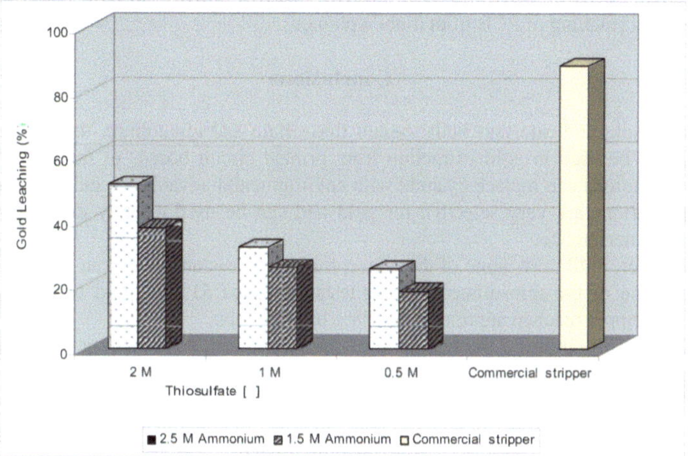

Figure 4: Results of PCBs leaching using ammonium thiosulfate in 24 hours

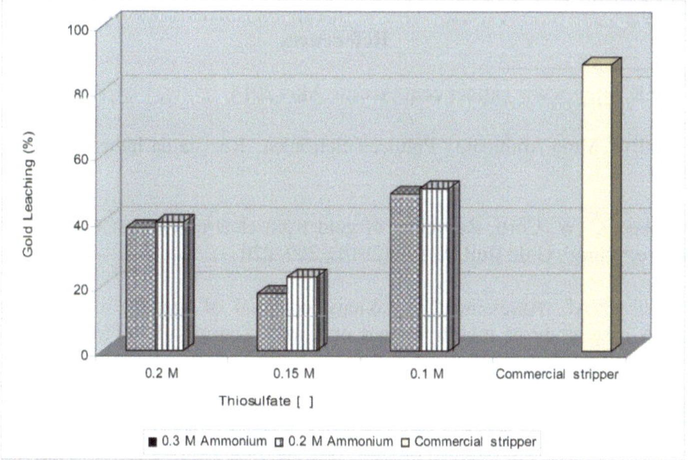

Figure 5: Results of PCBs leaching using ammonium thiosulfate in 4 hours

The figure 5 shows the results obtained in the test with ammonium thiosulfate from 4 hours of leaching. In this case, unlike of the results obtained in figure 4, the best results were obtained when lower concentrations of thiosulfate and ammonium were used.

The results obtained in experiments with the ammonium thiosulfate followed the pattern observed in experiments with sodium thiosulfate, where good results were obtained with low concentrations of reactants. In this case further experiments will also be performed to improve the rate of gold leaching of PCB from mobile phones.

Conclusions

The results indicate that both sodium thiosulfate and ammonium thiosulfate have good perspective to be used in gold extraction from printed circuit boards of mobile phones. These thiosulfate solutions can replace cyanide with environmental advantages and also due the fact of the leach solutions are very selective for gold and can be used directly for electrowinning of gold without interference.

However, until now, none of the concentration of reactants used can be considered ideal, given that none of the thiosulfate leaching tests exceeded 51% of gold extraction, while the commercial stripper reached approximately 88% of extraction.

These results indicate the necessity of further studies, with other reagent concentration, and/or test conditions (extraction time, temperature, pH, etc.).

Acknowledgements

The authors would like to thank FAPERGS, CAPES and CNPq for financial supports.

References

1. GARTNER, http://www.gartner.com. Access: May 2014.

2. A. Mawakdiye, Meio Ambiente - Poluição eletrônica. Revista da Indústria, 129 (7), (2007), 50-53.

3. C. Hageluken, C. W. Corti, Recycling of gold from electronics: cost effective use through 'Design for Recycling'. Gold Bull. 43 (3), (2010), 209–220.

4. B. Y. Wu et. Al., Assessment of toxicity potential of metallic elements in discarded electronics: A case study of mobile phones en China. Journal of Environmental Science, 20 (2008), 1403 –1408.

5. Angela C. Kasper et. Al., Printed wiring boards for mobile phones: Characterization and recycling of copper. Waste Management, 31 (2011), 2536–2545.

6. A. C. Grosse, et. al., Leaching and recovery of gold using ammoniacal thiosulfate leach liquors (a review). Hydrometallurgy, 69 (2003), 1 – 21.

7. M. G. Aylmore, D. M. Muir, Thiosulfate leaching of gold - A review. Minerals Engineering, 14 (2) (2001), 135-174.

8. X. Zhang, "The dissolution of gold colloids in aqueous thiosulfate solutions" (Ph.D. thesis Murdoch University, 2008).

9. J. Li et al. The ammoniacal thiosulfate system for precious metal recovery, XIX International Mineral Processing Congress, v.4, p.37, San Francisco, 1995.

10 Breuer, P. L., Jeffrey, M. I. Thiosulfate leaching kinetics of gold in the presence of copper and ammonia, Minerals Engineering, 13 (2000), p. 1071 .

11. G. Senanayake, Analysis of reaction kinetics, speciation and mechanism of gold leaching and thiosulfate oxidation by ammoniacal copper(II) solutions. Hydrometallurgy 75 (2004), 55–75.

12. S. Zhang, J. M. Nicol, An electrochemical study of the dissolution of gold in thiosulfate solutions. Part I. Alkaline solutions. J. Appl. Electrochem. 33 (2003), 767–775.

13. S. Zhang, J. M. Nicol, An electrochemical study of the dissolution of gold in thiosulfate solutions. Part II. Effect of copper. J. Appl. Electrochem. 35 (2005), 339–345.

14. S. Zhang, "Oxidation of refractory gold Concentrates and simultaneous Dissolution of gold in aerated Alkaline solutions" (Ph.D. thesis Murdoch University, 2004).

Energy Technology 2015
Carbon Dioxide Management and Other Technologies

Magnetic Materials for Energy Applications V

Energy Technology 2015
Carbon Dioxide Management and Other Technologies

MAGNETIC MATERIALS
FOR ENERGY APPLICATIONS V

Soft Magnetic Materials I

Session Chairs:

Paul Ohodnicki
Katie Jo Sunday

Energy Technology 2015: Carbon Dioxide Management and Other Technologies
Edited by: Animesh Jha, Cong Wang, Neale R. Neelameggham, Donna P. Guillen, Li Li, Cynthia K. Belt, Randolph Kirchain, Jeffrey S. Spangenberger, Frank Johnson, Andrew J. Gomes, Amit Pandey, and Peter Hosemann
TMS (The Minerals, Metals & Materials Society), 2015

EFFECT OF ELECTRIC CURRENT PULSE ON GRAIN BOUNDARY OF GRAIN ORIENTED SILICON STEEL DURING PRIMARY RECRYSTALLIZATION ANNEALING

Zheng Lu[1], Lijuan Li[1], Wenming Nan[1], Xiang Jiang[1], Qijie Zhai[1]

[1]Shanghai University;149 Yanchang Road;Shanghai 200072, P.R. China

Keywords: Grain-oriented silicon steel, Primary recrystallization, Grain boundary, Electrical current pulse

Abstract

Electrical current pulse（ECP）treatment is an advanced technology in material processing fields. The 70% rolling reduction cold-worked grain-oriented Fe-3%Si steel was selected as the experimental materials in this study. Electrical current pulse and furnace annealing were applied on the samples respectively. Microstructure of the treated samples were investigated by EBSD analysis. Based on the experimental results and comparison with the different treated samples, the effects of ECP on grain boundary evolution of grain-oriented silicon steel were studied during primary recrystallization annealing. The proportion of low angle grain boundary (2-15°) in specimens treated by ECP is reduced while the high energy grain boundary(>15°) is increased . And the proportion of high energy grain boundary is more than low angle grain boundary. Since the marked coincident site lattice (CSL) grain boundary is few around the Goss grains, the low sigma value CSL grain boundary can't promote the development of Goss grains during primary recrystallization.

Introduction

Grain-oriented silicon steel is one of the most important soft magnetic alloy, which is mainly used for manufacturing the iron core of transformer. With the excellent magnetic performance, the steel is widely used in the field of Electric power, electronics and military industry. According to statistics, power loss caused by loss core of silicon steel accounted for 2.5% -4.5% of annual generating capacity in China. In order to go with the current times of energy saving, it is necessary to improve magnetic properties of electrical steel as much as possible.

The main texture in the primary recrystallization is γ. While the main texture in finished oriented electrical steel is Goss, which goes after a process of abnormal grain grows called second

[1] Corresponding author. Tel./fax: +86-188-1826-4788. Email: neullj@163.com

recrystallization. Excellent magnetic properties of silicon steel owes to the highly concentrated Goss texture. The abnormal growth of Goss texture depends on appropriate γ texture. ECP is a fast non-equilibrium process, which causes large quantity of electron flow hit the nuclear to elevate the temperature of material. Electricity, heat, thermal stress will be instantly imposed on material, which changes the microstructure of the material[1]. Xu et al.[2] proposed that with the coupling of the thermal and electromigration effects, electropulsing assisted dislocation movement and vacancy diffusion. Jian et al.[3,4] thought that the rapid recrystallization process during EPT was proposed based on the enhancement of nucleation rate and atomic diffusion resulting from the coupling of the thermal and athermal effects. Hu et al.[5,6] showed that the electropulsing strengthened the recrystallization of cold-rolled Fe-3Si alloy strip and lowered recrystallization temperature. Thanks to these scholars' studies, a novel method was proposed to explore the way of improving quality of silicon steel strip products. The present work is to study the effect of ECP on grain boundary of grain-oriented silicon steel during primary recrystallization annealing.

Experimental process

Commercially oriented electrical steel is an industrially 0.3-mm thickness cold-rolled electrical steel strip which is rolled twice. Table.1 lists the steel's chemical composition. MnS and Cu2S are employed as inhibitors for the production of common grain-oriented electrical steel (CGO).

Table.1 Chemical composition(in wt%) of CGO used in the experiment

Element	C	Si	N	S	P	Als	Mn	Fe
Content	0.027	2.85	0.030	0.010	0.03	0.014	0.18	remain

The ends of polished oriented silicon steel were connected to the positive and negative of intermediate frequency pulse power(Fig.1) with copper and copper foil tape respectively. Center of the sample was welded with a thermocouple to control temperature. Specimens of this sheet (30×100mm2) were annealed in a peak current of 950A with the current density 1.06×108A/m2. The current frequency of the ECP was 200Hz and the time of ECP treatment was 300μs(Fig.2). After annealing in ECP, heating temperature curve of samples annealed without ECP participation were determined. Commercial annealing samples were heated and adjusted temperature to coincide with the heating curve.

The specimens were cut into 10 12mm size to do experiments in different treat methods(Table.2). They were additionally electrolytically polished to remove 10% total thickness in a solution of 80ml CH3OH and 20ml HCLO4 for the EBSD measurements. The mixed solution was kept in mixed ice water bath to stay 0℃ when polished. Microstructure were magnified by the APOLLO 300 scanning electric microscopic.

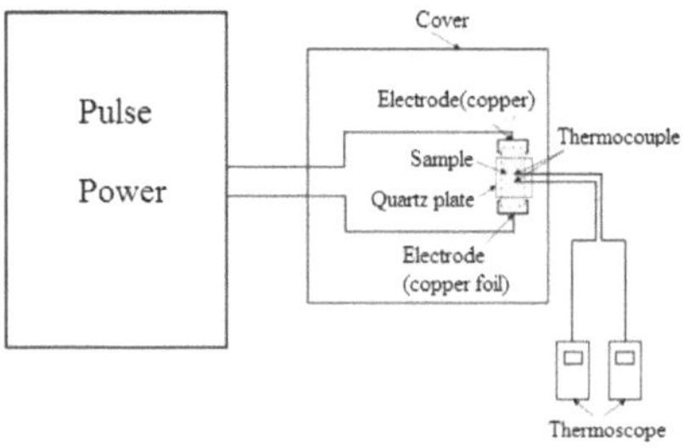

Fig. 1. Schematic of pulse current experimental installation.

Table. 2. Experimental craft

Specimens	Parameters			
	Temperature (℃)	Annealed time (min)	Peak current (A)	Frequency (Hz)
ECP-5	552	5	950	200
Furnace-5	553	5	-	-
ECP-7	604	7	950	200
Furnace-7	606	7	-	-
ECP-9	629	9	950	200
Furnace-9	634	9	-	-
ECP-12	645	12	950	200
Furnace-12	649	12	-	-
ECP-15	652	15	950	200
Furnace-15	654	15	-	-
ECP-20	656	20	950	200
Furnace-20	658	20	-	-

Results and discussion

Misorientation angle distribution

Misorientation angle distribution and relative percentage of different angle grain boundaries in specimens treated by furnace and ECP during crystallization(Fig. 2,Table. 2). Compared with samples annealed 5minutes in ECP , those specimens annealed without ECP have obvious high angle grain boundaries. Further heat treatment resulted in no distinctly difference between the annealing samples in 7min and 9 min. The content of low angle grain boundary in grain-oriented silicon steel decreased gradually after annealing in 12 minutes. While the high angle grain boundaries increased. It indicated that low angle grains transformed into high angle grains when deforming grains disappeared.

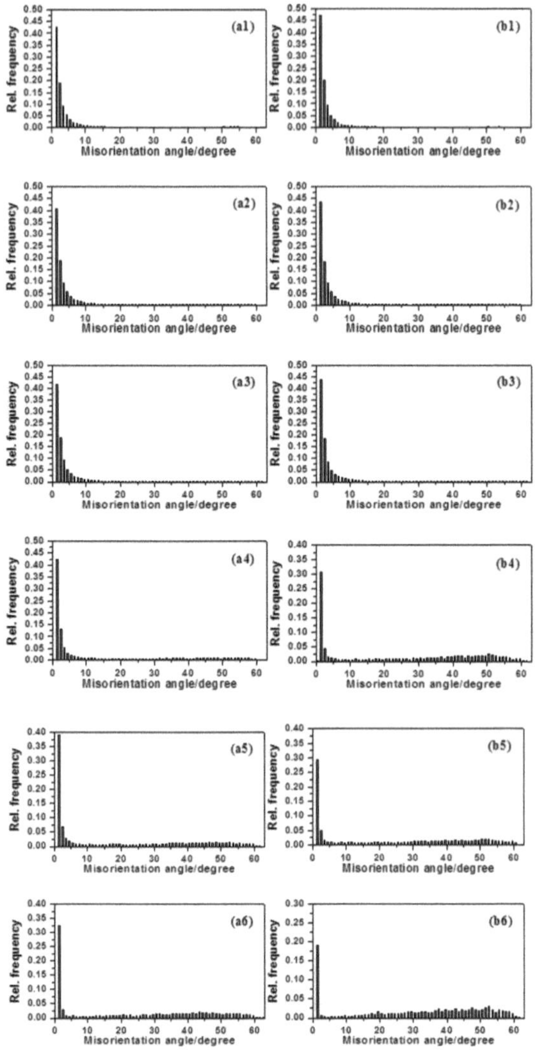

Fig. 2. Misorientation angle distribution of specimen treated by furnace and ECP (a1) Furnace-5; (a2) Furnace-7; (a3) Furnace-9; (a4) Furnace-12; (a5) Furnace-15; (a6) Furnace-20; (b1) ECP-5; (b2) ECP-7; (b3) ECP-9; (b4) ECP-12; (b5) ECP-15; (b6) ECP-20

Table. 3. The proportion of low angle grain boundaries and high angle grain boundaries in specimens treated by ECP and furnace.

Misorientation angle		5min	7min	9min	12min	15min	20min
Low angle grain boundaries(<15°)	ECP	92%	90.7%	88.4%	44.4%	43.9%	25.4%
	Furnace	88.8%	88.6%	88.2%	72.5%	56.7%	42.4%
High angle grain boundaries(>15°)	ECP	8.0%	9.3%	11.6%	55.6%	56.1%	74.6%
	Furnace	11.2%	11.4%	11.8%	27.5%	43.3%	57.6%

According to the statistics, the small angle grain boundaries' conversion rate of ECP annealing is faster than furnace. ECP annealing for more than 12 minutes resulted in distinctly less low angle grain boundaries and more high angle grain boundaries. Not until the 20th minute, the gap is even greater.

<u>Coincident site lattice(CSL) distribution</u>

CSL is used to describe the crystal lattice of orientation relationship between two grains, namely a grain rotates a certain angle around an axis angle to overlap with the adjacent grains. A large number of experiments indicated that CSL could change performance of materials such as improving ductility, impeding grain boundary fracture, improving the corrosion resistance performance and so on. Nowadays, the concept of Grain Boundary Engineering(GBE) is put forward to increase the content of unique grain boundaries in materials by heat treatment. CSL boundaries are usually in low energy state with less impurity segregation and high mobility, except for special CSL grain boundary types such as $\Sigma 1$ and $\Sigma 3$[8]. $\Sigma 3$ is twin grain boundary, whose migration rate is so low that can not move easily. For the body-centered cubic structure of oriented silicon steel, CSL grain boundaries $\Sigma 5$[7], $\Sigma 7$[8], $\Sigma 9$[9] conducive to the growth of Goss. Thus, here makes the Statistical comparisons on the total content of $\Sigma 5$, 7, 9.

According to the statistic and classification of the rotational axis in CSL ,$\Sigma<100>$ in low CSL index involves $\Sigma 5$,13a, 17a, 25a, 29a. $\Sigma<110>$ involves $\Sigma 9$, 11, 19a, 27a.$\Sigma<111>$ involves $\Sigma 3$, 7, 13b, 19b, 21a. It was shown on Table.4 that contents of CSL in specimens treated in two methods were almost not as much as 0.5 in the initial stage of recrystallization(5min to 9min). The CSL content elevates obviously in later recrystalliztion(9 min - 20 min).

Table. 4 Comparison of CSL boundaries between specimens treated by ECP and furnace.

Types of CSL grain boundary		5min	7min	9min	12min	15min	20min
Σ3	ECP	0.25	0.26	0.35	1.78	1.83	2.54
	Furnace	0.45	0.33	0.38	0.82	1.27	1.76
Σ 5+Σ7+Σ9	ECP	0.15	0.20	0.18	0.90	1.08	1.32
	Furnace	0.21	0.20	0.23	0.46	0.98	1.12
Σ<100>	ECP	0.05	0.03	0.04	0.20	0.17	0.32
	Furnace	0.02	0.03	0.03	0.09	0.29	0.43
Σ<110>	ECP	0.24	0.30	0.25	1.11	1.5	1.64
	Furnace	0.28	0.24	0.32	0.70	0.86	1.43
Σ<111>	ECP	0.61	0.46	0.54	1.19	1.95	2.41
	Furnace	0.33	0.36	0.50	2.43	2.54	3.56

The content of twin boundaries Σ3 in steels treated by ECP is slightly lower than the treated by furnace during 5-9 minutes. During the followed 12 to 20 minutes, the ECP involves much more value than the ordinary annealing. It indicates that ECP promotes the formation of twin boundaries.

The content of Σ5+Σ7+Σ9 is almost not change in the initial recrystallization. When treated by ECP, the specimens get more Σ5+Σ7+Σ9 in the later recrystallization. There is no Σ<100> in the initial recrystallization. In the middle stage, furnace annealing generates more Σ<100> than ECP . While annealed in the last stage, specimens treated by ECP promote the content of Σ<100>. It is similar for the Σ<110>. But for the Σ<111>, the situation seems to be opposite. At the beginning of recrystallization, the ECP has a larger number than the furnace. With the recrystallization continues, the ordinary annealing owns a larger number of Σ<111> boundaries. It is clear that ECP Inhibits the formation of Σ <111> grain boundaries.

Misorientation angle distribution around Goss grain

The data in Table.5 are calculated from the EBSD results like the Fig.5. The distribution of misorientation angle of grain boundaries in ordinary annealing accounts the maximum amount in 25-45° , the lesser in 45-60° and the least in 2-25°. While the misorientation angle in ECP experienced some changes. The 45-60° accounts the most, the 25-45° accounts the lesser and 2-25° is still in the minimum volume. Assuming distribution of the grain boundary is uniform, the proportion of grain boundaries around the Goss grain boundary which are in the range 25-45° would be in the maximum. While the fact is that the maximum in specimens treated by ECP is 45-60°.

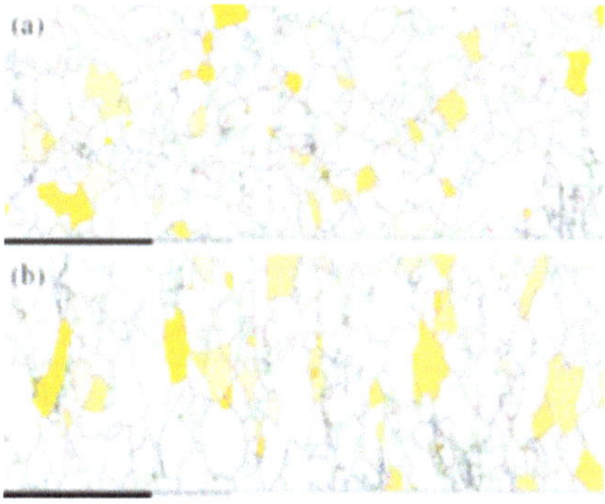

Fig. 5 Boundaries around Goss texture in specimens treated by furnace and ECP in 20 minutes.(a) Furnace-20; (b) ECP-20 Grain boundaries in different angles are marked in different colours:

— 2-15° — 15-25° — **25-45°** — 45-60° — {110}<001>

represents the Goss texture.

We can infer from the fact that ECP reduces the small-angle(2-15°) grain boundaries and improves the high-energy grain boundaries (>15°). It is known that the process of recrystallization is small-angle grain boundaries transform to large angle grain boundaries.

Table. 5 The content of boundaries in specimens treated by furnace and ECP

range	2-15°	15-25°	25-45°	45-60°
Furnace 20min	10.1%	8.6%	27.1%	21.7%
ECP 20min	6.2%	10.6%	23.8%	30.5%

Conclusion

The amount of small-angle boundaries in specimens treated by ECP is higher than ordinary annealing in initial recrystallization. At the same time, the amount of high angle boundaries in ECP has increased and is higher than the ordinary in later recrystallization. ECP has little effect on the development of CSL boundaries in initial recrystallization. But ECP promote the formation of $\Sigma3$ twin boundaries、$\Sigma5+\Sigma7+\Sigma9$ boundaries、Σ<100> and Σ<110> boundaries and reduce the Σ<111> content.ECP reduces the proportion of small boundaries（2-15°）and increase the high-energy boundaries (>15°) proportion, especially for the scale of 45-60°increased substantially.

Acknowledgments

The authors wish to express their thanks to Instrumental Analysis and Research Center of Shanghai University for EBSD technology, and Wuhan Iron and Steel (Group) Corporation for providing experimental materials.

References

1. LiuXiao ping et al, "Effect of Electric Current Pulse on Metal Materials," Research Studies on Foundry Equipment， "4 (2005) 51-54.
2. Qing Xu et al , "Thermal and electromigration effects of electropulsing on dynamic recrystallization in Mg-3Al-1Zn alloy," Materials Science and Engineering A, 528 (2011) 4431-4436.
3. Yanbin Jiang et al, "Influence of electropulsing treatment on microstructure and mechanical properties of cold-rolled Mg-9Al-1Zn alloy strip," Materials Science and Engineering A, 528 (2011) 5627-5635.
4. Yanbin Jiang et al , "Microstructure and texture evolution of the cold-rolled AZ91 magnesium alloy strip under electropulsing treatment," Journal of Alloys and Compounds, 509 (2011) 4308-4313.
5. Guoliang Hu et al, "Electropulsing induced texture evolution in the recrystallization of Fe-3 Pct Si alloy strip," Metallurgical and Materials Transactions A,42A (2011) 3484-3490.
6. Guoliang Hu et al, "Electropulsing induced G-texture evolution in a deformed Fe-3%Si alloy strip,"Journal of Materials Research, 26 (2011) 917-922.
7. J. Harase et al, "Texture evolution in the presence of precipitates in Fe-3%Si alloy," Acta metal mater, 39 (1991) 763-770.
8. J. Harase, "Texture evolution in the presence of precipitates in Fe-3% Si alloy, " Materials Science Forum, 94-96 (1992) 419-424.
9. Y. Yoshitomi, "Influence growth of primary recrystallized grains on secondary recrystallization texture in Fe-3% Si alloy, " Materials Science Forum, 113-115 (1993) 715-720

Energy Technology 2015
Carbon Dioxide Management and Other Technologies

MAGNETIC MATERIALS
FOR ENERGY APPLICATIONS V

Soft Magnetic Materials II

Session Chairs:

Thomas G. Woodcock
Orlando Rios

MAGNETOSTRICTION OF Co-Fe-BASED AMORPHOUS SOFT MAGNETIC MICROWIRES

A. Zhukov[1,2,3], M. Churyukanova[4], S. Kaloshkin[4],
V. Sudarchikova[4], S. Gudoshnikov[4],M. Ipatov[1,3], A. Talaat[1,3], J.M. Blanco[3] and V. Zhukova[1,3]

[1]Dpto. de Fís. Mater., UPV/EHU San Sebastián 20018, Spain
[2] IKERBASQUE, Basque Foundation for Science, 48011 Bilbao, Spain
[3]Dpto. de Física Aplicada, EUPDS, UPV/EHU, 20018, San Sebastian, Spain
[4]National University of Science and Technology «MISIS», Moscow, 119049, Russia

Keywords: Magnetostriction coefficient, magnetoelastic anisotropy, glass-coated microwires

Abstract

We present the results on comparative studies of soft magnetic properties and magnetostriction coefficient of Co-Fe-rich microwires. We found that the hysteresis loop as well as the magnetostriction coefficient of Co and Fe-rich microwires depend on internal stresses and can be tailored by annealing. Varying the time and the temperature of annealing we observed changes of the character of hysteresis loops. In the case of Co-rich microwires drastic changes of the hysteresis loop are correlated with change of the sign of the magnetostriction coefficient.

Introduction

Amorphous magnetically soft glass-coated microwires attract considerable attention of researchers and engineers owing to excellent soft magnetic properties, GMI (giant magnetoimpedance) effect and fast domain wall propagation suitable for prospective applications related to electronic surveillance, magnetic field sensors using the GMI effect and magnetic memories [1-3]. It is widely accepted that the character of hysteresis loop is closely related to the magnetic anisotropy and domain structure of amorphous microwires [1,2,4]. Domain structure of such microwires is defined by distribution and value of the internal stresses inside metallic nucleus of microwire and by the value and sign of the magnetostriction coefficient [1,2,4]. The origin of the internal stresses is related to the fabrication method of glass-coated microwires involving rapid solidification of composite microwire consisting of metallic nucleus surrounded by the glass coating from the melt [5-7]. Moreover the strength of the internal stresses inside the metallic nucleus is determined by the ρ-ratio between the metallic nucleus diameter, d, and total microwire diameter, D ($\rho=d/D$) [5-7] increasing with decreasing the ρ-ratio. Therefore changing the ρ-ratio modifying fabrication parameters we can tailor the strength on internal stresses and hence the magnetoelastic anisotropy and various magnetic properties like switching field which depends on internal stresses [1, 2].

Generally microwires with positive magnetostriction coefficient presents magnetically bistable character, while with negative magnetostriction coefficient exhibit linear hysteresis loops with low coercivity [1-5]. In general for conventional amorphous materials is it assumed that the magnetostriction coefficient depends on the composition of the metallic alloy [8]. But recent studies pointed out that the internal stress dependence of the magnetostriction coefficient of glass-coated microwires must be taken into account especially for the compositions with vanishing magnetostriction coefficient [9,10].

As earlier has been demonstrated presence of strong internal stresses can affect the magnetostriction coefficient due to previously reported stress dependence of magnetostriction [11]. This dependence must be relevant for low-magnetostrictive composition [9, 10].

Consequently measurements of the magnetostriction coefficient and understanding of mechanisms of the influence of internal and applied stresses on the magnetostriction coefficient is essentially important for tailoring of the magnetic properties of amorphous ferromagnetic microwires.

The most useful method for evaluation of the magnetostriction coefficient of amorphous materials is so-called small angle magnetization rotation (SAMR) method introduced in 1980-th [12, 13]. Initially this method has been developed only for amorphous materials where the magnetization rotation presents the determining role [4, 12- 15]. Consequently first attempt to use this method for evaluation of the magnetostriction coefficient of glass-coated microwires have been employed only for the for Co-rich glass-coated microwires [14,15]. Afterwards the possibility to extend SAMR method for the case of Fe-rich microwires presenting important contribution of domain wall propagation has been demonstrated [16,17]. The magnetic permeability corresponding to the magnetization rotation of the outer domain shell of Fe-rich microwires is rather low. Consequently the accuracy for the magnetostriction coefficient evaluation using SAMR method in Fe-rich microwires is lower. Recently novel experimental set-up with improved resolution has been successfully realized for measurements of the magnetosrtriction coefficient, λ_s, of any metallic nucleus composition (either Fe-rich or Co-rich) microwires [18].

Table 1. Compositions and properties of studied microwires

No	Metallic nucleus composition	d, μm	D, μm	$\rho = d/D$	$\lambda_s \cdot 10^6$
1	$Fe_{47.42}Ni_{26.6}Si_{11}B_{12.99}C_{1.99}$	29	32.2	0.9	24
2	$Fe_{6.1}Co_{57}Ni_{10}B_{15.9}Si_{11}$	20.4	24	0.85	-0.6
3	$Fe_{6.1}Co_{57}Ni_{10}B_{15.9}Si_{11}$	30	32.2	0.93	-0.5
4	$Fe_{6.1}Co_{57}Ni_{10}B_{15.9}Si_{11}$	22.8	25.6	0.89	-1.2
5	$Fe_{3.9}Co_{67}Ni_{1.5}B_{11.5}Si_{14.5}Mo_{1.6}$	25.6	26.6	0.96	-0,1
6	$Fe_{3.83}Co_{66.94}Ni_{1.44}B_{11.51}Si_{14.59}Mo_{1.69}$	19.4	22	0.88	-0,1

Consequently we present our last results on comparative studies of the evolution of the magnetostriction coefficient values and hysteretic properties of Co-Fe rich microwires.

We tried to understand the effect of internal stresses relaxation induced by annealing on the magnetostriction coefficient of amorphous ferromagnetic FeCoNi, FeCo, Fe- based glass coated microwires.

Experimental technique

We studied FeCoNi, FeCo, FeNi- based glass-coated microwires produced by Taylor-Ulitovsky technique [1, 2]. The microwires compositions, metallic nucleus diameters, *d*, and ρ-ratios are presented in the Table 1.

Magnetic properties of the samples were measured by induction method described elsewhere [9,10].

The magnetostriction coefficient of studied microwires has been measured using aforementioned SAMR method. During the measurements the initial tensile stress, σ_0, is created by the weigh m_0 attached to the microwire end. Axial magnetic field, H_z, is created along the microwire z -axis using the solenoid. In this method the sample is saturated by an axial magnetic field, H_z, while applying simultaneously a small *ac* transverse field, H_y, created by an *ac* electric current flowing along the sample. The combination of these fields leads to a reversible rotation of the magnetization within a small angle, θ, out of the axial direction. The induction voltage, $V(2\omega)$, due to the magnetization rotation is detected by a pick-up coil wound around the microwire. The magnetostriction constant is determined from the measurement of the dependence on axial magnetic field, H_z, versus on applied stress σ at fixed value of induction voltage $V(2\omega)$ in according to the expression (1). The $\mu_o M_s$ values of the investigated microwires obtained from room temperature measurements of the magnetization curves at high magnetic field. The AC current value, $i_{\sim}$, flowing through the wire is selected to avoid possible Joule heating of the sample: the current amplitude does not exceed 10-30 mA.

As has been shown [18,19] under conditions θ=*const* and H_f=const the value of magnetostriction coefficient, λ_s, is given by

$$\lambda_s = -(\mu_o M_s/3)(dH/d\sigma), \quad (1)$$

where $\mu_o M_s$ is the saturation magnetization.

Additional details on SAMR method and set-up features designed for evaluation of magnetostriction coefficient, λ_s, can be found elsewhere [12-18].

For every studied sample of microwire we measured the magnetostriction coefficient several times and then found the average value.

Additionally we removed mechanically the glass-coating in $Fe_{3.83}Co_{66.94}Ni_{1.44}B_{11.51}Si_{14.59}Mo_{1.69}$ and $Fe_{47.42}Ni_{26.6}Si_{11}B_{12.99}C_{1.99}$ microwires in order to evaluate the effect of internal stresses induced by the glass caoting on the magnetostriction coefficient value.

Results and discussion

Using described above SAMR method we evaluated the saturation magnetostriction values λ_s from the experimental dependences $[\Delta H_z/\Delta\sigma]_{V(2\omega)\text{-}const}$ using equation (1). The example of experimentally observed dependences of change of ΔH_z on applied stress, $\Delta\sigma$, for $Fe_{47.42}Ni_{26.6}Si_{11}B_{12.99}C_{1.99}$ microwire is shown in Fig.1.

As expected studied microwires in as-prepared state exhibit completely different hysteresis loops: Fe-rich sample with positive magnetostriction constant shows rectangular hysteresis loops, while Co-rich sample with low and negative magnetostriction constant exhibits inclined hysteresis loop with low coercivity (see Fig.2).

As can be appreciated form the table 1 and from Fig.3, the magnetostriction coefficient values increase with increasing of the Fe content. This compositional dependence is consistent with previous experimental results reported for amorphous Fe-Co- based alloys [4,20].

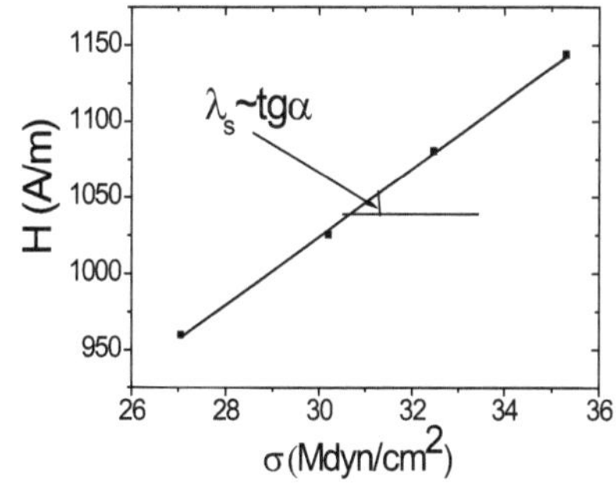

Fig.1. Experimentally measured $\Delta H_z(\Delta\sigma)$ dependence for the sample 6.

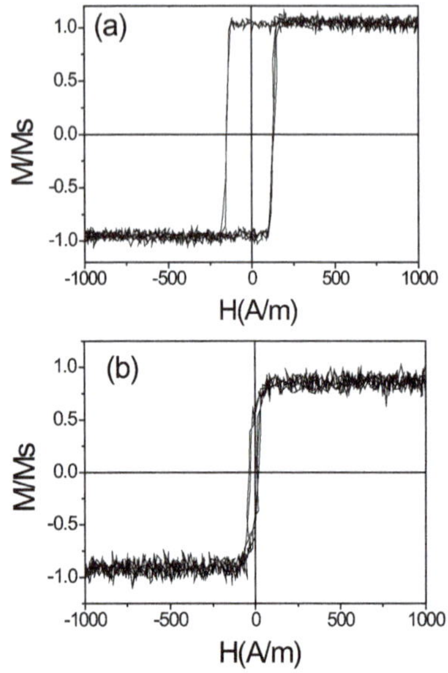

Fig. 2. Hysteresis loops of as-prepared microwires with different composition: $Fe_{47.42}Ni_{26.6}Si_{11}B_{12.99}C_{1.99}$ (a), $Fe_{3.83}Co_{66.94}Ni_{1.44}B_{11.51}Si_{14.59}Mo_{1.69}$ (b).

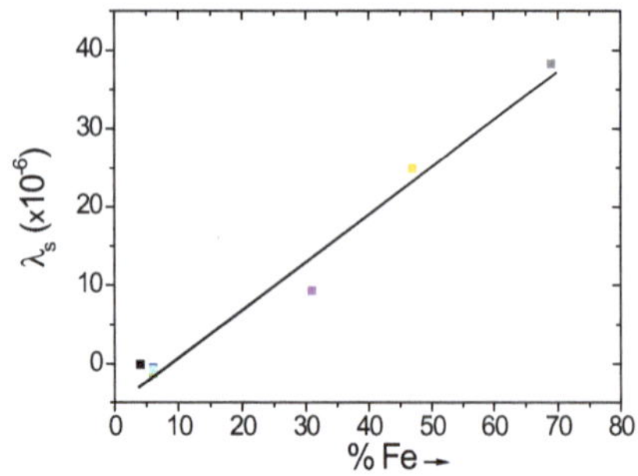

Fig.3. Effect of Fe content on λ_s–values of Fe-Co based as-prepared microwire.

On the other hand the peculiarity of studied materials is the existence of non-magnetic glass coating inducing strong internal stresses. As can be appreciated from Fig.4 (where dependence of λ_s -values on measurements number) substantial increasing of λ_s after glass coating removal is observed.

On the other hand microwire with the same chemical composition of metallic nucleus but with different ρ-ratio exhibit considerable effect of the ρ–ratio on hysteresis loops. As can be appreciated from Fig. 5a all $Fe_{6.1}Co_{57}Ni_{10}B_{15.9}Si_{11}$ microwires with the same chemical composition of the metallic nucleus but different ratio exhibit low coercivity and liner hysteresis loops. But the magnetic anisotropy field, H_k, decreases with increasing of the ρ-ratio. This increasing of the ρ-ratio correlates with decreasing of the magnetostriction coefficient value λ_s. This influence of ρ-ratio on magnetostriction coefficient value λ_s (Fig.5b) as well as the effect of the glass removal on λ_s–values must be attributed to the effect of internal stresses

Aforementioned stress dependence of the magnetostriction coefficient must be associated with both applied, σ_{appl}, and internal, σ_i, stresses ($\sigma = \sigma_{appl} + \sigma_i$).

Stress dependence of the magnetostriction has been expressed as:

$$\lambda_{s,\sigma} = \lambda_{s,0} - B\sigma \quad (2)$$

where $\lambda_{s,\sigma}$ is the magnetostriction coefficient under stress, $\lambda_{s,0}$ is the zero-stress magnetostriction coefficient and B is a positive coefficient of order 10^{-10} MPa and σ– stresses [11].

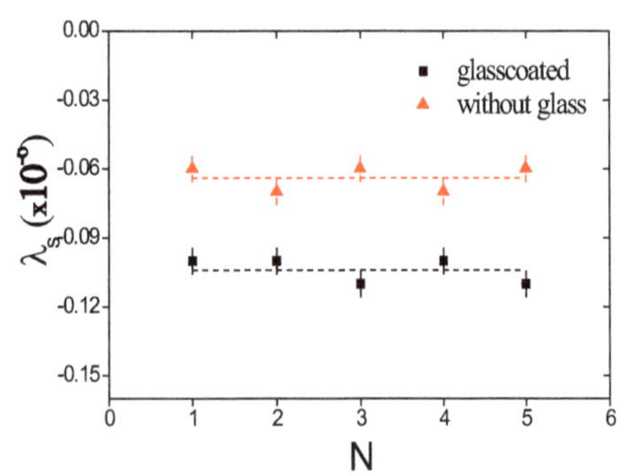

Fig.4. Effect of glass removal on λ_s–values of $Fe_{3.83}Co_{66.94}Ni_{1.44}B_{11.51}Si_{14.59}Mo_{1.69}$ as-prepared microwire.

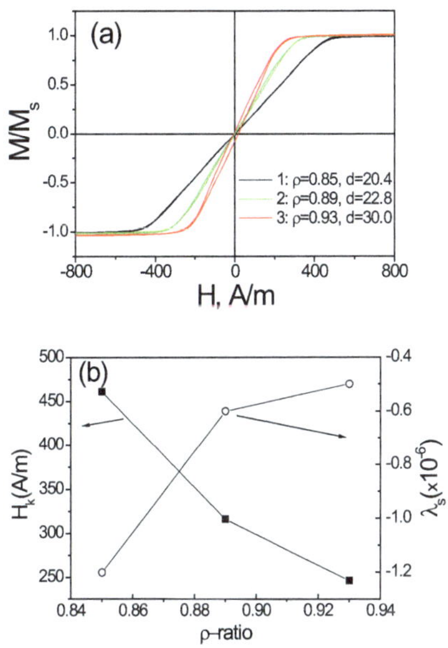

Fig. 5 Effect of ρ-ratio on hysteresis loops (a) magnetic anisotropy field, H_k, and magnetostriction coefficient, λ_s, of $Fe_{6.1}Co_{57}Ni_{10}B_{15.9}Si_{11}$ (b) microwires.

Consequently changes of the hysteresis loop shape from almost linear and inclined (typical for Co-rich compositions with negative magnetostriction coefficient) to almost rectangular (typical for microwires with positive magnetostriction coefficient, see Fig.7) must be related not only with the stress relaxation itself but also with the magnetostriction coefficient sign.

In $Fe_{47.42}Ni_{26.6}Si_{11}B_{12.99}C_{1.99}$ microwire we observed significant decreasing of the coercivity, H_c, after annealing (Fig.7). This H_c decreasing with increasing of annealing time correlates with decreasing of λ_s with annealing time (Fig.5a).

As already discussed before [9,10] in low-magnetostrictive compositions (with $\lambda_{s,0} \approx 10^{-7}$) and internal stresses of the order of 1000 MPa the second term of eq. (2) is almost of the same order as the first one. Therefore (as we observed in Fig.5) drastic changes of the magnetostriction coefficient after stress relaxation induced by

Consequently from obtained results we can conclude that the internal stresses induced by the glass coating affect the magnetostriction constant.

Recently we reported on drastic change of the hysteresis loops of Co-rich microwires after annealing [9,10]. Such strong effect of even very short and relatively low-temperature annealing on hysteresis loop we attributed to the stress relaxation. Below we present comparative data on effect of annealing on both hysteresis loop and λ_s –values. As can be appreciated from Fig.6, annealing considerable affect the magnetostriction coefficient of studied Fe-Co-based alloys. In the case of $Fe_{3.83}Co_{66.94}Ni_{1.44}B_{11.51}Si_{14.59}Mo_{1.69}$ microwire after annealing for more than 4 min the magnetostriction coefficient, λ_s, changes the sign from low negative to positive (Fig.6).

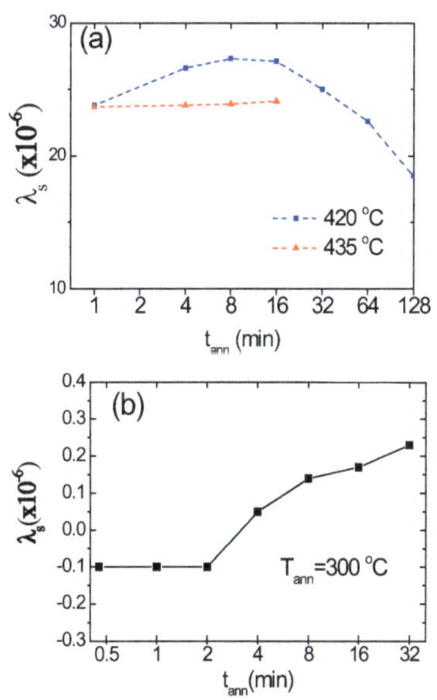

Fig. 6 Effect of annealing conditions on magnetostriction, λ_s of $Fe_{47.42}Ni_{26.6}Si_{11}B_{12.99}C_{1.99}$ (a), $Fe_{3.83}Co_{66.94}Ni_{1.44}B_{11.51}Si_{14.59}Mo_{1.69}$ (b) microwires

annealing can be expected. This can explain the observed changes of the hysteresis loops after annealing (Fig 7).

As discussed elsewhere [2,17], the magnetoelastic energy, K_{me}, plays the determining role in the formation of magnetic properties of amorphous microwires. K_{me} depends on the magnetiostriction coefficient, λ_s, as well as on the total stresses, $\sigma = \sigma_{appl} + \sigma_i$, ($\sigma_{appl}$ and σ_i are applied and internal stresses respectively) and given by [1,2,9]:

$$K_{me} \approx 3/2\ \lambda_s \sigma_i, \qquad (3)$$

The magnetostriction coefficient, λ_s, depends on the chemical composition of amorphous metallic alloy as well as on stresses (as we discussed above).

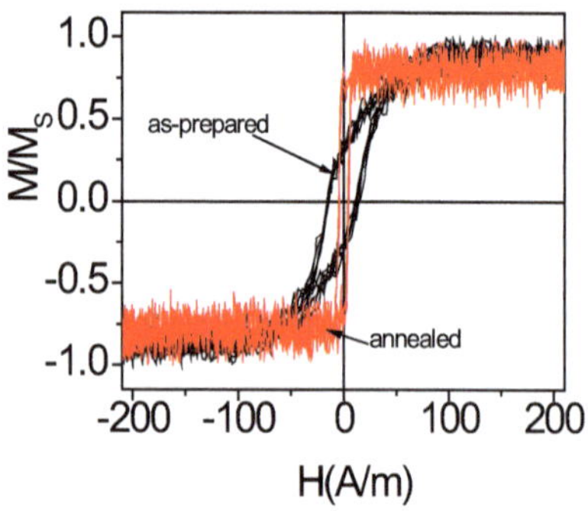

Fig.7. Hysteresis loops of as-prepared and annealed at 300 C for 60 min $Fe_{3.85}Co_{67.05}Ni_{1.44}B_{11.53}Si_{14.47}Mo_{1.66}$

Considering two contributions in the magnetoelastic energy given by eq. (3) we can assume that if the magnetostriction coefficient is low and negative we must consider two opposite effects of internal stresses. The first contribution is increasing of the total magnetoelastic energy (as usually discussed elsewhere, see [1]). The second one must be related to the stress dependence the magnetostriction coefficient described by eq. (2). This contribution is especially relevant for the case of the low zero-stress magnetostriction coefficient, $\lambda_{s,0}$. In according to the eq. (2) the magnetostriction constant under stress, $\lambda_{s,\sigma}$, must decrease.

Conclusions

We present the results on comparative studies of soft magnetic properties and magnetostriction coefficient of Co-Fe-rich microwires. We found that the hysteresis loop as well as the magnetostriction coefficient of Co and Fe-rich microwires depend on internal stresses and can be tailored by annealing. Varying the time and the temperature of annealing we observed changes of the character of hysteresis loops. In the case of Co-rich microwires drastic changes of the hysteresis loop are correlated with change of the sign of the magnetostriction coefficient.

Acknowledgements

This work was supported by Spanish MINECO under MAT2013-47231-C2-1-P, by the Basque Government under Saiotek 13 PROMAGMI (S-PE13UN014) and DURADMAG (S-PE13UN007) projects and by the Russian Ministry of Education and Science, the application № 2014-14-588-0010-8925. Technical and human support provided by SGIker (UPV/EHU, MICINN, GV/EJ, ERDF and ESF) is gratefully acknowledged.

References

[1] M. Vazquez, H. Chiriac, A. Zhukov, L. Panina, and T. Uchiyama, On the state-of-the-art in magnetic microwires and expected trends for scientific and technological studies, Phys. Status Solidi A, Vol. 208, (2011) 493-501.

[2] A. Zhukov and V. Zhukova, Magnetic Sensors and Applications Based on Thin Magnetically Soft Wires with Tuneable Magnetic Properties, International Frequency Sensor Association Publishing, Barcelona, 2013, 164p.

[3] F.Qin, H.-X. Peng, Ferromagnetic microwires enabled multifunctional composite materials, Progress in Materials Science Vol. 58 (2013) 183–259.
[4] A. Zhukov, V. Zhukova, J. M. Blanco, A. F. Cobeño, M. Vazquez, J. Gonzalez, Magnetostriction in glass-coated magnetic microwires J. Magn. Magn. Mater. Vol. 258-259, (2003) 151-157.
[5] H. Chiriac, T-A. Ovari, A. Zhukov, Magnetoelastic anisotropy of amorphous microwires, J. Magn. Magn. Mater. vol. 254–255 (2003) 469–471.
[6] J. Velázquez, M. Vazquez, A. Zhukov, Magnetoelastic anisotropy distribution in glass-coated microwires, J. Mater. Res. Vol. 11, (1996) 2499-2505.
[7] A. S. Antonov, V. T. Borisov, O. V. Borisov, A. F. Prokoshin, N. A. Usov, Residual quenching stresses in glass-coated amorphous ferromagnetic microwires, J. Phys. D: Appl. Phys. Vol. 33, (2000) 1161-1168.
[8] Y. Konno, K. Mohri, Magnetostriction measurements for amorphous wires, IEEE Trans Magn Vol. 25 No.5 (1989) 3623-3625.
[9] A. Zhukov, A. Talaat, M. Ipatov, J.M. Blanco, V. Zhukova, Tailoring of magnetic properties and GMI effect of Co-rich amorphous microwires by heat treatment, J. Alloys Compounds 615 (2014) 610–615, DOI: http://dx.doi.org/10.1016/j.jallcom.2014.07.079
[10] A. Zhukov, A. Talaat, J. M. Blanco, M. Ipatov and V. Zhukova, Tuning of magnetic properties and GMI effect of Co-based amorphous microwires by annealing, Journal of ELECTRONIC MATERIALS, 2014, DOI: 10.1007/s11664-014-3348-2
[11] M. Barandiaran, A. Hernando, V. Madurga, O.V. Nielsen, M. Vazquez and M. Vazquez-Lopez, Temprature, stress, and structural relaxtion dependece of the magnetostriction in metallic glasses, Phys. Rev.B, Vol. 35 (1987) 5066- 5071.
[12] K. Narita, J. Yamasaki and H. Fukunaga, Measurement of magnetostriction of a thin amorphous ribbon by means of small-angle magnetization rotation, IEEE Trans.Magn., Mag-16 (1980) 435-439.
[13] A. Siemko and H. Lachowicz, "On indirect measurements of magnetostriction in low-magnetostrtctive metallic glasses" IEEE Trans.Magn., Mag-23 (1987) 2563-2365.
[14] V. Zhukova, J.M. Blanco,A. Zhukov, J. Gonzalez, A. Torcunov and V. Larin, Magnetostriction of glass-coated Co-rich amorphous microwires and its dependence on current annealing, J. Magn. Magn. Mater. Vol. 254-255 (2003) 94-96.
[15] V. Zhukova, J.M. Blanco, A. Zhukov, J. Gonzalez, Studies of the magnetostriction of as-prepared and annealed glass coated Co-rich amorphous microwires by SAMR method, J. Phys. D: Appl. Phys., Vol. 34 (2001) L113-L116.
[16] J. Gonzalez, A. P. Zhukov, J. M. Blanco, A. F. Cobeño, M. Vázquez and K. Kulakowski, Evaluation of the magnetostriction in nearly-zero magnetostrictive glass-coated amorphous microwires, J. Appl. Phys. Vol. 87 No 9 (2000) 5950-5952.
[17] A.F. Cobeño, A. Zhukov, A.R. de Arellano - Lopez, F. Elías. J.M. Blanco, V. Larin and J. González, Physical properties of nearly zero magnetostriction Co-rich glass-coated amorphous microwires, J. Mater. Res. Vol. 14 (1999) 3775-3783.
[18] M. Churyukanova, V. Zhukova, A. Talaat, S. Kaloshkin, E. Kostitcyna, E. Shuvaeva, S. Gudoshnikov, V. Sudarchikova and A. Zhukov, Correlation Between Thermal and Magnetic properties of Glass Coated Microwires, J. Alloys Comp. (2013), DOI. http://dx.doi.org/10.1016/j.jallcom.2013.11.191
[19] S. Kaloshkin, A. Talaat, M. Ipatov, V. Zhukova, J. M. Blanco, M.Churyukanova, K. Chichay, and A.Zhukov, Correlation between the magnetostriction coefficient and thermal propertie of soft magnetic microwires, Phys. Status Solidi A, (2014) DOI 10.1002/pssa.201300723
[20] H. Fujimori, K.I. Arai, H. Shirae, H. Saito, T. Masumoto and N.Tsuya, Magnetostriction of Fe-Co amorphous alloys, Japan. J. Appl. Phys. 15 No4, (1976) 705.

Energy Technology 2015
Carbon Dioxide Management and Other Technologies

MAGNETIC MATERIALS
FOR ENERGY APPLICATIONS V

Poster Session

Energy Technology 2015: Carbon Dioxide Management and Other Technologies
Edited by: Animesh Jha, Cong Wang, Neale R. Neelameggham, Donna P. Guillen, Li Li, Cynthia K. Belt, Randolph Kirchain, Jeffrey S. Spangenberger, Frank Johnson, Andrew J. Gomes, Amit Pandey, and Peter Hosemann
TMS (The Minerals, Metals & Materials Society), 2015

MAGNETOSTRUCTURAL TRANSITION IN HEUSLER Mn-Ni-In MELT-SPUN RIBBONS

Hongwei Li[1], Jian Ren[1], Jinke Yu[1], Hongxing Zheng[*]

[1] *Laboratory for Microstructures, Shanghai University, Shanghai 200072, China*

Keywords: Heusler Mn-Ni-In; Melt spinning; Martensitic transformation; Magnetization

[*] Corresponding author. *E-mail address*: hxzheng@shu.edu.cn (H.X. Zheng).

Abstract

The magnetostructural transition of Heusler Mn-Ni-In melt-spun ribbons was investigated in the present work. Experimental results show that the characteristic temperatures of martensitic transformation enhance with substituting Mn with Ni for constant indium levels in $Mn_{50\text{-}x}Ni_{40+x}In_{10}$ (at.%, x=0,2,4) melt-spun ribbons. Meanwhile, the ribbons evolve from the bcc cubic $L2_1$-type structured austenite to fourteen-layered modulated (14*M*) monoclinic martensite at room temperature. The martensitic transformation happens prior to the magnetic transition of austenite in $Mn_{46}Ni_{44}In_{10}$ and lead to a sharp decrease of magnetization discrepancy across the martensitic transformation. It is beneficial for the magnetic enhancement to keep high Mn content.

1. Introduction

In past decades, plenty of investigations have been concentrated on the Ni-Mn based Heusler alloys due to their potential applications in the fields of magnetic actuator and refrigerator [1-4]. Those functional effects are closely linked to a first-order solid-state matensitic transformation. The low-temperature martensite demonstrates diversity in different alloy systems. For example, four martensitic crystal structures, modulated four- and five-layered orthorhombic, seven-layered monoclinic (4*O*, 10*M* and 14*M*), and unmodulated double tetragonal ($L1_0$) martensites were revealed in Ni–Mn–Sn [5-7]. 10*M*, 14*M* and 2*M* martensites have been reported in Ni-Mn-Ga [8,9]. Meanwhile, the transformation coupled with a magnetic transition. In case of Ni-Mn-In, the martensitic transformation takes place from a ferromagnetic austenite to a weak-magnetic martensite, and the magnetization discrepancy in the vicinity of the martensitic transformation is about 60 emu/g under a magnetic field change of 60 kOe [10]. Very recently, researchers put more stress on the Mn-rich Heusler alloys because it is expected that higher Mn content would effectively enhance magnetic properties [11-14]. The refrigerant capacity of $Mn_{50}Ni_{39.75}In_{10.25}$ alloy has been reported up to 54 J/kg under a low magnetic field of 12 kOe [15]. In this present work, a series of Mn-rich Mn-Ni-In melt-spun ribbons were prepared and the compositional dependence of magnetostructural transition behavior was studied.

2. Experimental procedure

Three samples with nominal compositions of $Mn_{50\text{-}x}Ni_{40+x}In_{10}$ (at.%, x=0,2,4) were investigated in the present work. The as-cast ingots about 8 g were first prepared using high-frequency induction melting in quartz tubes from Mn (99.5 wt.%), Ni (99.99 wt.%) and In (99.99 wt.%) elements under argon atmosphere. Then the ribbons were fabricated using melt-spinning technique at a linear speed of 10 m/s. The resultant ribbons were ~20 mm long, 4~6 mm wide and 30~35 μm thick.

X-ray diffraction technique (XRD DLMAX–2200) was employed to detect the phases and crystal structures. Differential scanning calorimetric measurement (NETZSCH DSC 404C) was carried out at heating/cooling rates of 10 K/min. The magnetic transition curves were measured on a physical property measurement system (PPMS-9).

3. Results and discussion

The X-ray diffraction patterns of Heusler Mn-Ni–In melt-spun ribbons at room temperature are shown in Figure 1. The $Mn_{50}Ni_{40}In_{10}$ is well indexed as $L2_1$ austenitic phase with cubic structure. With substituting 2at.% Mn with Ni, the $L2_1(220)$ peak of austenite completely diminished and splited into four 14*M* martensitic peaks, marked by a rectangular frame. A mixture of $L2_1$ austenite and fourteen-layered monoclinic (14*M*) martensitic phase was identified. Further increasing Ni to 44 at.%, the ribbon yields a pure 14*M* martensitic phase. Miller indices (*hkl*) for the $L2_1$ and 14*M* structures are indexed in Figure 1.

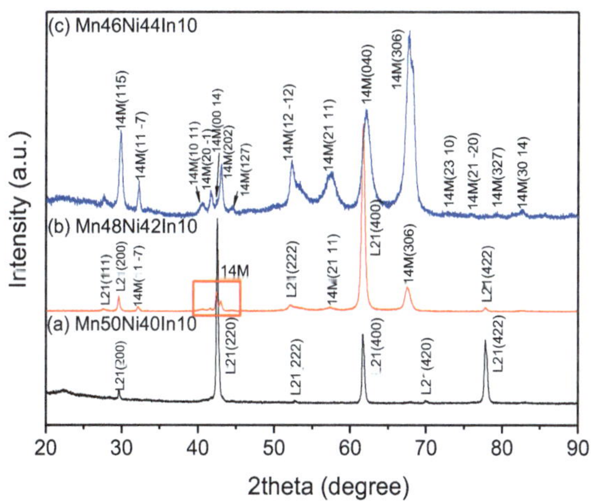

Figure 1: Room temperature X–ray diffraction patterns of $Mn_{50-x}Ni_{40+x}In_{10}$ melt-spun ribbons.

The DSC results are presented in Figure 2, where one can see that large exothermic and endothermic peaks during cooling and heating corresponding to direct and reverse martensitic transformations. Taking $Mn_{50}Ni_{40}In_{10}$ as an example, the characteristic transition temperatures, including the austenite start and finish (A_s and A_f), and martensite start and finish (M_s and M_f) temperatures can be determined, as shown in Figure 2. Keeping In levels constant, the transformation peaks clearly shift towards the high-temperature regime with Ni replacing Mn on the basis of $Mn_{50}Ni_{40}In_{10}$, indicating an increase of martensitic transformation temperature. The change is in well agreement with the well-accepted valence electron concentration principle

[6,15]. The increase of martensitic transformation temperature also agrees with the X-ray diffraction results shown in Figure 1. On the other hand, it should be noted that in case of $Mn_{50}Ni_{40}In_{10}$, weak exothermic and endothermic peaks can be identified at about 335 K upon cooling and heating, respectively, which correspond to the magnetic transition of austenite, T_C^A. The magnetic transition of austenite for the $Mn_{48}Ni_{42}In_{10}$ occurs in 312 K, while no clear peak is visible in the $Mn_{46}Ni_{44}In_{10}$.

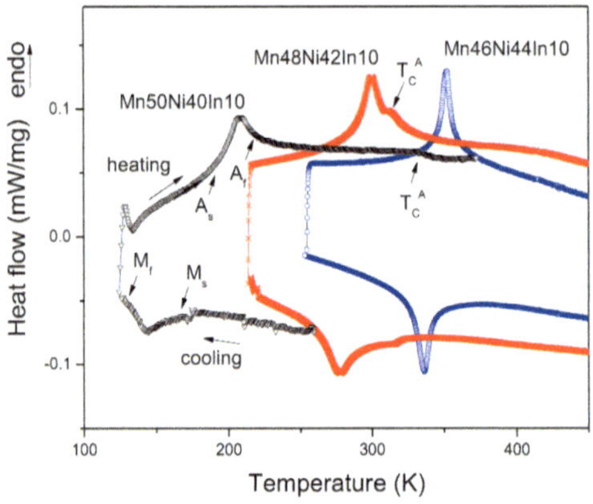

Figure 2: DSC charts of $Mn_{50-x}Ni_{40+x}In_{10}$ melt-spun ribbons.

Figure 3 presents the thermomagnetic curves under a magnetic field of 100 Oe for these three ribbons. The martensitic transition and the magnetic temperatures of austenite obtained from thermomagnetic measurements are almost in accordance with those obtained from the DSC curves (only the characteristic temperatures were marked in Figure 3a for clarity). Both $Mn_{50}Ni_{40}In_{10}$ (Figure 3a) and $Mn_{48}Ni_{42}In_{10}$ (Figure 3b) show a similar temperature dependence of the magnetization within the measured temperature range. Upon heating, magnetization initially increases slowly and then a sharp rise takes place which is associated with the reverse first-order martensitic transformation from martensite with weak magnetization to ferromagnetic austenite. Further heating, the magnetization decreases rapidly due to the occurrence of a ferromagnetic-paramagnetic transition in austenite. However, in case of $Mn_{46}Ni_{44}In_{10}$, the martensitic transformation takes place prior to the magnetic transition of austenite (see the square frame marked in Figure 3c by dashed line where shows the martensitic transformation). The martensitic transition temperature of $Mn_{46}Ni_{44}In_{10}$ is higher than the Curie temperature of austenite and that is why the magnetization property of $Mn_{46}Ni_{44}In_{10}$ is clearly lower than the other two alloys.

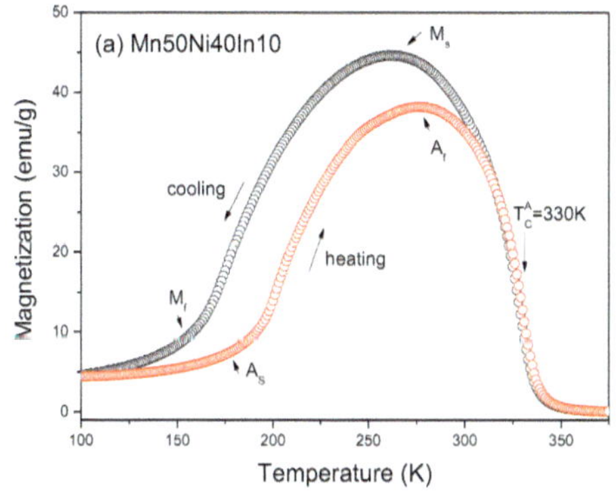
(a) Mn50Ni40In10
M_s
A_f
cooling
heating
T_C^A=330K
M_f
A_s
Magnetization (emu/g)
Temperature (K)

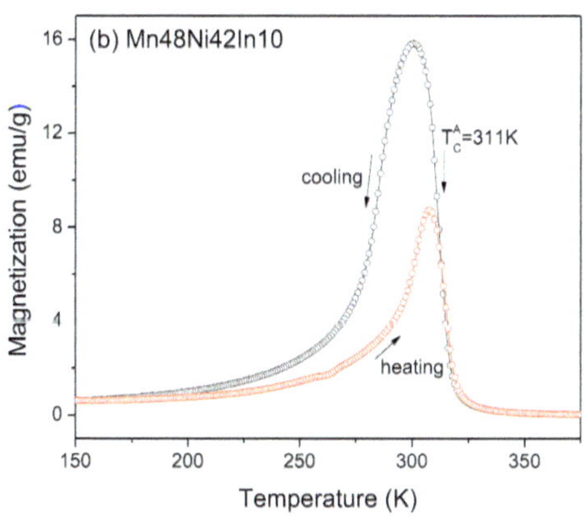
(b) Mn48Ni42In10
T_C^A=311K
cooling
heating
Magnetization (emu/g)
Temperature (K)

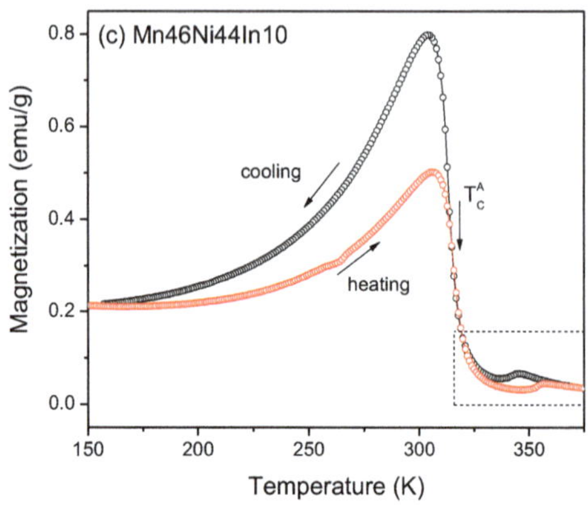

Figure 3: Temperature dependence of magnetization under a magnetic field of 100 Oe for (a) $Mn_{50}Ni_{40}In_{10}$, (b) $Mn_{48}Ni_{42}In_{10}$ and (c) $Mn_{46}Ni_{44}In_{10}$.

4. Conclusions

The magnetostructural transition of $Mn_{50-x}Ni_{40+x}In_{10}$ (at.%, x=0,2,4) melt-spun ribbons were investigated. With substituting Mn with Ni for constant In levels, the martensitic transition starting temperature increases from 170 K to 355 K. Enhanced magnetization properties can be obtained in $Mn_{50}Ni_{40}In_{10}$ and it is beneficial for the magnetization improvement by keeping high Mn level.

Acknowledgments

The authors gratefully acknowledge the supports from the National Natural Science Foundation of China (51474144, 51201096), the Specialized Research Fund for the Doctoral Program of Higher Education of China (20123108120019), and the Scientific Research Foundation for the Returned Overseas Chinese Scholars, State Education Ministry.

References

1. X. Moya et al., "Martensitic transition and magnetic properties in Ni–Mn–X alloys," *Materials Science and Engineering A*, 911 (2006) 438-440.
2. R. Kainuma et al., "Metamagnetic shape memory effect in NiMn-based Heusler-type alloys," *Journal of Materials Chemistry*, 18 (2008) 1837-1842.
3. Y. Sutou et al., "Magnetic and martensitic transformations of NiMnX(X=In, Sn, Sb) ferromagnetic shape memory alloys," *Applied Physics Letters*, 85 (2004) 4358-4360.
4. L. Mañosa et al., "Ni–Mn-based magnetic shape memory alloys: Magnetic properties and martensitic transition," *Materials Science and Engineering A* , 49 (2008) 481-482.
5. H.X. Zheng et al., "Composition-dependent crystal structure and martensitic transformation in Heusler Ni–Mn–Sn alloys," *Acta Materialia*, 61 (2013) 4648-4656.
6. T. Krenke et al., "Martensitic transitions and the nature of ferromagnetism in the austenitic and martensitic states of Ni-Mn-Sn alloys," *Physical. Review B*, 72 (2005) 014412.
7. H.X. Zheng et al., "Martensitic transformation in rapidly solidified Heusler $Ni_{49}Mn_{39}Sn_{12}$ ribbons," *Acta Materialia*, 59 (2011) 5692-5699.
8. N. Okamoto, T. Fukuda, T. Kakeshita, "Temperature dependence of rearrangement of martensite variants by magnetic field in 10M, 14M and 2M martensites of Ni–Mn–Ga alloys," *Materials Science and Engineering A*, 481-482 (2008) 306-309.
9. J. Pons et al., "Long-period martensitic structures of Ni-Mn-Ga alloys studied by high-resolution transmission electron microscopy," *Journal of applied physics*, 97 (2005) 083516.
10. V. Recarte, J.I. Pérez-Landazábal, V. Sánchez-Alarcos, "Dependence of the relative stability between austenite and martensite phases on the atomic order in a Ni–Mn–In Metamagnetic Shape Memory Alloy ," *Journal of Alloys and Compounds*, 536S (2012) S308-S311.
11. G.D. Liu et al., "Martensitic transformation and shape memory effect in a ferromagnetic shape memory alloy: Mn2NiGa, " *Applied Physics Letters*, 87 (2005) 262504.
12. H.Z. Luo et al., "Effect of the main-group elements on the electronic structures and magnetic properties of Heusler alloys Mn_2NiZ (Z = In, Sn, Sb)," *Journal of Magnetism and Magnetic Materials*, 321 (2009) 4063-4066.
13. S. Dwevedi, "Magnetic and transport properties of Mn-rich $Ni_{35}Mn_{50}In_{15}$ alloy," *Journal of Alloys and Compounds*, 574 (2013) 188-190.
14. S. Paul and S. Ghosh, "First-principles investigations of the electronic structure and properties related to shape-memory behavior in Mn2NiX (X=Al,Ga,In,Sn) alloys," *Journal of Applied Physics*, 110 (2011) 063523.
15. H.C. Xuan et al., "Martensitic transformation and magnetic properties in high-Mn content $Mn_{50}Ni_{50-x}In_x$ ferromagnetic shape memory alloys," *Journal of Alloys and Compounds*, 509 (2011) 5761-5764.

Energy Technology 2015: Carbon Dioxide Management and Other Technologies
Edited by: Animesh Jha, Cong Wang, Neale R. Neelameggham, Donna P. Guillen, Li Li, Cynthia K. Belt, Randolph Kirchain, Jeffrey S. Spangenberger, Frank Johnson, Andrew J. Gomes, Amit Pandey, and Peter Hosemann
TMS (The Minerals, Metals & Materials Society), 2015

MAGNETIC PROPERTIES OF NANOCRYSTALLINE MICROWIRES

A. Talaat[1,2], V. Zhukova[1,2], M. Ipatov[1], J. M. Blanco[2], P. Klein[3], R. Varga[3], L. Gonzalez-Legarreta[4], B. Hernando[4], and A. Zhukov[1,5]

[1]Dpto. de Fís. Mater., UPV/EHU San Sebastián 20018, Spain
[2]Dpto. de Física Aplicada, EUPDS, UPV/EHU, 20018, San Sebastian, Spain
[3]Inst. Phys., Fac. Sci., UPJS, Park Angelinum 9, 041 54 Kosice, Slovakia.
[4]Dpto de Física, Universidad de Oviedo, Oviedo 33007, Spain
[5]IKERBASQUE, Basque Foundation for Science, 48011 Bilbao, Spain

Keywords: glass-coated microwires, nanocrystalline materials,

Abstract

We have studied magnetic properties of Finemet-type and Hitperm-type microwires. In the case of the Finemet-type we observed that GMI effect and magnetic softness of microwires can be considerably improved by annealing. A considerable GMI effect has been observed in annealed FeCuNbSiB microwires exhibiting nanocrystalline structure. From the dependence of magnetic properties and GMI effect on the annealing temperature we identified optimal annealing conditions.

For the Hitperm-type annealing results in an enhancement in the domain wall velocity as well as its mobility.

Introduction

Excellent soft magnetic properties exhibited by amorphous glass-coated microwires attractive for various technical application have focused considerable attention in the field of applied magnetism [1,2]. These materials can be prepared using the so-called Taylor-Ulitovsky method allowing the fabrication of a few km long metallic microwires covered by a glass coating. Low-cost and fast preparation process and reduced dimensionality are the other advantages of this family of amorphous materials [3]. Most interesting properties for the applications are Giant magnetoimpedance (GMI) effect and fast magnetization switching suitable for technical applications in electronic surveillance, magnetic field sensors and magnetic memories [1-4].
Recently, the technological process has been modified allowing to prepare not only amorphous microwires, but also either crystalline or granular glass-coated microwires exhibiting a number of functional properties such as magnetic shape memory effect, giant magnetoresistance, magnetocaloric effect etc [5-7].
On the other hand, in certain cases the nanocrystallization of Fe-rich amorphous alloys allows the improvement of soft magnetic properties and obtaining an enhanced GMI effect [8-10]. Excellent soft magnetic properties observed in FeSiBNbCu (Finemet) alloys are usually attributed to vanishing magnetocrystalline anisotropy as well as to vanishing magnetostriction value of the material consisting of nano-sized grains, with average size about 10 nm, embedded in an amorphous matrix [8,9].
Consequently a great enhancement of GMI effect has been reported after nanocrystallization of amorphous ribbons and conventional wires [10].

Few papers reported on low coercivity values obtained for Finemet-type microwires [11-13]. On the other hand, only quite recently few successful attempts to enhance poor GMI effect

exhibited by amorphous Fe-rich glass-coated microwires through devitrification have been reported [14-16].
Alternative group of Cu-free nanocrystalline materials - HITPERM alloys with improved soft magnetic properties have been discovered at the end of 90-s [17,18]. For the Taylor –Ulitovsky technique involving drawing of the metallic alloy surrounding by the glass from the melt at high temperatures Zr- containing alloys are not suitable for the glass-coated microwires fabrication. Therefore, first attempts for HITPERM-type microwires fabrication using Hf instead of Zr have been reported [19]. Recently, a quite high domain wall (DW) velocity in HITPERM –type (Fe40Co38Mo4B18) microwires with DW velocities up to 3 km/s, and very stable almost independent on temperature DW dynamics have been reported [20].

As mentioned above, although Co-rich amorphous microwires can present quite soft magnetic properties and high GMI effect, substitution of Co by less expensive Fe can be essentially important from the point of view of microwires applications.

Therefore, in this work we present the study of the devitrification influence on magnetic properties, structure and GMI effect in nanocrystalline microwires.

Experimental details

Studied glass-coated microwires were produced by the modified Taylor-Ulitovsky method as has been described elsewhere [1,3]. Essentially, in the laboratory process the molten metal fills the glass capillary and a microwire is thus formed where the metal core is completely coated by a glass shell. We prepared Hitperm-type glass-coated microwires with the nominal composition $Fe_{38.5}Co_{38.5}B_{18}Mo_4Cu_1$ with different metallic nucleus diameter, d, total microwire diameter, D, and ρ= d/D ratio (D= 22.5μm, d= 9.4 μm and ρ= 0.41; D= 16.6 μm, d= 10 μm and ρ= 0.60) and Finemet-type $Fe_{70.8}Cu_1Nb_{3.1}Si_{16}B_{9.1}$ glass-coated microwires (D=19.2 μm, d=11.4 μm and ρ= 0.59, D=23 μm, d=14.2 μm and ρ= 0.62; D=15.8 μm, d=12.8 μm and ρ= 0.81; D=16.5 μm, d=14 μm and ρ= 0.85) [1-3]. Hysteresis loops have been determined by a flux-metric method described elsewhere [1, 3]. We measured magnetic field dependences of impedance, Z, and GMI ratio, $\Delta Z/Z$, for as-prepared samples and after heat treatments. We used a specially designed micro-strip sample holder. The sample holder was placed inside a sufficiently long solenoid that creates a homogeneous magnetic field, *H*. The sample impedance *Z* was measured using vector network analyzer from reflection coefficient S_{11}. The frequency range for GMI measurements is from 10 MHz up to 4 GHz. More details on experimental technique can be found in ref. (1).

The magneto impedance ratio, $\Delta Z/Z$, has been defined as:

$$\Delta Z/Z = [Z(H) - Z(H\max)] \cdot 100 / Z(H\max), \qquad (1)$$

An axial *DC*-field with maximum value, H_{max}, up to 8 kA/m was supplied by a magnetization coils.

The technique allowing measuring the DW velocity in thin wires is well described elsewhere [20,21]. In order to activate the DW propagation always from the other wire end in our experiment we placed one end of the sample outside the magnetization solenoid. We used 3 pick-up coils, mounted along the length of the wire and propagating DW induces electromotive force (*emf*) in the coils, as described in ref. [21]. These *emf* sharp peaks are picked up at an oscilloscope upon passing the propagating wall.

Then, DW velocity, *v*, is estimated as:

$$v = \frac{l}{\Delta t} \qquad (2)$$

where l is the distance between pick-up coils and Δt is the time difference between the maximum in the induced *emf*.

The microwire is placed coaxially inside of the primary and pick-up coils so, that one end is inside of the primary coil. Magnetic field, H, is generated by a solenoid applying rectangular shaped voltage.

Structure and phase composition have been checked using a BRUKER (D8 Advance) X-ray diffractometer with Cu K_{α} (λ=1.54 Å) radiation.

Results and discussion

Hysteresis loops of all as-prepared and even annealed microwires present rectangular shapes typical for Fe-rich amorphous microwires (see Fig.1). As-prepared Hitperm-type microwires present much higher coercivity values as –compared with Finemet-type microwires (Fig.1). For comparison we selected the microwires with similar ρ–ratio, i.e., with similar value of the internal stresses.

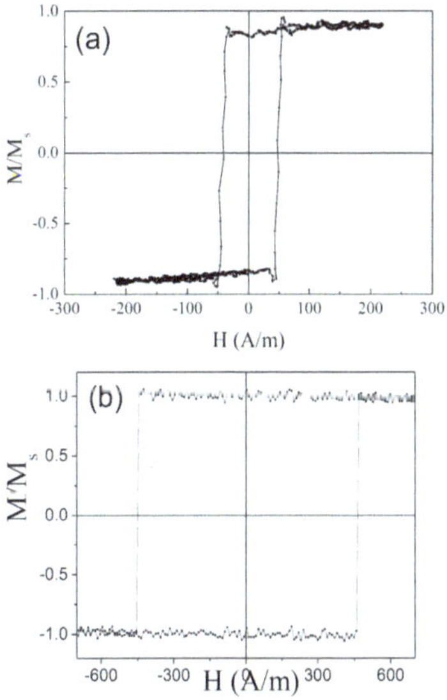

Fig. 1. Hysteresis loops of as-prepared $Fe_{70.8}Cu_1Nb_{3.1}Si_{16}B_{9.1}$ microwires with ρ=0.62 (a) and $Fe_{39.5}Co_{37.7}Mo_{3.7}B_{18}Cu_{1.1}$ with ρ= 0.60

XRD studies reveal that as-prepared microwires $Fe_{70.8}Cu_1Nb_{3.1}Si_{14.5}B_{10.6}$ microwires present amorphous structure. As has been reported recently, annealing of Finemet-type samples allows the improvement of magnetic softness and decreasing of coercivity [14-16]. Magnetic softness depends on the annealing temperature, T_{ann}. At T_{ann} ≥550 °C we observed the devitrification process. We observed the first crystallization process at T_{ann}≥ 823K, when the onset and noticeable growth of the crystalline peak is observed in studied sample (see Fig. 2).

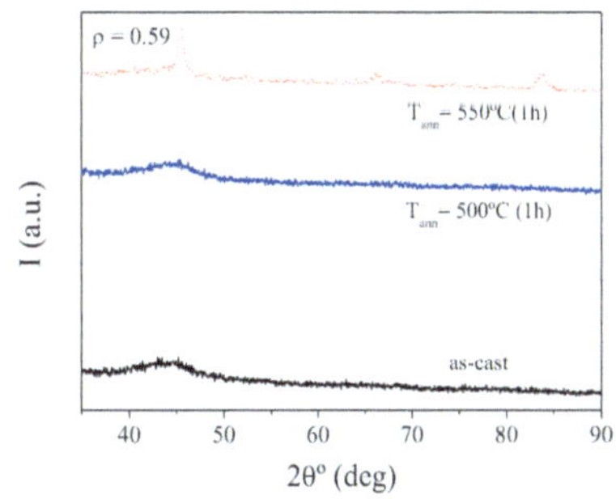

Fig.2. XRD patterns of $Fe_{70.8}Cu_1Nb_{3.1}Si_{14.5}B_{10.6}$ microwires with – ρ= 0.59, after annealing at different temperatures.

The first crystallization process corresponds to the precipitation of α-Fe (Si) BCC crystal structure, similar to devitrification of amorphous ribbons [8-10]. Similar dependences have been obtained for other $Fe_{70.8}Cu_1Nb_{3.1}Si_{14.5}B_{10.6}$ microwires with various ρ-ratios.

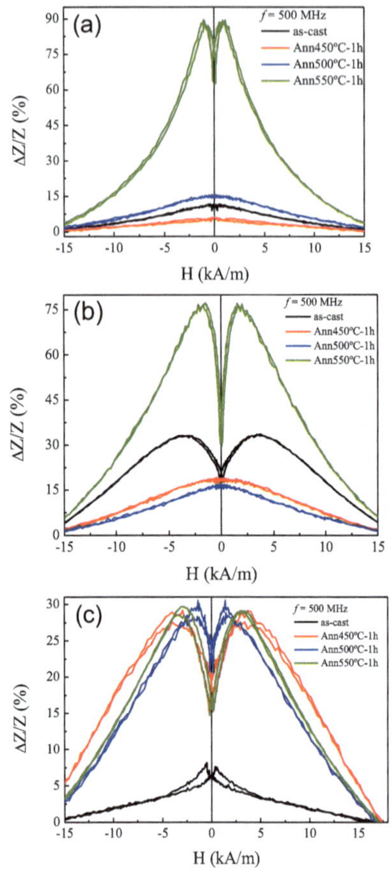

Fig. 3. *ΔZ/Z (H)* dependences of $Fe_{70.8}Cu_1Nb_{3.1}Si_{14.5}B_{10.6}$ microwires with geometric ratio $\rho = 0.59$ (a) $\rho = 0.0.81$ (b) and $\rho = 0.85$ (c) measured at 500 Mz in as-prepared and annealed at different temperatures samples.

Aforementioned magnetic softening achieved through the nanocrystallization allows improving considerably the GMI ratio (see Fig.3). It is worth mentioning that the highest GMI ratio has been observed at the same annealing temperature as we observed the samples devitrification.

We must assume, that the enhancement of the *ΔZ/Z* ratio is related with magnetic softening of studied microwires after annealing related with the internal stress relaxation and especially with the nanocrystallization.

Hitperm samples present higher coercivity but the hysteresis loops exhibit a perfectly rectangular shape typical for Fe-rich samples (see Fig.4). After current annealing we observed a considerable decreasing of coercivity, although the hysteresis loop shape remains rectangular.

We measured the DW dynamics in as-prepared and current annealed Hitperm microwire. For as-prepared and current annealed samples we observed a single DW propagation similarly as reported for amorphous microwires [21]. Observed domain wall (DW) velocities are rather high and are comparable with that observed in amorphous as well in nanocrystalline

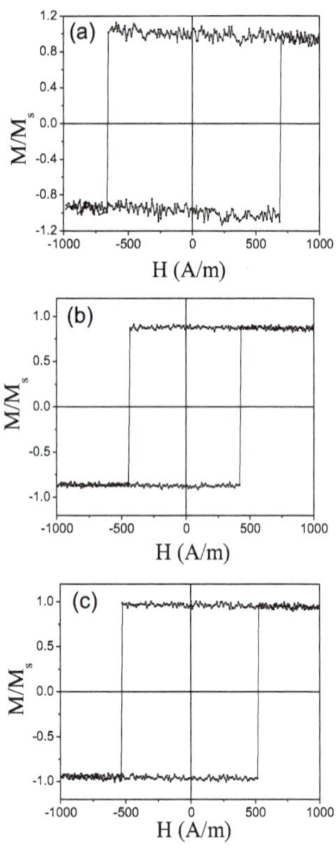

Fig. 4. Hysteresis loops of as-prepared (a) and current annealed for 10 min at 30 mA (b) and 45 mA (c) $Fe_{39.5}Co_{37.7}Mo_{3.7}B_{18}Cu_{1.1}$ with ρ= 0.41

microwires [20,21]. Performed annealing did not affect significantly the DW velocities for the $Fe_{38.5}Co_{38.5}B_{18}Mo_4Cu_1$,(ρ= 0.41) although the magnetic field range for the single DW propagation regime in annealed samples is more extended (Fig.5a). This extension of the magnetic field region correlates with the decreasing of the coercivities observed after current annealing. But for the $Fe_{38.5}Co_{38.5}B_{18}Mo_4Cu_1$,(ρ= 0.6) we observed a considerable increasing of

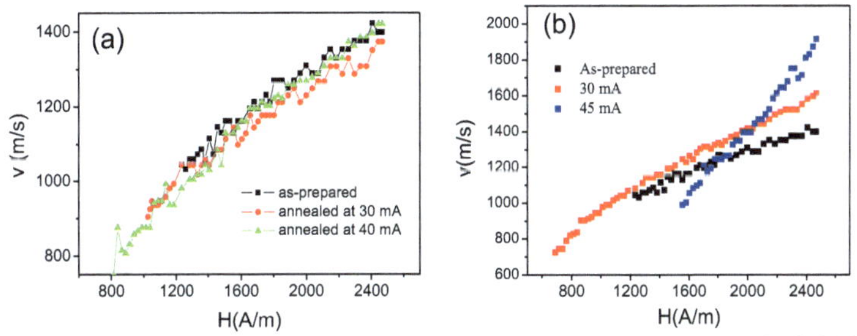

Fig.5. Experimentally measured dependence of DW velocity on magnetic field for $Fe_{38.5}Co_{38.5}B_{18}Mo_4Cu_1$ microwires with ρ= 0.41 (a) and ρ= 0.60 (b).

the DW velocity after annealing (see Fig.5b).

Consequently, we can assume that Hitperm-type and Finemet-type glass-coated microwires can exhibit interesting magnetic properties as either considerable GMI effect (achieved after devitrification of Finemet-type glass-coated microwires) and fast DW dynamics observed in Hitperm-type glass-coated microwires.

Conclusions

We studied magnetic properties of Finemet-type and Hitperm-type microwires. In the case of Finemet-type we observed that GMI effect and magnetic softness of microwires can be considerably improved by annealing. A considerable GMI effect has been observed in annealed FeCuNbSiB microwires exhibiting nanocrystalline structure. From dependence of magnetic properties and GMI effect on annealing temperature we have identified optimal annealing conditions.

For the Hitperm-type we observed a fast domain wall velocity as well as a high mobility. Current annealing allows the enhancement of DW velocity in some cases.

Acknowledgements

This work was supported by by Spanish MINECO under MAT2013-47231-C2-1-P and MAT2013-48054-C2-2-R, by the Basque Government under Saiotek 13 PROMAGMI (S-PE13UN014), DURADMAG (S-PE13UN007) projects and Slovak VEGA grants No. 1/0060/13, APVV-0027-11, APVV-0266-10. Technical and human support provided by SGIker (UPV/EHU, MICINN, GV/EJ, ERDF and ESF) is gratefully acknowledged.

References

[1] M. Vazquez, H. Chiriac, A. Zhukov, L. Panina, and T. Uchiyama, On the state-of-the-art in magnetic microwires and expected trends for scientific and technological studies, Phys. Status Solidi A, Vol. 208, (2011) 493–501.

[2] F.Qin, H.-X. Peng, "Ferromagnetic microwires enabled multifunctional composite materials",Progress in Materials Science 58 (2013) 183–259
[3] A. Zhukov and V. Zhukova, "Magnetic Sensors and Applications Based on Thin Magnetically Soft Wires with Tuneable Magnetic Properties", International Frequency Sensor Association Publishing, Barcelona, 2013, 156p.
[4] A. Zhukov, M. Ipatov, M. Churyukanova, S. Kaloshkin, V. Zhukova, Giant magnetoimpedance in thin amorphous wires: From manipulation of magnetic field dependence to industrial applications, J. Alloys Comp. 2012, http://dx.doi.org/10.1016/j.jallcom.2012.10.082
[5] A. Zhukov, J. Gonzalez and V. Zhukova, "Magnetoresistance in thin wires with granular structure", J. Magn. and Magn., Mater. 294 (2005) 165-173
[6] A. Zhukov, C. Garcia, M. Ilyn, R.Varga, J. J. del Val, A. Granovsky, V. Rodionova, M. Ipatov, V. Zhukova, "Magnetic and transport properties of granular and Heusler-type glass-coated microwires", J. Magn. Magn. Mater. 324 (2012) 3558–3562
[7] A. Zhukov, V. Rodionova, M. Ilyn, A.M. Aliev, R. Varga, S. Michalik, A. Aronin, G. Abrosimova, A. Kiselev, M. Ipatov, V. Zhukova, "Magnetic properties and magnetocaloric effect in Heusler-type glass-coated NiMnGa microwires", Journal of Alloys and Compounds 575 (2013) 73–79
[8] G. Herzer, Anisotropies in soft magnetic nanocrystalline alloys, J. Magn. Magn. Mater. Vol. 294 (2005) 99
[9] M. E. McHenry, M. A. Willard, D. E. Laughlin, Progress in Materials Science 44 (1999) 291
[10] H. Q. Guo, H. Kronmüller, T. Dragon, Z. H. Cheng, and B. G. Shen, J. Appl. Phys. 89, 514 (2001).
[11] J. Arcas, C. Gómez-Polo, A. Zhukov, M. Vázquez, V. Larin and A. Hernando, Magnetic properties of amorphous and devitrified FeSiBCuNb glass-coated microwires, Nanostructured Materials Vol. 7, No 8 (1996) 823-834.
[12] H. Chiriac, T.A. Ovari, C.S. Marinescu, Giant magneto-impedance effect in nanocrystalline glass-covered wires, J. Appl. Phys. Vol. 83 (1998) 6584–6586.
[13] C. Dudek, A. L. Adenot-Engelvin, F. Bertin, O. Acher, Engineering of the magnetic properties of Finemet based nanocrystalline glass-coated microwires, J Non-Cryst Solids 353 (2007) 925-927.
[14] A. Zhukov, A. Talaat, M. Ipatov, J. J. del Val, L. Gonzalez-Legarreta, B. Hernando and V. Zhukova, " Effect of nanocrystallization on magnetic properties and GMI effect of Fe-rich microwires", Journal of ELECTRONIC MATERIALS, 2014, DOI: 10.1007/s11664-014-3370-4
[15] A. Talaat, V. Zhukova, M. Ipatov, J. M. Blanco, L. Gonzalez-Legarreta, B. Hernando, J. J. del Val, J. Gonzalez, and A. Zhukov, "Optimization of the giant magnetoimpedance effect of Finemet-type microwires through the nanocrystallization", J. Appl. Phys. 115, 17A313 (2014)
[16] A. Zhukov, A. Talaat, M. Ipatov, J. M. Blanco, L. Gonzalez-Legarreta, B. Hernando, and V. Zhukova, "Effect of Nanocrystallization on Magnetic Properties and GMI Effect of Microwires" IEEE Trans. Magn. (2014) DOI: 10.1109/TMAG.2014.2303396
[17] M. A. Willard, M.-Q. Huang, D. E. Laughlin, M. E. McHenry, J. O. Cross, V. G. Harris, C. Franchetti, "Magnetic properties of HITPERM $(Fe,Co)_{88}Zr_7B_4Cu_1$ magnets", J. Appl. Phys. 85 (1999) 4421- 4423
[18] I. Škorvánek, P. Švec, J. Marcin, J. Kováč, T. Krenický and M. Deanko, " Nanocrystalline Cu-free HITPERM alloys with improved soft magnetic properties", phys. stat. sol. (a) 196, No. 1, 217– 220 (2003) / DOI 10.1002/pssa.200306390

[19] C. García, A. Zhukov, J. Gonzalez, V. Zhukova, R. Varga, J.J. del Val, V. Larin and J.M. Blanco, “Studies of a structural and magnetic properties of glass-coated nanocrystalline $Fe_{79}Hf_7B_{12}Si_2$ microwires”, J. Alloys and compounds, 423 (2006)pp.116-119
[20] P.Klein., R. Varga, and Ma. Vazquez, **“**Domain wall dynamics in nanocrystalline microwires**”,** P hys. Status Solidi C 11, No. 5–6, 1139–1143 (2014) / DOI 10.1002/pssc.201300707
[21] A. Zhukov, J. M. Blanco, M. Ipatov, A. Chizhik and V. Zhukova, Nanoscale Research Letters, 7 (2012) 223, doi:10.1186/1556-276X-7-223

Energy Technology 2015: Carbon Dioxide Management and Other Technologies
Edited by: Animesh Jha, Cong Wang, Neale R. Neelameggham, Donna P. Guillen, Li Li, Cynthia K. Belt, Randolph Kirchain, Jeffrey S. Spangenberger, Frank Johnson, Andrew J. Gomes, Amit Pandey, and Peter Hosemann
TMS (The Minerals, Metals & Materials Society), 2015

On Magnetocaloric Properties of (Pr,Dy)Fe9 Alloys

R. Guetari[1,2,3], C. B. Cizmas[3], K. Zehani[1], L. Bessais [1] , and N. Mliki[2]

[1]CMTR, ICMPE, UMR7182, CNRS-UPEC, 2 rue Henri Dunant F-94320 Thiais, France
[2]LMOP, Faculté des Sciences de Tunis, Campus Universitaire, 2092 Tunis, Tunisia
[3]DEEAP, Transilvania University of Brasov, Brasov, Romania

Keywords: Nanocrystalline metastable material, Rare earth compounds, Magnetic materials, Magnetocaloric effect.

Abstract

Pr_2Fe_{17} intermetallics are considered as very attractive systems because of their interesting magnetic properties and their low cost. In order to increase the Curie temperature (T_C), it is possible to mix rare earth elements such as Pr and Dy. In this work, we report the influence of Dy substitution on structural and magnetocaloric properties of Pr_2Fe_{17} and its out of equilibrium phase; (Pr,Dy)Fe_9. Various (Pr,Dy)Fe_9 alloys with nominal compositions were elaborated by high-energy ball milling technique and annealed at different temperatures. X-ray diffraction coupled to Rietveld analysis was used for the structural study. The $P6/mmm$ hexagonal phase, derived from $CaCu_5$ was found more efficient than the $R\bar{3}m$ one. We show that T_C increases with Dy content, raising of around 40 K compared to the $(Pr,Dy)_2Fe_{17}$ equilibrium phase. The temperature dependence of the magnetization, the magnetic entropy change and the RCP around the second order magnetic transition were determined.

Introduction

Rare earth transition ($4f$) metal intermetallic ($3d$) compounds receive significant attention due to their important magnetic characteristics. In the last decade, there has been an increased interest in magnetocaloric materials. Among many of the applications, the current interest has been on magnetic refrigeration near room temperature because that was being energetically more favorable [1, 2].

Most of the current prototypes for room temperature magnetic refrigeration have used rare-earth (R) based materials [2, 3]. However, some Fe-rich R_2Fe_{17} compounds have shown values of RCP comparable with those of Gd-based magnetic materials. In addition, these alloys are of interest due to lower cost of the main component (Fe) and easy fabrication procedures[4, 5, 6, 7, 8, 9, 10].

It is possible to modify the Curie temperature T_C of R_2Fe_{17} intermetallics by mixing different rare earth elements. In this way, we have mixed Pr with Dy.

In this paper, a structural, magnetic and magnetocaloric properties of the new out-of-equilibrium $Pr_{1-x}Dy_xFe_9$ phase ($x \leq 0.5$) has been investigated.

Experiment

A polycrystalline $Pr_{2-x}Dy_xFe_{17}$ samples, for $x = 0.125$, 0.25, 0.375 and 0.5, have been prepared by melting high-purity starting elements (99.9%) in an induction furnace under an atmosphere of high purity [11]. An excess of Praseodymium powder was used in order to maintain an overpressure of Pr on the sample. After milling, the powder mixtures was wrapped in tantalum foil and sealed in silica tubes, then annealed for one week at 1373 K followed by quenching in water. The samples's structure was checked by means of a Brucker X-ray diffractometer (XRD) (CuKα radiation $\lambda = 1.54178$ Å). The Rietveld method was used to analyze the crystal structure and calculate the lattice parameters. The magnetic and magnetocaloric properties were determined using a Physical Property Measurement System (PPMS) magnetometer from Quantum Design.

Results and discussion

Structure analysis

XRD pattern evolution of nanostructured $Pr_{2-x}Dy_xFe_{17}$ annealed at two different temperatures 600 and 1000 °C is shown in Fig. 1. The pattern of the sample annealed at 600 °C is representative of the hexagonal sub-cell $P6/mmm$ derived from $CaCu_5$. While the sample annealed at higher temperature above 900 °C is characterized by the full transformation of the alloys into the Th_2Zn_{17} structure type with the expected intensities of the superstructure extra lines typical of the $R\bar{3}m$ space group (104), (211), (112), (024) [12]

In the $Pr_{2-x}Dy_xFe_{17}$ compounds, the Fe atoms occupy four crystallographic nonequivalent sites, denoted as Fe(6*c*), Fe(9*d*), Fe(18*f*) and Fe(18*h*).

The Rietveld refinement for the XRD patterns corresponding to (Pr,Dy)Fe_9 (x = 0.25) compounds annealed at 600 ° reveals that the s value 0.22 which corresponds to the 1/7 ratio is not in agreement with the x-ray intensity data, s was found to be equal to 0.360.38, representative of the stoichiometry 1/9 with a $P6/mmm$ space group.

It appears that the unit cell parameters a decrease and c increase with increasing x. However, the c/a ratio is constant and the volume decreases slightly. Table 1 summarizes the values of the crystallographic parameters as well as the reliability factors of the refinements.

The partial substitution of praseodymium by dysprosium resulted in a variation of the lattice parameters showing that the cell volume is constant. The atomic radius of Dy is small compared to the Pr one, so we can expect this evolution of crystallographic parameters.

Magnetic properties

Thermomagnetic measurements for different Dy content have been carried out. The Curie temperature T_C was deduced from the maximum slope in dM/dT. Dy doping causes an increase of Curie temperature , which reaches a maximum value for $x = 0.5$ ($T_C = 341$ K).

Figure 2 presents the magnetization isotherm curves of the $Pr_{0.5}Dy_{0.5}Fe_9$ compound, for various temperatures with a temperature step of 5 K in order to include the Curie temperature. No magnetic hysteresis effect appears upon decreasing and increasing the magnetic field.

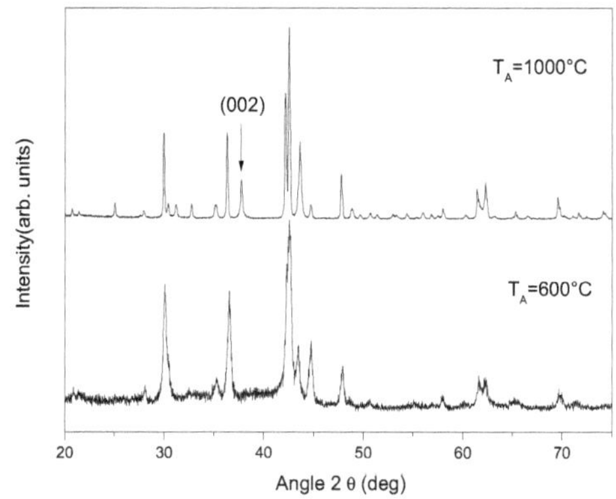

Figure 1: XRD pattern evolution of nanostructured $Pr_{1.75}Dy_{0.25}Fe_{17}$ annealed at two different temperatures.

Table 1: Atom Wyckoff positions, number of subsequent atom in the hexagonal $P6/mmm$ structure as a function of the s parameter atom occupancy.

Atomic Position	Occupancy	s=0	s=0.03 [4]	s=0.33
		$CaCu_5$	$Sm_{1-s}Co_{5+2s}$	2/17
Pr/Dy (1a) : (0, 0, 0)	1-s	1	0.97	0.66
Fe (2c) : (1/3, 2/3, 0)	2(1-s)	2	1.82	0
Fe (6l) : (x, 2x, 0)	6s(6)	0	0.18	2
Fe (3g) : (1/2, 0, 1/2)	3	3	3	3
Fe (2e) : (0, 0, z)	2s	0	0.06	0.66

The magnetization vs. applied magnetic field, $M(H)$ curves, show a similar trend for the different compounds although the magnetization values are different.

Magnetocaloric properties

It is well known that, for an isothermal process, ΔS_m can be evaluated by integrating the Maxwell relation [8]:

$$\Delta S_m(T,H) = S_m(T,H) - S_m(T,0) = \int_0^H \left(\frac{\partial M}{\partial T}\right)_H dH$$

where, at a fixed temperature T, S_m (T,H) and S_m $(T,0)$ are the magnetic entropy under an applied magnetic field H and in the absence of the magnetic field, respectively.

The $|\Delta S_m|$ value was found enhanced for the $Pr_{1-x}Dy_xFe_9$ compounds: For all compounds, $|\Delta S_m|$ increases with applied magnetic field; One maximum value of $|\Delta S_m|$ was observed for $x = 0$ and 0.125, coresponding to Curie temperature, which slightly increases with applied magnetic field.

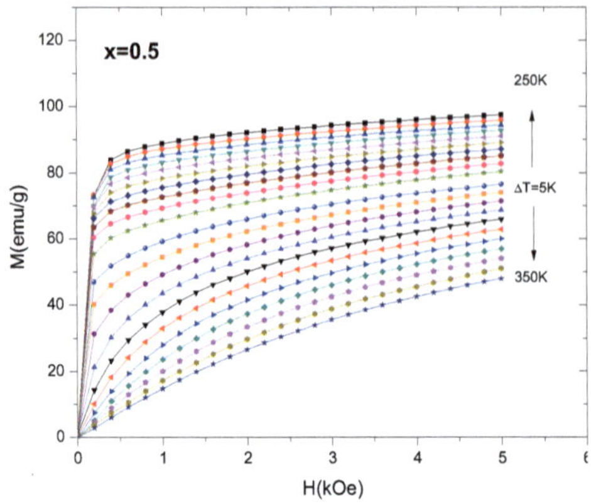

Figure 2: Magnetization isotherms of the $Pr_{0.5}Dy_{0.5}Fe_9$ compound. The curves were measured by increasing and decreasing the magnetic field in the vicinity of T_C.

The magnetic entropy change decreases from 5.51 $Jkg^{-1}K^{-1}$ (x=0) to 4.71 $Jkg^{-1}K^{-1}$ (x=0.5) under a magnetic field change of 0 - 5T.

Summary and Conclusion

High energy ball milling and subsequent annealing at $T_A = 600$ ° show an hexagonal $P6/mmm$ $CaCu_5$-type phase, precursor of $R\bar{3}m$ $Pr_{2-x}Dy_xFe_{17}$ phase, is detected as the main phase. The Rietveld analysis gives a phase stoechiometry 1/9 with three crystallographic transition metal sites and one Pr site $1-s$ ($1a$). Upon Dy substitution, a considerable increase of T_C was observed. The values of Curie temperature are between 306 and 341 K. The entropy change ΔS_M between 0 and 5 T was calculated.

Acknowledgments

This work was mainly supported by the CNRS and the "Ministère de l'Enseignement Supérieur, de la Recherche Scientifique et de la Technologie" (LAB MA03) (Tunisia) and AUF (PCSI type Project 16PS005).

The authors acknowledge the French-Tunisian Cooperation PHC-Utique (project 11/G 1301).

References

[1] K. A. G. Jr, V. K. Pecharsky, and A. O. Tsokol, Rep. Prog. Phys. **68**, 1479 (2005).

[2] A. Tishin and Y. Spichkin, IOP Publishing (2003).

[3] M. Phejar, V. Paul-Boncour, and L. Bessais, Intermetallics **18**, 2301 (2010).

[4] D. Givord, J. Laforest, J. Schweizer, and F. Tasset, J. Appl. Phys. **50**, 2008 (1979).

[5] K. H. J. Buschow, Rep. Prog. Phys. **54**, 1123 (1991).

[6] C. Djega-Mariadassou, L. Bessais, A. Nandra, and E. Burzo, Phys. Rev. B **68**, 24406 (2003).

[7] L. Bessais *et al.*, Phys. Rev. B **69**, 64402 (2004).

[8] K. Mandal *et al.*, J. Phys. D **37**, 2628 (2004).

[9] L. Bessais, E. Dorolti, and C. Djega-Mariadassou, **87**, 192503 (2005).

[10] S. Khazzan, N. Mliki, L. Bessais, and C. Djega-Mariadassou, J. Magn. Magn. Mater. **322**, 224 (2010).

[11] K. Younsi, V. Russier, and L. Bessais, J. Appl. Phys. **107**, 083916 (2010).

[12] C. Djéga-Mariadassou *et al.*, Phys. Rev. B **65**, 14419 (2001).

Sustainable Energy and Layered Double Hydroxides

Energy Technology 2015
Carbon Dioxide Management and Other Technologies

SUSTAINABLE ENERGY AND LAYERED DOUBLE HYDROXIDES

Sustainable Energy and Layered Double Hydroxides

Session Chairs:

Andrew Gomes
Christian Ruby

Energy Technology 2015: Carbon Dioxide Management and Other Technologies
Edited by: Animesh Jha, Cong Wang, Neale R. Neelameggham, Donna P. Guillen, Li Li, Cynthia K. Belt, Randolph Kirchain, Jeffrey S. Spangenberger, Frank Johnson, Andrew J. Gomes, Amit Pandey, and Peter Hosemann
TMS (The Minerals, Metals & Materials Society), 2015

TECHNICAL INNOVATION AND ENTREPRENEURIAL POTENTIAL OF "HYDROTALCITE LIKE" MATERIALS

David Cocke[1,2], Paul Latiolais[2], Andrew Gomes[3], Gary Beall[4]

[1]Center for Innovation and Commercialization, Lamar University, Beaumont, TX
[2]Gill Chair of Chemical Engineering, Lamar University, Beaumont, TX
[3]Materials Instrumentation Center, Lamar University, Beaumont, TX
[4]Texas State University, San Marcos, TX

Keywords: Hydrotalcite, Layered Double Hydroxides, Innovation, Entrepreneurship

Abstract

The process of technological innovation embraces a sequence of activities that translates technical and scientific knowledge into physical reality to be used on a scale having substantial societal impact. This goes beyond invention to accumulation of technical data, expansion of necessary knowledge, transformation into product and its diffusion and adaptation for significant impact. Hydrotalcite like materials or double layered hydroxides in general have been a playground for discovery, have shown a history of innovation, and have the potential for an even greater surge in innovation in the future that can drive an enormous expansion in technology entrepreneurship. The versatility for use in energy materials, composites, bioactives, electrochemicals, environmentals, anticorrosives, catalysts, drugs, coatings, and yet to be discovered uses are enormous. The pathways and barriers to new innovations and entrepreneurial activities are discussed in this overview in context of their science and technology.

Introduction

The concept in science and engineering that structure particularly at the molecular level determines properties, structure-property relationships, SPRs, is a powerful concept impacting, material science, environmental science, biology, biochemistry, medicine, and energy. Layered double hydroxides exemplified by hydrotalcite are a prime example of this. Not only are these materials examples of SPRs but they are fodder for discussions of Pasteur's quadrant, translational research, innovation, commercialization, and the materials genome. Materials Entrepreneurship involves processes beyond the SPRs to other relationships as shown in Figure 1 that help define materials innovation.

Pasteur's Quadrant

Donald Stokes, in Pasteur's Quadrant [1] argues that the standard view that science progresses from pure to applied research to engineering to adaptation to societal benefit is often wrong. Stokes argues that this path has been the exception rather than the rule even though it has provided examples of exceptional success and is more often the result of research at the intersection of basic and applied technology. This is what Stokes famously called Pasteur's

Quadrant. The recent surge in expectation from the government and from society that research be a source of innovation and commercialization is increasingly in line with Pasteur's quadrant and increasingly referred to as translational research.

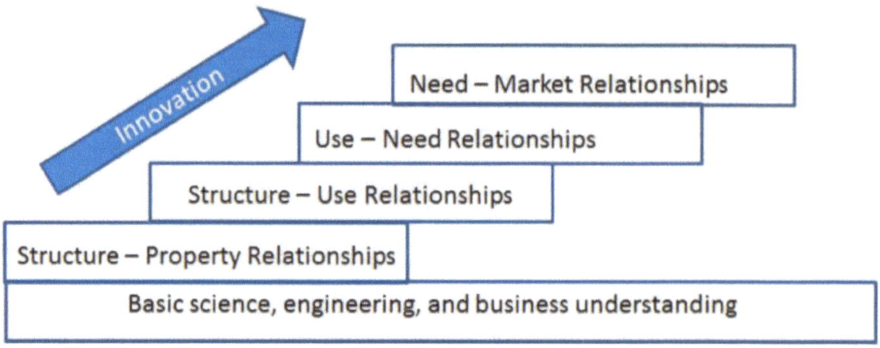

Figure 1. Processes of Materials Technology Entrepreneurship [1]

Translational Research

Translational research is becoming well established in biomedical, biotechnology, and medical research [2]. Translational research programs are busy formulating specific program objectives, and delineating the knowledge and skills that researchers are expected to develop. Efforts are underway to create an appropriate curricula, and track outcomes. Journals have been launched as exemplified by Translational Research, The Journal of Laboratory and Clinical Medicine, American Journal of Transitional Research, and Drug Delivery and Translational Research, the latter being particularly relevant to materials science discussed in this paper [3-5], as it is published by the Controlled Release Society.

Transitional Materials Research is also moving forward as exemplified by the launch in 2014 of the journal Transitional Materials Research by IOPscience [6]. The process was summarized in a "lab-to -market" road map in academe shown in in Figure 2 in the first issue by Windheim and Meyers [7]. A much shorter road map or developing a product that is brought to market by licensing as opposed to company building.

If resources are not significantly flowing in, a disciplined approach can play the central role to advance the most promising technologies. This kind of program can be applied according to the strengths of the university. This will move univerisities away from an unsolicited approach to an environment where promising innovations are cultivated, and transformed into a long-term success areas that will generate meaningful return at small risk to the universities and their communities including the inventors [8].

Before starting patenting of one's invention, the inventor should check whether the invention is novel, and unique, and when the future profit of the invention commercialization will start to come in. In a country, there are special departments where the university employees can proceed to with their inventions. In USA, Office of Technology Transfer is the proper departments where the inventors need to proceed for commercialization of their inventions. This Office has the following jobs: it receives the disclosures of inventions, assists in the registration of copyrights and trademarks, helps with the filing of patent applications typically performed by appointing law firm, assesses who owns intellectual property, and handles the commercialization/licensing of intellectual property and the distribution of net royalties received. "Office of Technology Transfer at the university receives hundreds of invention disclosure forms (IDFs). It typically chooses to pursue about 1/3 of those and licenses 1/2 of these. Out of all licensed patents: about 10% break even, 5% make some money and 0.01% generates a million dollars of revenue. Typical license fee includes: reimbursement and ongoing payment of patent prosecution costs, milestone payment minimum annual royalties, and a percentage royalty on sales. The revenue is typically split between the inventor(s), university and inventor's lab. For example, it could be split as follows: 34% +33%+33%; inventor share varies from 25% to 70%. In some Universities (e.g. Iowa) first $100,000.00 of net revenue is given to the inventor(s). In case of many inventors the revenue for them is typically share equally unless inventors agree otherwise in writing. The economic impact of U.S. academia licensing results in the investments in $20 billion per year range, thus creating over 150,000 jobs per year" [8].

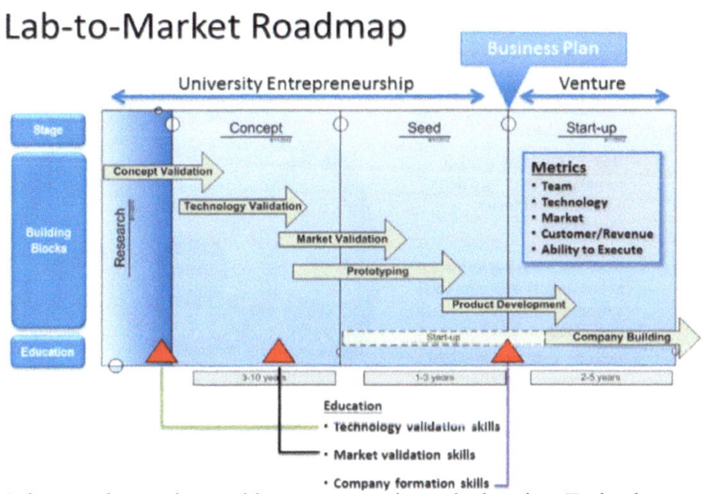

Figure 2. Lab-to-market roadmap with process, metrics and education. Technology moves from lab to market [7].

Innovation

For anyone, it is possible to learn, practice, and master the skills to become upsetting innovator. For that, one needs to develop associating thinking. That means the ability to connect previously unconnected ideas. Questioning, observing, networking, and experimenting are a few examples that will enhance the process. They can be mastered by concentrating more on these behaviors, by allocating time to them, and by practicing them. After achieving mastery, one will be able to constantly produce innovative new ideas [9].

"As defined in The Innovator's Manifesto: Deliberate Disruption for Transformational Growth, an innovation is "any combination of activities or technologies that breaks existing performance trade-offs in the attainment of an outcome in a manner that expands the realm of the possible."3

DTTL's Global Manufacturing Industry group's Advanced Materials Systems (AMS) framework provides a new approach for companies to target these opportunities for innovation and invigorate value creation (see Figure 3). The approach urges companies to move beyond the frontier of invention of new molecules and materials into the realm of functional solutions as a new source of growth and innovation.

Achieving the desired outcomes/results, however, will require additional and different types of capabilities and approaches beyond those required for the invention of new materials. For example, system-level design and engineering will be a core capability necessary to leverage combinations of both new and existing materials and process technologies. In addition, new business models, together with new approaches to partnerships and collaborations that extend well beyond current concepts of "open innovation," will likely be essential to create and grow new businesses based on functional solutions. The AMS concept is therefore a call to action for players in the AMS ecosystem to rethink conventional strategies and approaches for creating innovation and growth." [10]

Drivers of performance	Challenges to address
Demand	• How to determine which market opportunities should be targeted • How to translate market requirements to functional solutions and focus innovation
Functional solutions	• How to define performance requirements and approach system-level design, optimization, and testing • How to approach material and process engineering (i.e., technology selection, development, and integration)
Value capture	• How technical and commercial development activities reinforce one another • How to assess and pursue business model alternatives • How best to develop and access a network of required capabilities

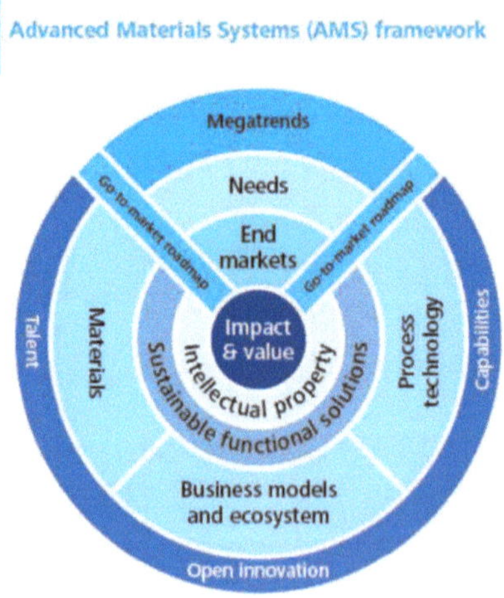

Figure 3: The AMS framework can help answer key questions along the path to innovation [11].

Effect on Patenting on Science

When it comes to commercialization of the technology or material or process or device disclosed in the patent, and then the inventor has a right to request a monetary reward for using the patented invention. So the positive sides of university patenting are: (1) recognition of the scientist as an inventor; (2) another publication; (3) it is beneficial for the university and therefore improves the scientist's position there; (4) and the last, but maybe one of the most important – is monetary reward collected through licensing, which is important both for the scientist and for the university. Many scientists have pointed out negative effect of patenting on science, which is in refusal to share information, in its contradicting to open science norm. This delay in the results publication may affect areas where advances are cumulative. This thesis is mostly related to European countries, where the patent law names "inventor" a person who first filed the patent application, in contrast with US patent law where the inventor is the first who made the invention. US patent law does not preclude publishing the results and later their patenting. In European countries, the invention needs to be patented first, and then published. European scientists often prefer to publish results and not to file a patent, since patenting requires keeping the results secret for quite a long time.

Materials Genome (Materials Innovation Infrastructure

The Materials Genome Initiative (MGI) is a multi-agency multi-stakeholder effort to develop the infrastructure needed to enable the materials science community to design, develop, manufacture, and deploy advanced materials at least twice as fast as possible today, at a fraction of the cost. This new Materials Innovation Infrastructure (MII) will leverage advances in materials modeling, computing, and communications to accelerate advanced materials design and deployment in the United States across many fields. More widespread implementation of advanced materials will contribute to new products with enhanced functionality, and potentially enhance U.S. global competitiveness. Major advances in theory and modeling have led to a remarkable opportunity for the use of computational simulation in predicting the behavior of material systems. However, such computational tools are not in widespread use today due to limitations in their capabilities, a lack of expertise needed to employ them, and a general lack of confidence in accepting conclusions that are not empirically based. Similarly, advances in networked communications have led to remarkable opportunities for the sharing of technical information, such as materials property data. This too has had limited use due to the lack of suitable data repositories, standards, and incentives for sharing [12].

Layered Double Hydroxides and Hydrotalcite Like Materials

Layered double hydroxides (LDH), also known as anionic clays or hydrotalcites, consist of two types of metallic cations accommodated with the aid of a close packed configuration of OH - groups in a positively charged brucite-like layer. The charge typically results from the substitution of the lower valence metal in the brucite structure by a metal of higher valence. The interlayer space in the LDH is, typically, occupied by water and various anions for charge compensation [13]. The general LDH chemical structure can be represented by the general formula: [MII1-x MIIIx (OH)2]x+ [Amx/m •nH2O]x where MII is a divalent (e.g. Mg, Co, Mn,Ni, Cu, Fe, Zn) and MIII a trivalent metal cation (e.g. Al, V, Fe, Mn), Am - is the intercalating anion, such as Cl-, SO42-, NO3-and n represents the water content [14]. An extensive development of these materials for various applications such as anion exchangers host structures for the synthesis of nanocomposite materials and precursors of catalysts [15]. In this latter case the attractive interest comes from the control of the crystallinity and texture, as well as the intrinsic acid–base and redox properties, which can be fine tuned by the composition, the synthesis method and the activation process [16].

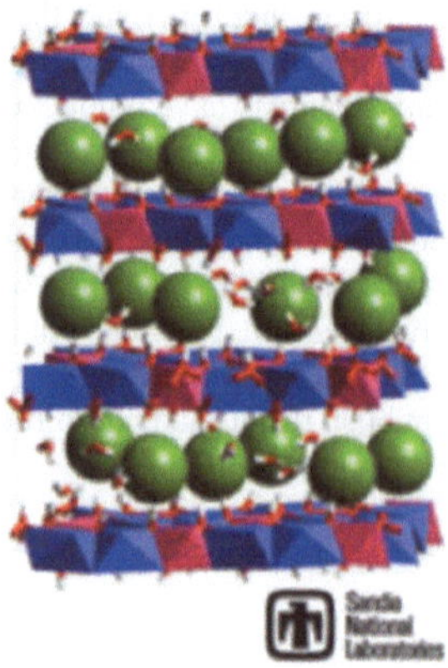

Figure 4. Schematic of hydrotacite (courtesy; Sandia National laboratories) [17].

The DNA of Innovators

- Associating: Knowing all about LDHs (e.g., reviews, papers, and experiments)
- Table 1 shows a general picture of associating the known parameters, application on hydrotalcites.
- Questioning: Formulate questions. Figrue 4 shows the overview how to associate hydrotalcite with different areas of science and engineering.
- Observing: Broad look at the technology area, needs, opportunities, markets, and others.
- Networking: Get to know the best individuals, companies, commercial potentials, and others.
- Experimenting: proof of principle, prototype test, apply, and get feedback.

Table 1: Associating known parameters with application of hydrotalcites

Property	Application
Energy surface sites	Catalysis
Ion proton conduction	Fuel cells, batteries
Anion exchange	Purification, water purification
Neutralization capacity	Antiacid
Intercalation	Drug delivery, bioactive molecule stabilization
Low to null toxicity	Medication
Formability, structural diversity	Powders to film
Polymer/cement compatibility	Composites
Water content	Flame retardance
Interlayer reactions	Confined space chemistry

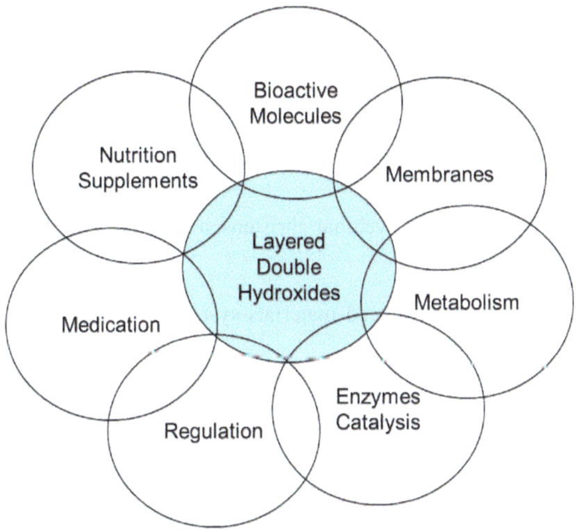

Figure 5: overview how to associate hydrotalcite with different areas of science and engineering.

Conclusions

Hydrotalcite-like materials have the tremendous potential to be utilized as an example for associating the methods of innovation, and make the innovative products be marketed.

References

1. M. Harvey, *Pasteur's Quadrant — Basic Science and Technological Innovation* (Washington DC, The Brookings Institute, 1997).

2. D. M. Rubio, et al., "Defining translational research: implications for training, "*Academic Medicine*, 85(3) (2010), 470–475.

3. M. Eili, et al., "Degradability Enhancement of Poly(Lactic Acid) by Stearte-Zn(3)Al LDH Nanolayers," *International Journal of Molecular Sciences*, 13(7) (2012), 7938-51.

4. S. Pranatharthiharan, et al., "Inorganic Nanovectors for Nucleic acid Delivery," *Drug Delivery and Translational Research*, 3(4) (2013), 446-470.

5. S. Mallakpour, and M. Dinari, "Structural Characterization and Thermal Properties of Chiral Poly(amide-imide)/Modified MgAl Layered Double Hydroxide Nanocomposites Prepared via Solution Intercalation, *Polymer-Plastics Technology and Engineering*, 53 (2014), 1047-1055.
6. http://iopscience.iop.org/2053-1613/

7. J. von Windheim, and B. Myers, "A lab-to-market roadmap for early-stage entrepreneurship ," *Translational Materials* Research, 1 (2014) 016001.

8. A.A. Goncalves Jr., G.H. Kaufmann, edi., "Speckle 2010: Optical Metrology," *Proceedings of SPIE*, 7387 (2010), 738712.

9. http://hbr.org/hbrg-main/resources/pdfs/marketing/webinar/christensen_innovator_dna_webinar_summary_101911.pdf.

10. http://dupress.com/articles/advanced-materials-systems/

11. http://d2mtr37y39tpbu.cloudfront.net/wp-content/uploads/2013/03/DUP286_fig1.jpg

12. http://www.nist.gov/mgi/upload/NIST-IR-7898.pdf

13. F. Cavani, et al., "Layered double hydroxides," *Catalysis Today,* 11 (1991), 173-209.

14. W. Jones, and M. Chibwe, *Pillared Layer Structures Current Trends andApplications*, (London, Elsevier, 1991), 67-81.

15. G. Carja, et al., "Textural properties of layered double hydroxides:effect of magnesium substitution by copper or iron," *Microporous and Mesoporous Materials*, 47 (2001), 275-284.

16. G. Carja et al., "MgVAl mixed oxides with mesoporous properties using LDHs as precursors," *Journal of Catalysis,* 218 (2003), 104-111.

17. http://www.sandia.gov/geo/research.html.

Energy Technology 2015: Carbon Dioxide Management and Other Technologies
Edited by: Animesh Jha, Cong Wang, Neale R. Neelameggham, Donna P. Guillen, Li Li, Cynthia K. Belt, Randolph Kirchain, Jeffrey S. Spangenberger, Frank Johnson, Andrew J. Gomes, Amit Pandey, and Peter Hosemann
TMS (The Minerals, Metals & Materials Society), 2015

LAYERED DOUBLE HYDROXIDES IN ENERGY RESEARCH: ADVANTAGES AND CHALLENGES

Andrew Gomes[1], David Cocke[2], Doanh Tran[3], Arnab Baksi[1]

[1]Chemical Engineering, Materials Instrumentation Center, Lamar University, Beaumont, TX
[2]Gill Chair of Chemical Engineering, Lamar University, Beaumont, TX
[3]GE Power and Water, Houston, TX
Keywords: Hydrotalcite, Layered Double Hydroxides, Energy, Challenges

Abstract

Layered Double Hydroxides (LDH) are being used in energy related research for last several years. Among others, research interests include synthesis of catalysts, electrocatalysts, photocatalysts, electrode materials for batteries and super capacitors, and luminescence materials. They can be also applied for the production of renewable energy sources, such as hydrogen and oxygen. LDH materials have several advantages in general, such as their low cost, versatility of their chemical composition, easily manipulated properties, a wide range of preparation variables, unique anion exchange and intercalation properties, chemical stability, and colloidal and thermal behavior. On the other hand, their use in particular energy research produces some challenging physical and chemical aspects that need to be addressed. These include lack of knowledge on exact structure, postulation of interlayer arrangement, and some selectivity issues in catalysis. In this work, we will revisit these topics with an overview of the historical background and current literature survey.

Introduction

Layered double hydroxides (LDHs) are one of the most promising materials families, widely used as catalyst, catalyst support, adsorbent, and favorable catalysts support with large anion adsorption capacity. They can be considered as anti-types of 2:1 (three-layer) clay minerals [1]. The general chemical formula is written as $[M^{II}_{1-x}M^{III}_{x}(OH)_2]^{x+}[A^{n-}_{x/n}]^{x-}\cdot yH_2O$, where M^{II} is a divalent cation (Mg^{2+} Zn^{2+}, Mn^{2+} , Mo^{2+}, Ni^{2+}, Cd^{2+}), M^{III} is a trivalent cation (Al^{3+}, Fe^{3+}, Cr^{3+}, Ga^{3+}), A^{n-} is a anion, such as carbonate, chloride, or sulfate. The cations by trivalent ones (e.g., Al^{3+}) generate positive charges on the layers, which are compensated by anions located in the interlayer region. Water molecules are also found in the interlayer space [2]. In a brucite layer, each Mg^{2+} ion is octahedrally surrounded by six OH^- ions, and the different octahedra share edges to form an infinite two-dimensional layer. Brucite crystals are then formed when the 2-D layers are stacked one on the top of another via O–H···O hydrogen bonding between the adjacent layers (Fig. 1) [3]. Partial replacement of Mg^{2+} ions by Al^{3+} gives the brucite-like layers a positive charge, which in hydrotalcite itself is balanced by carbonate anions, located in the interlayer region (gallery) between the two brucite-like layers. This gallery also contains water molecules, hydrogen bonded to layer OH^- ions and/or to the interlayer anions. The electrostatic interactions and hydrogen bonds between the brucite-like layers and the gallery species hold the layers together, forming the three-dimensional structure [4].

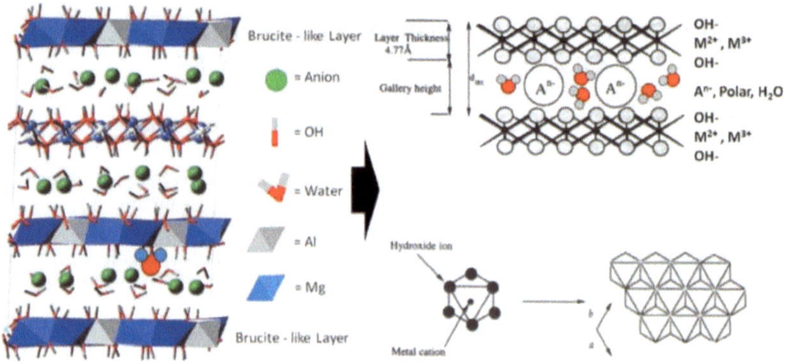

Figure 1: Schematic Representation of a Layered Double Hydroxide. After [3]

History

Hydrotalcite was first discovered in Sweden in 1842, and it was first regarded as mixed hydroxides but were later identified as magnesium aluminum and magnesium iron hydroxide carbonates in the 1920s. By the 1930s, Feitknecht conducted many studies on the synthesis, stability, solubility and structural determination of these materials and regarded them as double layered materials [5-6]. It was not until the 1960s that Allmann and Taylor used single crystal X-ray diffraction (XRD) studies on these mineral samples that correctly identified the layered structure features of layered double hydroxides (LDHs). LDHs are known as hydrotalcite-like compounds due to their structural similarity and are a family of natural and synthetic compounds having the general formula. Today, many of the fine details of the structure, the extent of ordering of metal cations within the layers, the stacking arrangement of the layers and the arrangement of anions and water molecules in the interlayer galleries are still not fully understood [6].

Application of LDH

Layered double hydroxides can be used for several applications due to their ability to vary the various components and different chemical properties of the compound in incipient or derived form. LDHs have extensive applications which include use as additives in polymers, as precursors to magnetic materials, in biology and medicine, anti-corrosion coatings, components in optical or magnetic devices, sensors, in catalysis and in environmental remediation [1-7].
For many applications in practical devices and applications and products such as solar cells and sensors, [Zn-Al] type LDHs can be in a uniformly aligned two-dimensional (2-D) arrays or films on various substrates [6-7]. A [Zn-Al-Cl] LDH compound via screen printing process fabricated a small room temperature humidity sensor [6].
LDHs can be used as precursors of catalysts or catalyst supports because of specific features of mixed oxides obtained by controlled thermal activation [8]. The features that makes LDHs promising as catalysts are: i) high surface area, ii) homogeneous and thermally stable interdispersion of the metal cations, that by reduction can form small and stable metal crystallites, iii) synergetic effects between the elements which favor unusual basic or hydrogenating properties, and iv) memory effect, with structure reconstruction under mild

conditions [1, 8]. With a deeper understanding of LDHs parameters, more active LDH-based catalysts can be produce for applications in $DeNO_x$, $DeSO_x$, full oxidation of VOCs and steam reforming reactions for H_2 production [8].
Complex impedance method studies on ionic conduction mechanisms of such anionic clays show a strong dependence on conductivity in relation to hydration state, on both the nature of intercalated anions and hydroxylated framework [3]. This shows the potential of LDHs for applications in battery and fuel cell materials to reduce cost due to high surface areas and transition metal oxide content along with good thermal stability [3].
Powdered LDHs have uses for medical applications. LDHs have a prolonged buffering effect and bile-binding activity and can act as antacids for the treatment of gastric or duodenal ulcers [9]. Powdered CaAl-LDHs are also used in the construction industry as one types of additives in concrete as hardening accelerators. Additionally, powered ZnMgAl-LDHs are used as additive in flame retardants due to the low toxicity emission and high thermal stability. Since the interlayer may host very diverse chemicals such as contaminants LDH is a very attractive sorbent for water treatment of inorganic contaminants, such as NO_3^-, SO_4^{2-}, PO_4^{3-}, AsO_4^{3-}, and CrO_4^{2-}, as well as monoatomic anions, F^-, Cl^-, Br^-, and I^- [7].

Energy Research and LDH

Energy research with LDH includes synthesis of catalysts, electrocatalysts, photocatalysts, electrode materials for batteries and super capacitors, and luminescence materials. They can be also applied for the production of renewable energy sources, such as hydrogen and oxygen.

Catalysts

Catalysts can be prepared from LDH through calcination and rehydration, calcination and reduction, delamination, intercalation, change of assembly, and uniform distribution of metal ions [10]. Alves et al. [11] used LDH as heterogeneous catalysts for conversion of sun flower seeds to biodiesel. LDH materials containing transition metal cations, particularly Ni- and Co, have been widely used in electrocatalytic reactions. Zou et al. [12] used ZnCo-LDH as electrocatlysts for water and alcohol oxidation. This electrocatalyst retains a lower overpotential by 100 mV compared to $Co(OH)_2$ and Co_3O_4 system for electrochemical oxidation of water, and the turn over frequency is about 10 times higher than the latter system. It was hypothesized that Zn^{2+} acts as a structural support assisting the reactant-catalyst interactions. Semiconductive photocatalysts are used to split water by solar energy. LDH containing Ti^{4+}, Nb^{5+}, or Ta^{5+} cations possess photocatalytic capability for water splitting, since they may provide spatially separate reduction and oxidation reaction sites [10].

Electrode Materials/Super Capacitors

There are three major types of electrode materials widely used for electrochemical supercapacitors: metal oxides/hydroxides, carbon materials and conducting polymers [1-6,Warsi]. Due to their low cost, high redox activity, flexible ion exchangeability, tunable composition and environmentally friendly nature, LDH has higher potential in this area. CoNi-LDH has been widely studied as a high-performance electrode material for supercapacitors due to its ability to offer richer redox reactions with better electrochemical activity and electrical conductivity as compared to the to the both corresponding single component hydroxides ($Ni(OH)_2$ and $Co(OH)_2$) [10].

Batteries

Dai et al. [13] synthesized NiAlCo-LDH nanoplates attached to few-walled carbon nanotubes (NiAlCo/CNT) as the cathode material of rechargeable NiZn battery in aqueous alkaline electrolytes. Tsuneishi et al. [14] applied Mg–Al layered double hydroxide (LDH) as anion conductive solid electrolytes in Fe–air secondary battery. The ion conductivity of LDH was markedly enhanced upon addition of KOH.

Luminescence Materials

The main requirements for developing luminescence materials are high quantum efficiencies, the repeatability of synthetic procedures and the stability of light output [15]. They investigated the synthesis of Tb- and Eu-containing LDH-based luminescent hybrid materials with quantum yields up to 45%. Energy transfer studies in corresponding mixed Tb^{3+}/Eu^{3+} systems revealed highly efficient ligand→Tb^{3+}/Eu^{3+} and Tb^{3+}→Eu^{3+} communication channels.

Production of H_2 and O_2

Carbonate-intercalated ZnCr–LDH can act as a visible-light-active photocatalyst for water splitting to generate hydrogen. The carbonate radicals originating from the oxidation of intercalated carbonate anions by photogenerated holes were found to improve the light-induced electron–hole separation. Replacing Zn^{2+} with Ni^{2+} also results in metal to metal charge transfer transitions of Zn(II)/Ni(II)–O–Cr(III) to Zn(I)/Ni(I)–O–Cr(IV) accounting for the enhanced photocatalytic activity for hydrogen evolution. In addition, the high dispersion of TiO_6 octahedral confined in the 2D matrix of a series of MTi–LDH (M = Ni, Zn, Mg) can suppress the carrier recombination under visible-light excitation, by taking advantage of the large number of surface Ti^{3+}–O defects serving as trapping sites to improve the electron–hole separation. The modified band gap of about 2.1 eV obtained by the formation of the M^{2+}–O–Ti network greatly facilitated water splitting, accounting for the 1.23 eV thermodynamic overpotential and other kinetic overpotentials for both hydrogen and oxygen evolution. As a result, Ti-containing LDHs showed an extremely high photocatalytic H_2-production rate of 31.4 mmol h−1, which was 18 times higher than that of the control $K_2Ti_4O_9$ sample [10].

Advantages and Challenges

LDH materials have several advantages in general, such as their low cost, environment-friendly, versatility of their chemical composition, easily manipulated properties, a wide range of preparation variables, unique anion exchange and intercalation properties, chemical stability, and colloidal and thermal behavior. On the other hand, their use in particular energy research produces some challenging physical and chemical aspects that need to be addressed. These include lack of knowledge on exact structure, postulation of interlayer arrangement, and some selectivity issues in catalysis. Table I shows a comparative study of eight cases of LDH uses regarding their advantages, challenges, and future aspects.

Table I. Literature Survey on Energy-Related LDH Research

LDH Type	Application	Advantages	Challenges	Research Prospect
Fe-/Co-/Ni-containing Mg-Al-LDH [16]	Nuclear fuel storage.	High co-precipitation yield.	Corrosion protection. Overcome layer stacking faults. Maintaining Fe(II) in the structure.	Different humidity level study during XRD analysis. Thermodynamic study of the crystal lattice formation.
Cobalt–nickel-LDH [17]	Redox-super capacitor design.	Highest Specific capacitance value. High power density. Very high theoretical specific capacitances.	Agglomeration of particles during preparation. Large particle size and requirement of a binder	Achieve nanomaterial composites with high specific capacitances.
20% Ni/Al_2O_3-LDH [18]	Biofuel	Inexpensive Ni-based catalysts.	Intermediate fatty acid formation.	Reaction kinetics study to optimize the rate limiting step.
α-$Ni(OH)_2$-LDH [19]	Rechargeable batteries.	Improved crystallinity of a-$Ni(OH)_2$, and anion exchange of NO_3^- by OH^-, better electrochemical performance with high-rate discharge ability.	Negative effect on the tap-density. Unstable α-$Ni(OH)_2$, easily transformed to β-$Ni(OH)_2$ upon aging at high pH. Decrease of discharge capacity with increasing amount of substituted Al.	Obtain a large enough size and high enough density of the particles co-precipitation of metals containing Al.
Co/Al -LDH [20]	Alternative electrical energy storage	Increase in capacitance with increase in amount of Co/Al-LDH. Simple synthesis process. Good rate capability, good coulombic efficiency, and electrochemical stability.	High electrode resistance.	Further fine-tuning by optimizing with anodic alumina of different pore sizes and densities.

Table I. Literature Survey on Energy-Related LDH Research (continued)

LDH Type	Application	Advantages	Challenges	Research Prospect
Mg–Al CO_3 type LDH [21]	Biomass-derived materials for Green & Sustainable Chemistry	Superior adsorption capacity; cation-exchange ability of the brucite layer; anion-exchange ability of the interlayer space; and tunable basicity of the surface	Adsorption capacity; cation-exchange ability of the brucite layer; anion-exchange ability of the interlayer space; tunable basicity of the surface.	Understanding of the structure and crystallization mechanism of LDH.
Neutral poly(vinylcarbazole) (PVK) and tris[2-(4,6-difluorophenyl) pyridinato-C2,N]iridium(III) ($Ir(F_2ppy)_3$) were assembled with LDH nanosheets forming ordered ultrathin films (UTFs) [22]	Chemical sensors	Low-cost and reliable sensors for detecting volatile organic compounds. Instantaneous detection, minimizing UV excitation.	Good homogeneity and continuity with only poly(vinylcarbazole) assembled LDH nanosheets to form ultrathin films.	Investigation on optical functionality of UTFs with 2D FRET process and interaction with VOC molecules, novel ideas of the design and preparation of LDH-based sensors for industrial/agricultural/biological fields.
Ni/Al LDHs [23]	Electrochemical capacitors	High theoretical specific capacitance and low fabrication cost.	Design of multiscale porous materials without surfactants or templates.	Design of hierarchical porous Ni/Al LDHs by self-assemble without any catalyst, template or surfactant. Study on the growth mechanism and more understanding on the flower like structure.

Conclusion

In conclusion, there are immense opportunities of layered double hydroxide (LDH) available in the energy related research in material science. It is a challenge to develop novel materials to make catalyst or energy storage device from a low cost, easy available material. Due to a high surface to volume ratio and ordered structure of LDH, it has a lot of potential in energy utilization and conservation.

References

1. G. Grünewald, K. Kaiser, and R. Jahn, "Hydrotalcite—A potentially significant sorbent of organic matter in calcareous alkaline soils," *Geoderma,* 147 (2008), 141–150.

2. Dongjin Wan et al., "Role of the Mg/Al atomic ratio in hydrotalcite-supported Pd/Sn catalysts for nitrate adsorption and hydrogenation reduction," *Journal of Colloid and Interface Science,* 332 (2009), 151–157.

3. X. Guo, F. Zhang, D.G. Evans, and X. Duan, "Layered double hydroxide films: synthesis, properties and applications". *Chemical Communications*, 46 (29) (2010), 5197-5210.

4. L P.S. Braterman, Z.P. Xu and F. Yarberry In: S.M. Auerbach, K.A. Carrado and P.K. Dutta, Editors, *Handbook of Layered Materials* (New York, Dekker, 2004), 373–474.

5. V. Rives, ed., *Layered Double Hydroxides: Present and Future* (New York, Nova Science Publishers, 2001).

6. D. Evans, and R. Slade, "Structural Aspects of Layered Double Hydroxides Layered Double Hydroxides", *Layered Double Hydroxides (Structure and Bonding)*, ed., X. Duan and D. Evans (Berlin/ Heidelberg, Springer. 2006), 1-87.

7. W. Canter, *Nitrates in Groundwater* (Boca Raton, FL, CRC Press, 1996), 217-256.

8. T.J. Pinnavaia, et al., "Organic chemical conversions catalyzed by intercalated layered double hydroxides (LDHs)", *Applied Clay Science*, 10 (1-2) (1995), 117-117.

9. A.D. Roy, and J.P. Besse, "Evolution of protonic conduction in some synthetic anionic clays", *Solid State Ionics*, 46 (1-2) (1991), 95-101.

10. G. Fan, F. Li, D.G. Evans, and X. Duan, "Catalytic applications of layered double hydroxides: recent advances and perspectives," *Chemical Society Reviews,* 43 (2014), 7040-7066.

11. O.L. Alves, et al., "Application Layered Double Hydroxides (LDH) as heterogeneous catalysts for synthesis of biodiesel," *Journal of Physics Conference Series,* 304 (012024) (2011),1-8.

12. X. Zou, A. Goswami and T. Asefa, "Efficient Noble Metal-Free (Electro)Catalysis of Water and Alcohol Oxidations by Zinc–Cobalt Layered Double Hydroxide", *Journal of American Chemical Society*, 135 (2013), 17242-45.

13. M. Gong, et al., "An Advanced Ni Fe Layered Double Hydroxide Electrocatalyst for Water Oxidation" *Journal of American Chemical Society*, 135 (2013), 8452-8455.

14. T. Tsuneishia, et al., "Preparation of hydroxide ion conductive KOH–layered double hydroxide electrolytes for an all-solid-state iron–air secondary battery", *Journal of Asian Ceramic Societies*, 2 (2014), 165–168.

15. N.. Zhuravleva, et al., "Energy transfer in luminescent Tb- and Eu-containing layered double hydroxides," *Mendeleev Communications*, 14 (2004), 176-178.

16. H. Curtius et al., "Preparation and characterization of Fe-, Co-, and Ni-containing Mg-Al-layered double hydroxides," *Clays and Clay Minerals*, 61 (2013), 424–439.

17. V. Gupta, S. Gupta, and N. Miura, "Potentiostatically deposited nanostructured CoxNi1-x layered double hydroxides as electrode materials for redox-supercapacitors," *Journal of Power Sources* 175 (2008) 680–685.

18. Jimenez et al., "Catalytic deoxygenation of triglycerides and fatty acids to hydrocarbons over Ni–Al layered double hydroxide," *Catalysis Today* 237 (2014) 136–144.

19. Li et al., "Synthesis, characterization and electrochemical performance of high-density aluminum substituted α-nickel hydroxide cathode material for nickel-based rechargeable batteries," *Journal of Power Sources* 270 (2014) 121-130.

20. Loh et al., "Co/Al layered double hydroxides nanostructures: A binderless electrode for electrochemical capacitor," *Electrochemistry Communications* 43 (2014) 9–12.

21. S. Nishimura, A. Takagaki, and K. Ebitani, "Characterization, synthesis and catalysis of hydrotalcite-related materials for highly efficient materials transformations," *Green Chem.*, 15 (2013) 2026.

22. Qin et al., "Phosphorescent Sensor Based on Iridium Complex/Poly(vinylcarbazole) Orderly Assembled with Layered Double Hydroxide Nanosheets: Two-Dimensional Föster Resonance Energy Transfer and Reversible Luminescence Response for VOCs, *J. Phys. Chem. C* 118 (2014) 20538–20544.

23. Song et al., "Self-assembled hierarchical porous layered double hydroxides by solvothermal method and their application for capacitors," *Microporous and Mesoporous Materials* 148 (2012) 159–165.

Energy Technology 2015: Carbon Dioxide Management and Other Technologies
Edited by: Animesh Jha, Cong Wang, Neale R. Neelameggham, Donna P. Guillen, Li Li, Cynthia K. Belt, Randolph Kirchain, Jeffrey S. Spangenberger, Frank Johnson, Andrew J. Gomes, Amit Pandey, and Peter Hosemann
TMS (The Minerals, Metals & Materials Society), 2015

MODELLING THE STRUCTURE AND VIBRATIONAL PROPERTIES OF LAYERED DOUBLE HYDROXIDES

Erwan Andre, Jean Fahel, Cedric Carteret

Laboratoire de Chimie Physique et Microbiologie pour l'Environnement (LCPME) UMR7564
CNRS – Lorraine University; 405 rue de Vandoeuvre; Villers-lès-Nancy, 54602, France

Keywords: Layered Double Hydroxides, DFT, Infra-Red, Raman, XRD

Abstract

A model structure is proposed for the layered double hydroxide $[Mg_2Al(OH)_6]^+ \cdot Cl^-$ (Mg/Al ratio R = 2). The structural parameters of this model have been optimized by DFT calculations and both PXRD and vibrational spectra have been simulated. The comparison with experimental data confirms the validity of the model. Thanks to the synergy between experiment and theory it is now possible to analyze in detail the vibrational signals and understand the interactions between the different components of this material.

Introduction

Layered Double Hydroxides (LDHs) also known as anionic clays are well known for their proprieties of anion exchanger, pollutant scavenger, or catalyst precursors [1,2]. They can be represented by their general formula $[M^{II}_{(1-x)}M^{III}_{x}(OH)_2]^{x+}[A^{n-}_{x/n}\ , z\ H_2O]^{x-}$.In this formula, one can see the different factors which rule the composition of this material: MII and MIII form a couple of divalent and trivalent cations constituting sheets of face sharing hydroxyl octahedra, An- is a charge balancing anion, x is the layer charge and z is the amount of intercalated water molecules. One can also define R = (x-1)/x which is the M^{II}/M^{III} ratio. LDHs form a complex system in which different factors such as: The layers composition, the stacking sequence or the structuration of the interlayer domain can influence their properties of anion exchanger or nano-reactor [3].

If the X-Ray Diffraction (XDR) is the technique of choice to characterize the structure of these materials at the crystal scale (stacking sequence, interlayer spacing...) through vibrational spectroscopy (Infra-Red and Raman) it is possible to access in situ information, at molecular scale, of both anionic and cationic parts of these systems.

The aim of this work is to show how quantum mechanical simulations (realized at DFT level through a periodic approach) can rely on experimental data brought by XRD and vibrational spectroscopies to validate a model structure. And in return, how computed data obtained from this model can help to assign the experimental signatures of LDHs as well as they give access to various properties depending on the electronic distribution.

To illustrate this methodology, a focus is made on the compound $[Mg_2Al(OH)_6]^+ \cdot Cl^-$ (named HTC hereafter), that possess the same layer structure as Hydrotalcite (natural mineral) but a different intercalated anion (CO_3^{2-}). This choice is motivated by the fact that chloride is an atom and thus does not possess any vibrational signature, making this compound the simplest system that can be modelled.

Experimental Details

Sample preparation

The synthesis of the carbonate LDH precursor was performed through the method of precipitation at low supersaturation [4]. Typically, a 0.4 M solution of $MgCl_2$ and $AlCl_3$ salts with R = 2 was added slowly (0.3 mL/min) to an anionic solution of Na_2CO_3 (0.25 M). The pH was kept constant at 11 by simultaneous addition of NaOH (1M) solution using an automatic titrator device. After complete addition of the cationic solution, the slurry was then placed in autoclave and heated at 100°C for 20h. The final solid was washed several times with deionized water and dried in oven at 60°C over a night.
The conversion of the carbonate precursor to the chloride LDH was performed using the method proposed by Iyi [5]. An ethanolic solution prepared by diluting a 1M HCl solution in 5 mL of ethanol was slowly added to a vigorously stirred CO_3^{2-} LDH solution (0.662 mmol) in 45 mL ethanol solution under Argon flow. The factor f, defined by Iyi *et al.* as f = [HX] / (2*[CO_3^{2-}]) where [HX] represents the monoprotic acid concentration, was set to 1.5. The solution was stirred for 1 hour at room temperature under Argon flow. Then, the resulting suspension was filtered with a polycarbonate membrane with 0.2µm pore size, washed thoroughly with absolute ethanol and dried in oven at 60°C.

Characterization

The ratio R, as well as the chloride content of both precursor and exchanged LDH was determined by Inductively Coupled Plasma equipped with atomic emission spectrometry detection (ICP-AES).
Infrared measurements, recorded in the range 4000-400 cm^{-1} on a Nicolet 8700 apparatus continuously purged with ultrapure N_2, were carried out with a DTGS-KBr detector and a KBr beam splitter. The spectral resolution was 4 cm^{-1} and 100 scans were co-added for each spectrum. Samples were pellets made of a ground mixture of dried KBr (FTIR grade) and 1% weight of LDH powder pressed up to 5 tons. A pure KBr pellet was used as background. In the range 650-80 cm^{-1}, spectra were acquired with a DTGS-PE detector and a solid substrate beam splitter. Samples were prepared by depositing 170 µL of a 4 g/L LDH suspension on a 13 mm polyethylene (PE) pellet diameter. All suspensions were prepared with deionized water to avoid carbonate contamination of Cl-LDH samples. A pure PE pellet was used as background. For background and samples, two spectra were averaged to minimize random noise.
Raman spectra were obtained using the Argon laser excitation line at 514.5 nm of a Jobin-Yvon T64000 spectrometer equipped with a liquid nitrogen cooled Charged-Coupled device (CCD). The power of the excitation source was set to 200mW. The scattered light was collected in a backscattering configuration by a microscope using an X50 olympus objective (laser spot ~ 4 µm). The instrumental resolution was ~ 4 cm^{-1}.
X-ray diffraction patterns were recorded with a Phillips X'Pert Pro MPD diffractometer in reflection geometry using Cu $K_{\alpha 1}$ radiation (λ = 154,06 pm). Data were collected with a dedicated chamber from finely ground samples. The 2θ range was explored between 5 and 100° at a speed of 2°/min.

Computational details

Calculations were performed with the CRYSTAL14 periodic *ab initio* code [6,7], by using an all-electron Gaussian type basis set and various functionals. Most of the results will refer to the hybrid B3LYP functional [8,9], that has been shown to provide excellent results for the vibrational properties of ionic and semi-ionic compounds [10], but slightly overestimated distances and cell volumes. To overcome this issue a B3LYP-D* scheme have been adopted, where the hybrid B3LYP functional is coupled to a Grimme approach [11,12] with parameters specifically optimized for the solid state [13]; this permits to take into account dispersive forces, that are expected to play an important role in determining the equilibrium structure of the present systems. Magnesium, aluminum, chlorine, oxygen and hydrogen were described by (8s)-(511sp)-(1d), (8s)-(621sp)-(1d), (8s)-(6311sp)-(1d),(8s)-(411sp)-(1d) and (211s)-(1p) contractions, respectively. The level of accuracy in evaluating the Coulomb and Hartree-Fock exchange series is controlled by five parameters [7], for which the 7 7 7 8 18 values were used. The threshold on the SCF energy was set to 10^{-10} Ha for both the geometry optimization and the frequency calculation. The reciprocal space was sampled according to a sublattice with shrinking factor set to 4 (along the 3 lattice vectors), corresponding to 24 and 36 independent k vectors in the irreducible part of the Brillouin zone.

Construction and validation of the model

In order to model the vibrational signature of LDHs, the first step is to construct a proper model of its crystalline structure. LDHs possess the same layer structure as Brucite $Mg(OH)_2$ that crystallizes in P-3m1 space group (sequential number n°: 164) with Mg in site 1a and OH in site 2d (Wyckoff positions) [14]. In the case of HTC, the metal position can be occupied either by Mg or Al with a global ratio of R(Mg/Al) = 2. But periodic quantum calculations do not support partial occupations. One has to define a supercell to describe the cation distribution. In the case of a ratio R(Mg/Al) = 2 there is only one way to distribute the cations in order to avoid two consecutive $Al(OH)_6$ octahedra. This distribution can be described with a hexagonal supercell of a√3 length [15] as displayed in figure 1.

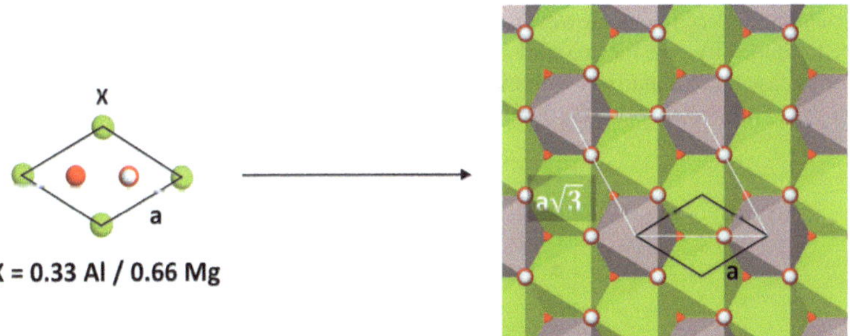

Figure 1: Primitive unit cell of HTC's brucitic layers with partial occupations (left) as well as the corresponding supercell used in calculations (right). Mg octahedra are green, Al octahedral are purple.

Depending on the way the layers stack, different polytypes can be formed [16]. The relative position of the layers hydroxides defines two types of crystallographic sites for the anion insertion: Octahedral or Prismatic. The nature of the preferred site (and thus polytype) depends on the interlayer content. One can cite the nature of the intercalated anion and its way to interact with the layer (hydrogen bonding versus electrostatic interaction) or the water content of the interlayer domain. All the stacking possibilities up to three layers have been tested. After optimization of their structures, the energy differences between the different models (a few $kcal.mol^{-1}$) were too small to draw a firm conclusion. To be able to go further, theoretical results had to be compared with experimental data. As powder X-Ray Diffraction (PXRD) is the technic of choice regarding the structure determination of LDH, this technic has been chosen to discriminate the different models. The PXRD patterns have been constructed using the theoretical structure factors obtained from the optimized structures. The patterns of the two most stable structures are compared to experimental data in figure 2.

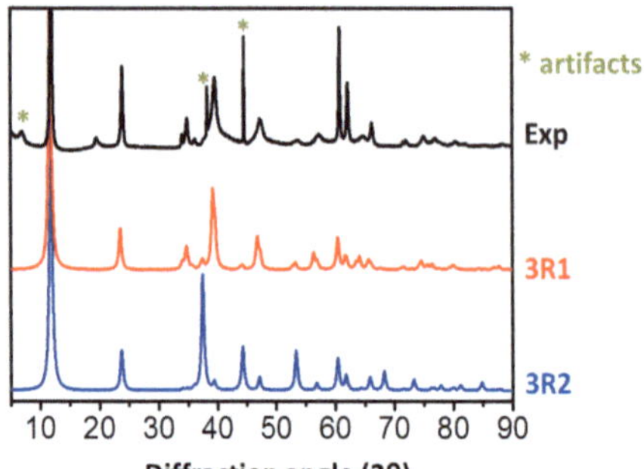

Figure 2: Comparison between experimental powder diffraction pattern of HTC and simulated patterns corresponding to 3R1 and 3R2 polytypes.

While the theoretical patterns are similar for low angles (corresponding to interlayer distance) they differ in the 30 – 55° (2θ) region, which is characteristic of the stacking sequence. One can clearly see that in that case, the 3R1 polytype fit better the experimental pattern. This stacking sequence can be described with an orthorhombic cell of R-3m symmetry (sequential number n°: 166) containing three layers However, for the sake of computational time, the size of the system has been reduced by considering a monoclinic cell of Cm symmetry that contain only one layer. In order to ease the comparison with experiment data, the crystallographic information displayed Table 1 are expressed in the orthorhombic cell.

HTC	Experiment	Theory
a	3.048	3.048
c	22.314	22.410
α	90.0	89.2
γ	120.0	119.9
V	179.5	180.4
Cz	7.438	7.469

Table 1: Calculated and experimental (XRD) in R-3m symmetry cell parameters (lengths [Å], angles [°]) and cell volume V [$Å^3$]. Cz is the interlayer spacing.

As it appeared in the XRD patterns, the agreement between theory and experiment is fairly good. The interlayer is slightly overestimated and, due to the symmetry lowering, the angles are distorted in the monoclinic structure although remaining very close to ideal values. These results confirm the validity of the current model. It is now possible to study its vibrational signature.

Vibrational spectra

The vibrational frequencies of the HTC's crystal have been calculated within the harmonic approximation at the Γ point. The computation of the dipole moment derivative gives access to

Infra-Red intensities. It is then possible to reconstruct a theoretical IR spectrum and to compare it to experimental data as shown in figure 3. Likewise, Raman intensities can be calculated from the third derivative of the total energy (twice with respect to an externally applied electric field and once with respect to the atomic displacements) and a theoretical spectrum can be constructed. Those theoretical spectra are compared to experimental data in figure 3.

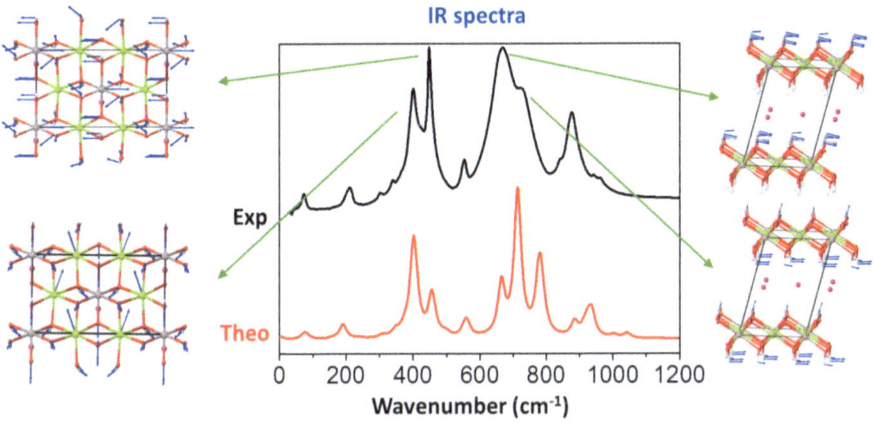

Figure 3: Comparison between experimental and simulated IR spectra of HTC. The normal vibration modes of the main pics are also displayed.

The Infra-Red spectrum of HTC is characterized by two main groups of pics. One around 400 cm^{-1} and the second around 700 cm^{-1} Each one of these pics can be assigned in the theoretical spectrum to its corresponding normal mode. For example, the two most intense pics in the first group correspond to deformations of the layer's octahedra. One can see clearly that the contributions of the divalent and trivalent cations are decoupled. The pic at low frequency expresses the contribution of the magnesium's framework while the one at 450 cm^{-1} correspond to the deformation of the aluminum's octahedra. One can note that even if the model reproduces correctly the positions of the pics, the modelling of intensities is less good as the relative intensities of these two modes is reversed. The second group of pics can be assigned to layers hydroxides bending modes. The splitting of these modes is due to differences of chemical environment of between those hydroxides. Indeed, the shift in layer stacking defined by the polytype, imposes that hydroxides interacting with the chloride anion are not the same in the "upper" layer than those of the "lower" layer.

The figure of the Raman spectra illustrate a limitation of the modeling process: The width of the different pics is not computed. To draw a spectrum a constant value is given to all the pics. The consequence is that the width of some pics is strongly underestimated. It can be seen for the pics around 700 cm^{-1} in the theoretical Raman spectrum. This is due to the high anharmonicity of these vibrations. A more realistic model for the width would require computing the coupling between each mode and all the others that is not possible in the current version of the crystal code.

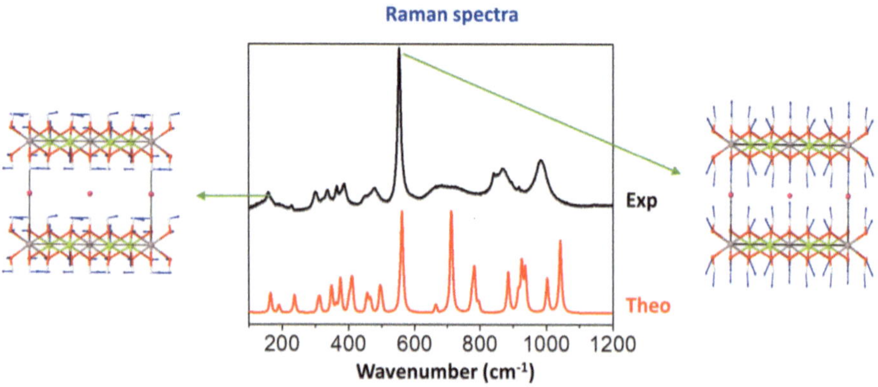

Figure 4: Comparison between experimental and simulated Raman spectra of HTC along with vibration modes of characteristic pics.

The main pic in the Raman spectrum of HTC corresponds to the symmetric elongation of the M-OH distances (breathing) of the octahedral layers. As it is easier to record to access to low frequencies with the Raman spectroscopy, we can highlight the pic at 160 cm^{-1}. It corresponds to an out of layer's plane motion of the magnesium atoms.

Conclusion

A model structure has been constructed for the LDH $[Mg_2Al(OH)_6]^+ \cdot Cl^-$. The validity of this model, including the stacking sequence, has been confirmed by the comparison to experimental XRD powder pattern. Using this model, it is possible to simulate the vibrational spectra and to compare with experimental Infra-Red and Raman data. The comparison between theoretical spectra and experimental ones serves two different goals: First, it is possible to detect in the experimental spectrum any signal that does not correspond to the model (amorphous phases, impurities…). But more important: thanks to the theoretical spectrum, it is possible to assign each pic to its corresponding normal mode. Then, it becomes easier afterwards to explain the evolution of the measured signals when a perturbation is applied to the material (hydration, anion exchange…).

References

1. V. Rives, *Layered Double Hydroxides: Present and Future* (Nova Publishers, 2001).
2. X. Duan and D. G. Evans, *Layered Double Hydroxides* (Springer Science & Business Media, 2006).
3. G. R. Williams and D. O'Hare, J. Mater. Chem. **16**, 3065 (2006).
4. B. Grégoire, C. Ruby, and C. Carteret, Cryst. Growth Des. **12**, 4324 (2012).
5. N. Iyi, H. Yamada, and T. Sasaki, Appl. Clay Sci. **54**, 132 (2011).
6. R. Dovesi, R. Orlando, A. Erba, C. M. Zicovich-Wilson, B. Civalleri, S. Casassa, L. Maschio, M. Ferrabone, M. De La Pierre, P. D'Arco, Y. Noël, M. Causà, M. Rérat, and B. Kirtman, Int. J. Quantum Chem. **114**, 1287 (2014).

7. R. Dovesi, V. R. Saunders, C. Roetti, R. Orlando, C. M. Zicovich-Wilson, F. Pascale, B. Civalleri, K. Doll, N. M. Harrison, P. D'Arco, M. Llunell, M. Causà, and Y. Noël, *CRYSTAL14 User's Manual* (2014).
8. A. D. Becke, J. Chem. Phys. **98**, 5648 (1993).
9. C. Lee, W. Yang, and R. G. Parr, Phys. Rev. B **37**, 785 (1988).
10. M. De La Pierre, C. Carteret, L. Maschio, E. André, R. Orlando, and R. Dovesi, J. Chem. Phys. **140**, 164509 (2014).
11. S. Grimme, J. Comput. Chem. **25**, 1463 (2004).
12. S. Grimme, J. Comput. Chem. **27**, 1787 (2006).
13. P. Ugliengo, C. M. Zicovich-Wilson, S. Tosoni, and B. Civalleri, J. Mater. Chem. **19**, 2564 (2009).
14. T. Nagai, T. Hattori, and T. Yamanaka, Am. Mineral. **85**, 760 (2000).
15. D. G. Costa, A. B. Rocha, R. Diniz, W. F. Souza, S. S. X. Chiaro, and A. A. Leitão, J. Phys. Chem. C **114**, 14133 (2010).
16. A. S. Bookin and V. A. Drits, Clays Clay Miner. **41**, 551 (1993).

Energy Technology 2015: Carbon Dioxide Management and Other Technologies
Edited by: Animesh Jha, Cong Wang, Neale R. Neelameggham, Donna P. Guillen, Li Li, Cynthia K. Belt, Randolph Kirchain, Jeffrey S. Spangenberger, Frank Johnson, Andrew J. Gomes, Amit Pandey, and Peter Hosemann
TMS (The Minerals, Metals & Materials Society), 2015

Synthesis of Hydrotalcite-like Compounds from Blast Furnace Slag: The Effect of Synthesis Parameters on Structure and Crystallinity

Maocheng He[1], Jianliang Zhang[1], Zhiwen Shi[1], Feng Liu[1], Xinyu Li[1]

[1]School of Metallurgical and Ecological Engineering, University of Science and Technology Beijing, 30 Xueyuan Rd; Haidian District, Beijing, 100083, China

Keywords: Hydrotalcite-like compounds, Blast furnace slag, Acid leaching, High saturation method

Abstract

Blast furnace slag (BFS) is a kind of by-product in iron-making process. The main components of BFS are CaO, SiO_2, Al_2O_3, MgO and a small deal of Fe, Mn, Cr, etc. In 2012, 450 million tons of BFS be generated in the world. In most cases, BFS is regarded as a kind of waste except when the water-granulated BFS was used for cement material and paving material. Therefore, it is extremely urgent to utilize BFS in a comprehensive way. Fortunately, The mole ratio of Ca and Mg to that of Al is about 3.5 in the BFS, which is an appropriate value in hydrotalcite-like compounds synthesizing, and $Si0_2$ in the BFS can be used as raw material of zeolite synthesis, researchers have synthesized the two products successfully and Explored the influence of synthesis parameters on the synthetic products deeply.

Introduction

Hydrotalcite-like compounds (HTLcs) is a layered double hydroxide, In the present, HTLcs successfully be used as exchange agent, adsorbent, catalyst, heat stabilizer and ultraviolet absorbing material [1-5]. However, most of HTLcs manufactured by chemically pure reagent, the production is high, and in china, the market prices of HTLcs as high as 1000$~18000$ per ton, therefore it is particularly important to use an inexpensive materials to synthesize such excellent performance product. Currently found in BFS is an appropriate HTLcs raw material.

BFS is a low-value secondary product produced during the extraction iron from iron ore. The major components of BFS are CaO, SiO_2, Al_2O_3, MgO which usually constitute 95% of the total weight, the main components and proportion of an ordinary BFS is relative stable (see Table I). Most of BFS is produced as water-granulated blast furnace slag, in this process, because high temperature BFS failed to convert all heat to chemical energy, this part of unreleased heat is converted to an amorphous state and vitreous storage, and thus cooled BFS have the potential of chemical activity. In the cement industry, the production of Portland cement is often necessary to add 5 to 20 percent of the active material, thus BFS- superfine powder is an excellent cement raw material [6-8]. In China, 210 million tons BFS is generated per year and huge quantity of blast furnace slag accounted for about 60% of the world, BFS is mainly applied in the cement industry and the civil engineering, the utilization rate of the BFS resource is only 75%.

According to the reference, the initial molar ratio of M^{2+} to M^{3+} for synthesizing HTLcs should be from 2 to 5 [9], and for most BFS, the molar ratio of $(Ca^{2+}+Mg^{2+})$ to Al^{3+} is around 3.5 (see Table I). Above all, the residual part mainly is silica gel after metal elements be extracted from BFS by acid leaching, while silica gel is a necessary material be used to synthesize zeolite [10].

This paper is committed to exploring a method, which is able to synthesize HTLcs from BFS conveniently, and this idea is conformity with the theme of developing green economy.

Table I. The sample main composition of blast furnace slag from different company

Components	BFS a	BFS b	BFS c
SiO_2	34.89	36.94	34.91
CaO	37.03	38.17	40.25
Al2O3	15.16	12.45	15.32
MgO	10.62	10.14	6.05
SO3	0.85	0.97	0.71
MnO	0.44	0.31	1.79
TiO2	0.65	0.61	0.41
Fe2O3	0.36	0.42	0.55
$(Ca^{2+}+Mg^{2+})/Al^{3+}$	3.11	3.83	2.90

*BFS a was provided by Jinan Group Co.,Ltd, BFS b was provided by Shougang Group Co.,Ltd, BFS c was provided by Baosteel Group Co.,Ltd.

The synthesis method of HTLcs by using chemically pure reagents is called coprecipitation method, other methods based on the coprecipitation principle are also developed such as High Saturation Method (also called Single Drop Method) [11-13], Low Saturation Method [14], Nucleation/Crystallization Isolation Method [15], and Urea Test [16]. In this paper, single drop method was chosen for the preparation of BFS HTLcs and the influence of the synthesis pH and aging time on the synthetic BFS HTLcs was investigated. The necessary procedures are shown in Figure 1.

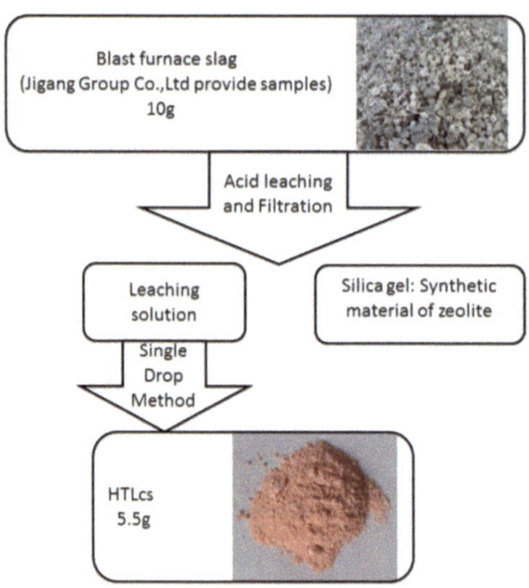

Figure 1. BFS HTLcs preparation process

Experiment

Materials

The BFS sample was provided by Jinan Group Co., Ltd, and the chemical composition of BFS used in this study(raw slag) is listed in Table 1. The chemical compositions were determined by chemical analysis. In order to promote the subsequent acid-leaching process, prior to synthesis, the raw slag was ball milled at 450 rpm for 20 min and screened using a 76 μm mesh to prepare for the next work. In addition, both HCl and NaOH were analytical grade (Sinopharm Chemical Reagent Co.,Ltd.), and deionized water was used to form aqueous solution.

HTLcs synthesis

The synthesis of BFS-HTLcs was performed as follows, adopting the single drop method according to the literature. First, 10g of the ball-milled blast furnace slag was completely dissolved in 200 mL of 3.6 M HCl aqueous solution, and then the solution was placed in the 100 °C thermostat water bath and stirred for 2 h to fully pectize the Si component. The gelation of the suspension easily occurs by heating, generating a hydrated silica gel. In this procedure, hydrochloric acid should be oversupplied to chloridize all elements, dosage is based on which BFS be selected in specific experiment. The obtained suspension was separated into hydrated silica gel and leaching solution containing other elements by filtration and washing with 300 mL of deionized water.

The silica gel can be used to synthesize zeolite and the leaching solution can be used to synthesize HTLcs, in this study, to the thus obtained leaching solution, 3 M NaOH aqueous solution was added dropwised at 60 °C and the final pH was adjusted to 11. The suspension was vigorously stirred for 1 h and then aged at 100 °C for another 9h to carry out crystallization, followed by filtration, washing with deionized water, and drying at 100 °C overnight to yield 5.5 g of the light pink product.In order to study the effect of synthesis parameters on structure and crystallinity, the samples were synthesized at different pHs (10, 11, 12, 13) and aging times (6h, 9h).

Characterization

The crystal structure was identified by powder X-ray diffraction (XRD) using a Ricon Ultima diffraction with CuKα radiation, and the tube voltage and current were set as 40 kV and 40 mA respectively. The XRD patterns were collected over the 2θ range of 10~90 ° using a step size of 0.02 ° (2θ). The scanning electron microscopy (SEM) analysis was performed by using FEI Quanta 250 to attain the characteristics of the crystalline phases such as the crystal morphology and the crystal size. Meanwhile, EDX was also applied to conduct elemental analysis.

Results and discussion

Synthesis of hydrotalcite-like compounds from BFS

To ordinary blast furnace slag, almost all of the components can be dissolved in hydrochloric acid, the free silicic acid will be aggregated and precipitated in amorphous silica, after acid leaching stage and fraction stage, obtained about 4.8g residual silica, removed most of Si from BFS. So the main elements were Ca, Mg, Al and Cl in the leaching solution. After the coprecipitation process, and washing, drying the products, by a range of detection, proved the successful synthesis of BFS-HTLcs. Figure 2 shows the XRD results of BFS HTLcs synthesized at pH=11 and age at 100 °C for 9 h, by contrast with standard PDF card, it proved that the synthetic sample was HTLcs, and a small amount of calcite ($CaCO_3$) was found in sample, this is due to the aging process conducted in the air. At the same time, combined with Figure 3a which shows the SEM results of synthetic HTLcs sample, it can be seen the layered plate structure clearly which is typical characteristics of HTLcs. Combined with Figure 3b which shows the EDX results of BFS HTLcs, it can be seen that the sample contain six major elements: magnesium, calcium, aluminum, chlorine, oxygen and hydrogen (hydrogen cannot be detected by EDX).

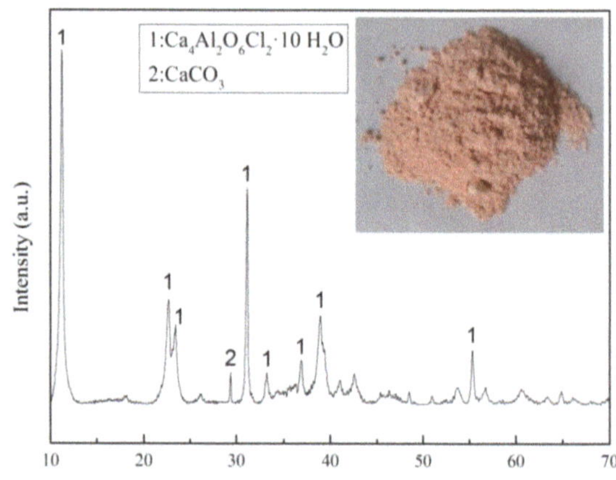

Figure 2. X-Ray diffraction patterns of the HTLcs sample synthesized at pH=11 at 9 h

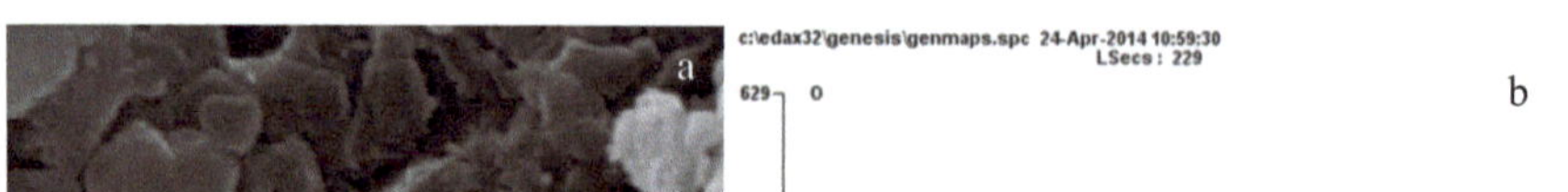

Figure 3. SEM images and EDX result of the HTLcs sample synthesized at pH=11 at 9 h: (a) SEM images, (b) EDX result

Effect of synthesis pH

The effect of synthesis pH on the crystallinity and structure was evaluated by means of XRD,

Figure 4 shows the XRD patterns of HTLcs samples synthesized at four different pHs at a fixed time of 9 h. All samples exhibited clear diffraction peaks assignable to the Ca-Al-Cl hydrocalumite phase, however, BFS-HTLcs synthesized at pH=11 and pH=12 (Figure 4b, Figure 4c) showed stronger peak intensities for basal reflections (002) and (004) than those at pH=10 and 13 (Figure 4a, Figure 4c). Relative crystallinity were calculated by means of the XRD patterns of samples, sample b was selected as the reference sample (Relative crystallinity is 100%), its crystallinity is similar to sample c, the relative crystallinity of sample a and d is 59% and 35%, respectively. Figure.4d indicated that too high pH will result in generate other phases, such as portlandite, katoite and calcite.

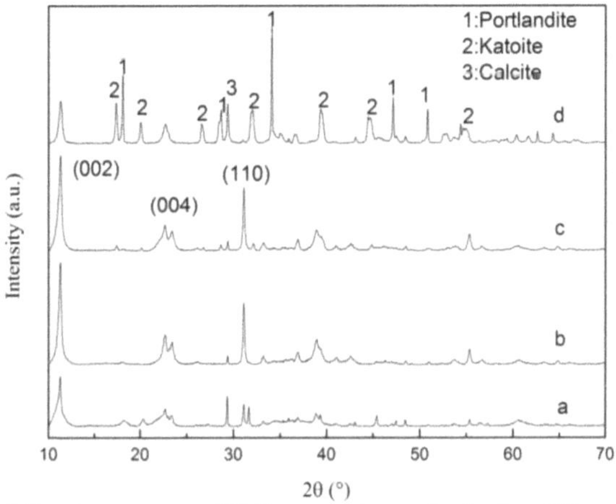

Figure 4. X-Ray diffraction patterns of HTLcs samples synthesized at different pHs: (a) 10, (b) 11, (c) 12, (d) 13

Effect of aging time

By chemical kinetics view, any chemical reactions need to consider the effect of time. Figure 5 shows the XRD result of HTLcs samples synthesized at two different aging times at a fixed pH of 11, and it indicates the effect of time in this study. The two samples have the same phase completely, the peak intensities at 9 h is stronger than at 6 h, and the former has a higher crystallinity. So it can be seen that an extension of time was capable of promoting the crystal growth.

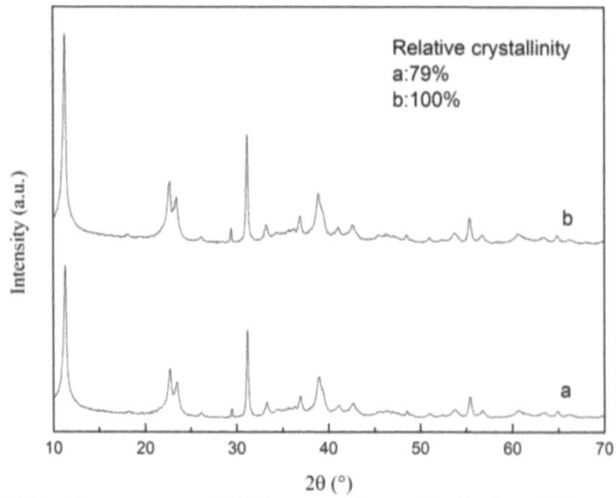

Figure 5. X-Ray diffraction patterns of HTLcs samples synthesized at different aging time: (a) 6 h, (b) 9 h

Conclusion

(1) After a simple preprocessing, blast furnace slag can be used as the material to synthesize hydrotalcite-like compounds.
(2) pH is a key factor in the process of synthesizing BFS HTLcs, in this study, pH=11~12 is considered to be an appropriate scope, too high or too low pH will result in lower crystallinity and generating other phases.
(3) The effect of aging time conform to the general rule: the longer the aging time, the higher the extent of crystal growth as well as the crystallinity.

Acknowledgments

The authors would like to acknowledge the help of The School of Metallurgical and Ecological Engineering, University of Science and Technology Beijing for providing experiment platform, this work supported by the National Key Technology R&D Program in the 12th Five year Plan of China (NO:2011BAC01B02).

References

1. D.L. Li et al., "Preparation of Ni-Cu/Mg/Al catalysts from hydrotalcite-like compounds for hydrogen production by steam reforming of biomass tar," *International Journal of Hydrogen Energy*, 39 (2014), 10959-10970.

2. R.G. Prado et al., "Multivariate Method for Transesterification Reaction of Soybean Oil Using Calcined Mg-Al Layered Double Hydroxide as Catalyst," *Catalysis Letters*, 144 (2014), 1062-1073.

3. Z.H. Li et al., "Ni-based catalyst derived from Ni/Mg/Al hydrotalcite-like compounds and its activity in the methanation of carbon monoxide," *Kinetics and Catalysis*, 55 (2014), 217-223.

4. C. Zhang et al., "Adsorption behavior and mechanism of reactive brilliant red X-3B in aqueous solution over three kinds of hydrotalcite-like LDHs," *Applied Surface Science*, 301 (2014), 329-337.

5. Q. He, S. Yin, T. Sato, "Synthesis and photochemical properties of zinc-aluminum layered double hydroxide/organic UV ray absorbing molecule/silica nanocomposites," *Journal of Physics and Chemistry of Solids*, 65 (2004), 395-402.

6. B. Das et al., "An overview of utilization of slag and sludge from steel industries," *Resources Conservation and Recycling*, 50 (2007), 40-57.

7. M. Maslehuddin et al., "Comparison of properties of steel slag and crushed limestone aggregate concretes," *Construction and Building Materials*, 17 (2003), 105-112.

8. M. Bohac et al., "Investigation on early hydration of ternary Portland cement-blast-furnace slag–metakaolin blends," *Construction and Building Materials*, 64 (2014), 333-341.

9. L. Zhou, "Preparation of hydrotalcite-like materials and research on structure" (M.S. thesis Beijing University of Chemical Technology (in Chinese), 2007), 10-11.

10. Y. Kuwahara et al., "Synthesis of zeolite from steel slag and its application as a support of nano-sized TiO2 photocatalyst," *Journal of Materials Science*, 43 (2008), 2407-2410.

11. S.H. Han et al., "Study on the preparation and structure of positive sol composed of mixed metal hydroxide," *Colloid and Polymer Science*, 274 (1996), 860-865.

12. G. Fornasari et al., "Structure and reactivity of high-surface-area Ni/Mg/Al mixed oxides," *Applied Clay Science*, 10 (1995), 69-82.

13. F. Millange, R.I. Walton, D. O'Hare, "Time-resolved in situ X-ray diffraction study of the liquid-phase reconstruction of Mg-Al-carbonate hydrotalcite-like compounds," *Journal of Materials Chemistry*, 10 (2000), 1713-1720.

14. A. Corma et al., "Determination of base properties of hydrotalcites: Condensation of benzaldehyde with ethyl acetoacetate," *Journal of Catalysis*, 134 (1992), 58-65.

15. L. Li, Q. Luo, X. Duan, "Clean route for the synthesis of hydrotalcites and their property of selective intercalation with benzenedicarboxylate anions," *Journal of Materials Science Letters*, 21 (2002), 439-441.

16. P. Adachi-Pagano, C. Forano, J. Besse, "Synthesis of Al-rich hydrotalcite-like compounds by using the urea hydrolysis reaction control of size and morphology," *Journal of Materials Chemistry*, 13 (2003), 1988.

Energy Technology 2015
Carbon Dioxide Management and Other Technologies

High-Temperature Systems for Energy Conversion and Storage

Energy Technology 2015
Carbon Dioxide Management and Other Technologies

HIGH-TEMPERATURE SYSTEMS FOR ENERGY CONVERSION AND STORAGE

Solid Oxide Fuel Cell: Recent Developments II

Session Chairs:

Paul Ohodnicki
Amit Pandey

Energy Technology 2015: Carbon Dioxide Management and Other Technologies
Edited by: Animesh Jha, Cong Wang, Neale R. Neelameggham, Donna P. Guillen, Li Li, Cynthia K. Belt, Randolph Kirchain, Jeffrey S. Spangenberger, Frank Johnson, Andrew J. Gomes, Amit Pandey, and Peter Hosemann
TMS (The Minerals, Metals & Materials Society), 2015

EVALUATING ELECTROPHORETICALLY DEPOSITED Cu-Mn-O SPINEL COATINGS ON STAINLESS STEEL SUBSTRATES USED IN SOLID OXIDE FUEL CELL INTERCONNECTS

Michael Galbo[1,3], Kyung Joong Yoon[1,4], Uday B. Pal[1,2], Srikanth Gopalan[1,2], Soumendra N. Basu[1,2]

[1] Div. of Materials Science and Engineering, Boston University, Brookline, MA, 02446, USA
[2] Department of Mechanical Engineering, Boston University, Boston, MA, 02215, USA
[3] Currently at Aidin, New York, NY 10013, USA
[4] Currently at Korea Institute of Science and Technology, Seoul, 136-791, Republic of Korea

Keywords: Fuel cells, Interconnect coatings, Electrophoretic deposition, Spinels

Abstract

Interconnects in solid oxide fuel cells (SOFCs) oxidize under SOFC operating conditions, leading to formation of Cr-rich oxides. This leads to the formation of chromium-containing vapor-phase species that poisons the SOFC cathodes and lead to long-term degradation in device performance. This necessitates the use of protective coatings on SOFC interconnects. Coatings of three Cu-Mn-O spinel compositions, $Cu_{1.3}Mn_{1.7}O_4$, $Cu_{1.2}Mn_{1.8}O_4$, and $CuMn_{1.8}O_4$, were deposited on Crofer 22 APU substrates by electrophoretic deposition (EPD). The effects of the different deposition parameters and the subsequent thermo-mechanical processing steps in the EPD deposition process were studied. The phase stability and the electrical conductivity of the three materials were measured. Finally, thermogravimetric measurements under oxidizing conditions were carried out to study the effectiveness of the coatings as diffusion barriers. The diffusion coefficients of chromium in the three coatings were calculated from the chromium concentration profile.

Introduction

The principle use of an interconnection in an SOFC stack is to provide electrical connectivity between the individual cells in a stack. In order to function properly, the interconnect must be stable in both oxidizing and reducing atmospheres found in fuel cells, be stable against reaction with the electrode materials, have a coefficient of thermal expansion (CTE) match with other cell components to avoid excessive stresses and fracture during thermal cycling, be impermeable to gaseous oxygen and hydrogen to prevent the direct combination of oxidant and fuel during operation, have high thermal conductivity to allow the heat generated at the cathode to be conducted to the anode where the endothermic fuel reforming reaction takes place, and be easy to fabricate [1,2]. Both ceramic and metallic interconnect materials have been explored for use in the fuel cell. Ceramic interconnects have been designed using complex ceramic oxides with perovskite structure [1,3,4]. However, issues with sintering prevent full densification of the layer [1]. While research has shown that dense layers can be produced [5], processing steps require high temperature manufacturing which increases overall cost. Another concern lies in the poor mechanical properties in reducing environments [6]. With the advent of anode supported planar SOFC's, the operating temperature of the fuel cell has been reduced to around 800°C, allowing the evolution of metallic interconnects. Metallic interconnects are lower cost than their ceramic counterparts, have a higher electronic and thermal conductivity, and are easy to manufacture [7]. Currently stainless steels containing chromium are the prime candidates for this application due to their low cost and close CTE match [8]. Also, a chromium oxide scale (Cr_2O_3) that forms on the alloy surface has good electrical conductivity [9]. However due to volatility of Cr_2O_3, over time vapor phase species containing Cr can poison the cathode, leading to rapid degradation of performance [10-12].

Copper manganese spinels of the form $Cu_xMn_{3-x}O_4$ have been used for this application since they demonstrate high conductivity at the fuel cell operating temperature and exhibit a CTE similar to the Croffer 22 steel being used as the interconnect material [13,14]. Electrophoretic deposition (EPD) has been explored as a way to apply ceramic material [15-18]. This method involves particles suspended in a liquid medium that migrate under the influence of an electric field and are deposited onto an electrode [19]. It is useful for applying materials to any electrically conductive surface. This process is ideal for interconnect applications because it has a simple setup, applies coatings which generally have a well controlled uniform coating thickness that are formed in a short time period, and can be used to coat objects with complex shapes [19].

Experimental

Three compositions of copper manganese spinel powders ($Cu_{1.3}Mn_{1.7}O_4$, $Cu_{1.2}Mn_{1.8}O_4$, and $CuMn_{1.8}O_4$) were synthesized using a solid-state reaction process. CuO and MnO_2 powders were mixed by ball milling in ethanol for 24 hours then dried and ball milled again before being calcined at 800°C in air for 10 hours. To achieve phase purity this process was repeated three times. Phase composition was verified by x-ray diffraction (XRD) analysis at room temperature using a Bruker D8 Advance XRD system with Cu Kα radiation (Figure 1) and the powders were refined through a 400 mesh sieve.

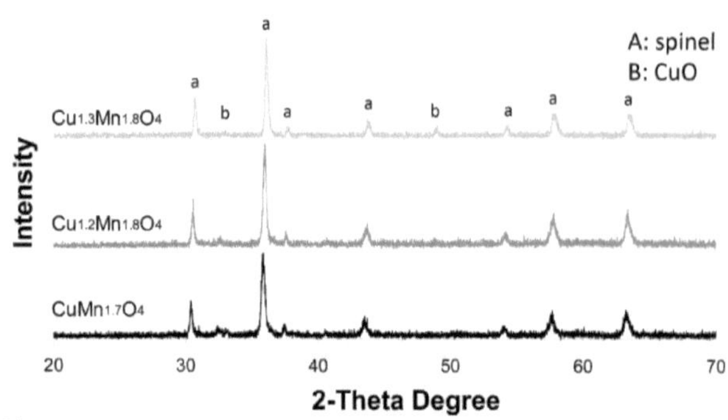

Figure 1. XRD scans of $Cu_{1.3}Mn_{1.7}O_4$, $Cu_{1.2}Mn_{1.8}O_4$, and $CuMn_{1.8}O_4$ spinel powders.

A process was developed to deposit dense and conductive spinel coatings to minimize oxide scale growth and chromium vaporization [20]. At the onset of coating application the Crofer 22 APU substrates (weight percent chemical composition of 22.8 Cr, 0.45 Mn, 0.08 Ti, 0.06 La, 0.005 C, ≤0.03 P, ≤0.03 S, and balance Fe) were polished to P4000 surface finish and cleaned ultrasonically in acetone. A combination of 25% ethanol and 75% acetone by volume was used as the suspension media, having a dielectric constant around 25 which is within the range of 12-25 where deposition is known to occur [15]. Iodine is added to act as a charge carrier along with the spinel powder. The solution is then mixed ultrasonically for 20 minutes to break apart any agglomerates in the solution. After waiting for any remaining large particles to settle, a current is applied across the two electrodes. A copper cathode is used and a DC current accelerates the spinel particles toward the substrate. Once the deposition is complete, the sample is pressed to densify the coating in the green state. The coating was subsequently reduced in forming gas and uniaxially pressed again to remove voids created during reduction. The coating was then annealed in air. A 4-probe conductivity apparatus using Ag wires was used to measure electrical conductivity on 5X0.5X0.7 cm^3 samples. Oxidation experiments were carried out in air at 800°C by thermogravimetry using a TA Q600 microbalance.

Results and Discussion

Figure 2 shows the results of the four-probe conductivity measurements in air. All three compositions are sufficiently conductive in the fuel cell operating temperature with conductivities easily exceeding 100 S/cm in the 750-800°C range.

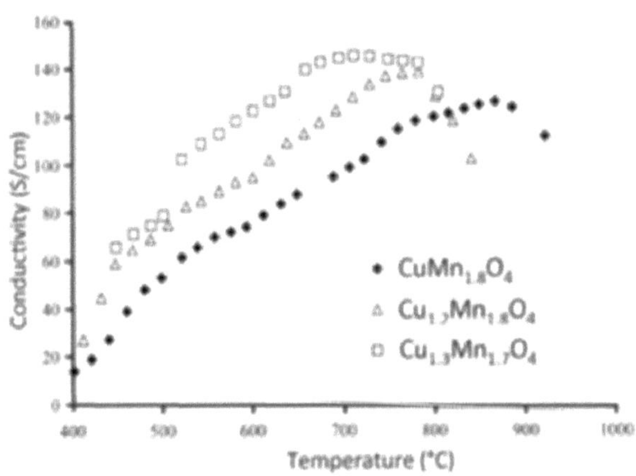

Figure 2. Conductivities of $Cu_{1.3}Mn_{1.7}O_4$, $Cu_{1.2}Mn_{1.8}O_4$, and $CuMn_{1.8}O_4$ as a function of temperature.

Iodine, the charge carrier in the EPD suspension reacts with acetone as [21]:

$$CH_3\text{-}CO\text{-}CH_3 + 2I_2 = ICH_2\text{-}CO\text{-}CH_2I + 2H^+ + 2I^- \quad (1)$$

These negatively charged ions are adsorbed onto the spinel particle surface carrying the particle towards the positive electrode when the current is applied. Varying the concentration (by weight) of spinel powder to iodine in the suspension indicated that there is a small band under which deposition can occur (Figure 3). There are two observed mechanisms that cause this result. With a high charge carrier concentration, the high negative particle charge causes the particles to repel. While this creates a well-dispersed suspension, it effectively slows particle kinetics by creating a charge boundary on the substrate wall. Evidence of this effect is seen with a very low weight gain and a thin uniform deposition layer on the substrate. Conversely, a low charge carrier concentration is not conductive enough to accelerate the particles to the substrate, or only the smallest particles will participate, again leading to a thin coating. In this case, the suspension tends to agglomerate and is not well dispersed, leading to non-uniform coatings.

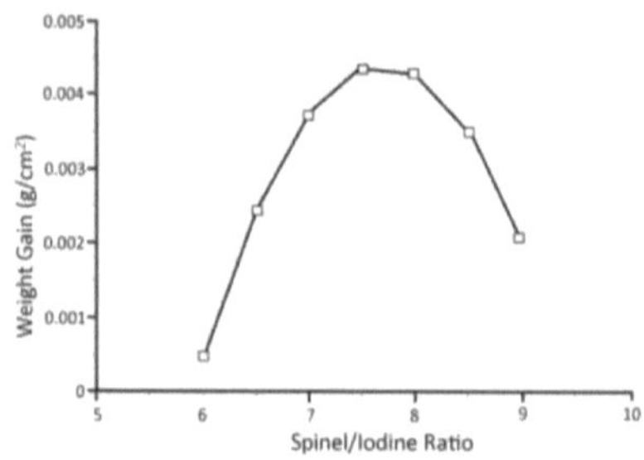

Figure 3. The weight gain after EPD deposition of $CuMn_{1.8}O_4$ powders as a function of the ratio of the weight of spinel powder to the weight of iodine in the suspension.

The voltage applied across the electrodes exponentially increased the amount of spinel particles deposited on the substrate. However, higher voltages resulted in significant non-uniformities in the coating. In the high voltage regime, the potential was high enough to mobilize larger particles at the onset of the EPD process. The large particles at the coating interface lead to poor coating adhesion and cracks at the coating interface after annealing. An applied voltage of 20V was chosen to increase the uniformity of the coating and decrease any possibility of delamination from the substrate surface.

After the deposition was complete the coated samples were placed in a uniaxial press to compact the coating before heat treatment. Once the samples were pressed in the green state it was reduced in forming gas for one hour at 850°C. After the reduction, the sample was pressed again before a 100h anneal at 850°C. For this investigation, all samples were processed identically, and only the pre- and post reduction anneal pressing pressure was varied. The effects of applied pressure on processed coatings can be seen in the micrographs in Figure 4.

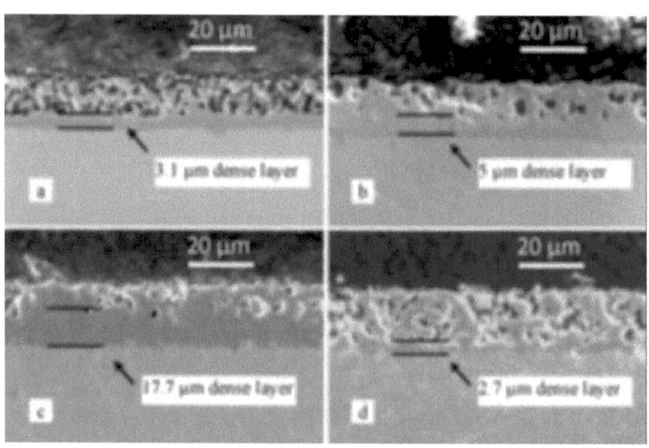

Figure 4. Scanning electron micrographs showing the effect of pressing pressure on $CuMn_{1.8}O_4$ coating density. a) 0 psi, b) 5000 psi, c) 10000 psi, d) 15000 psi.

A summary in Table 1 shows the porosity and thickness of the fully dense layer, as the pressure increases. Density trends upwards with increasing pressure, but it is presumed that for pressures above 10000 internal stresses caused the coating to crack and porosity to increase [21].

Table 1. Effect of pressure on thickness of dense layer and porosity of $CuMn_{1.8}O_4$ coatings

Pressure (psi)	Thickness of dense layer (μm)	Porosity (%)
0	3.7	35.0
5,000	5.0	10.5
10,000	17.7	2.3
15,000	2.7	11.8

The heat treatment process is the key step in this process to create a dense coating. The reduction step creates voids in the coating as the powder transitions to a higher density phase. These voids can be evacuated during pressing, allowing the coating to be annealed to high density. Figure 5 compares the XRD spectra of $Cu_{1.2}Mn_{1.8}O_4$ in the as-deposited coating in the powder form, after the reduction step and after 100hr anneal in air. The as-deposited peaks indicate phase-pure $Cu_{1.2}Mn_{1.8}O_4$ spinel, which is consistent with previous results reported by Martin and Petric [22]. During the reduction step, the spinel undergoes a phase transformation in the low oxygen partial pressure environment to form more stable MnO and Cu metal phases. This is a critical step in the densification process as the dense and malleable copper metal is easily compacted. After a 100hr anneal at 850 °C, the coated Crofer 22 APU transitioned back to the pure $Cu_{1.2}Mn_{1.8}O_4$ spinel phase.

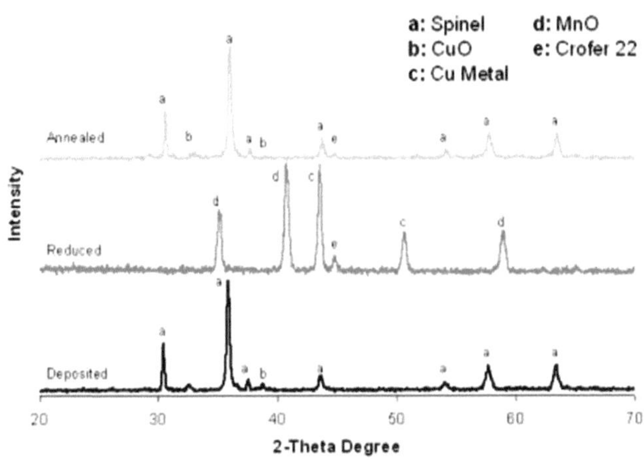

Figure 5. XRD patterns of $Cu_{1.2}Mn_{1.8}O_4$ as-deposited in the powder form, after the reduction step and after 100hr anneal in air.

Crofer 22 APU samples with $Cu_{1.3}Mn_{1.7}O_4$, $Cu_{1.2}Mn_{1.8}O_4$, and $CuMn_{1.8}O_4$ coatings were oxidized in air for 100 hours at 800°C and weight gains were measured using a TA Q600 microbalance. All three coatings maintained densities in excess of 95% after oxidation. Elemental distribution linescans of the spinel coatings after (100 hours anneal and) 100 hours of oxidation at 800 °C for all three compositions are shown in Figure 6. Consistent with previous analysis by Huang [23] in which there is evidence of a small layer of Cr_2O_3 and $MnCr_2O_4$ at the coating interface; the linescans shows a small peak in chromium concentration. Results indicate the coating acts well as a diffusion barrier, as chromium migration is negligible halfway through the layer thickness.

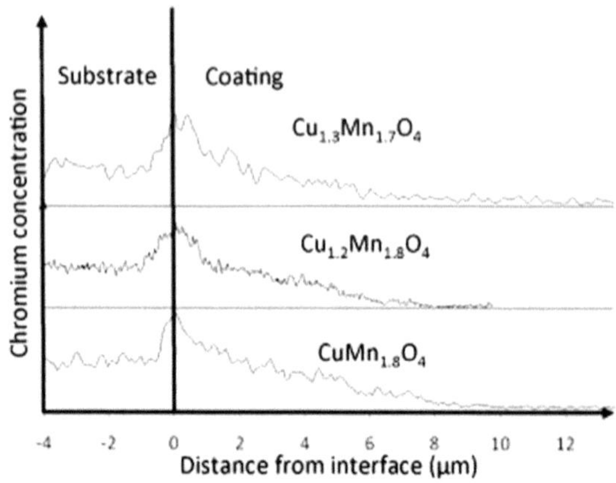

Figure 6. Elemental distribution linescans of the spinel coating after 100 hours anneal and 100 hours of oxidation at 800 °C for all three coating compositions.

An estimation of chromium diffusivity can be calculated using an error function solution model into the coating with the coating/substrate interface fixed at the chromium content in Cr_2O_3. The results are shown in Table 2.

Table 2. Calculated Cr diffusivities in the spinel coatings.

Composition	D (cm^2/s)
$Cu_{1.3}Mn_{1.7}O_4$	$1.27X10^{-15}$
$Cu_{1.2}Mn_{1.8}O_4$	$1.14X10^{-15}$
$CuMn_{1.8}O_4$	$1.79X10^{-15}$

Figure 7 shows the oxidation weight gain per unit area with time measured by TGA for the three coated samples as well as an uncoated Crofer 22 APU sample for comparison. The figure shows that the Crofer 22 APU sample coated with $CuMn_{1.8}O_4$ exhibited the smallest weight gain. Weight gain during oxidation is due to formation of oxides at the alloy/coating interface, In a coated sample, initially the rate controlling step is oxygen diffusion from the atmosphere to the alloy surface through the coating [23]. This implies that $CuMn_{1.8}O_4$ has the smallest oxygen diffusivity among the three spinel compositions. This is not surprising, since it is the most cation deficient spinel, implying that it has the highest concentration of cation vacancies, leading to the highest chromium diffusivity. By Schottky equilibrium, this would also lead to the lowest concentration of oxygen vacancies, which explains the smallest weight gain during the early stages of oxidation.

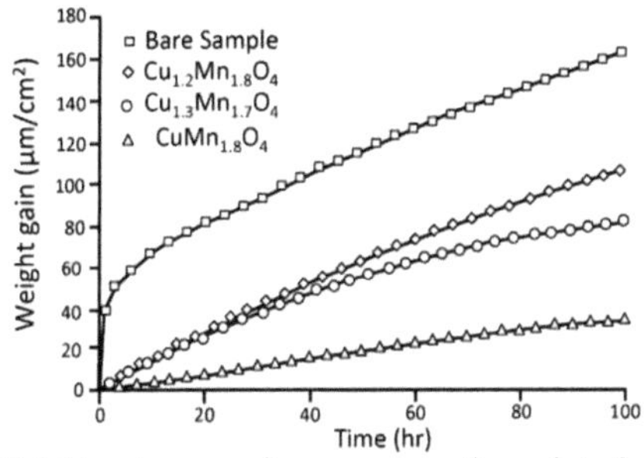

Figure 7. Weight gain per unit area versus time plots for uncoated and coated Crofer 22 APU samples.

Conclusions

Coatings of $Cu_{1.3}Mn_{1.7}O_4$, $Cu_{1.2}Mn_{1.8}O_4$, and $CuMn_{1.8}O_4$ were deposited on Crofer 22 APU substrates EPD. It was found that a spinel to iodine weight ratio of 7.5-8 led to maximum deposition rate of coatings during the EPD deposition process. A deposition voltage of 20V was found to be optimal for uniform coating deposition. To get optimal coating density, the following steps were necessary: i) a uniaxial press to compact the deposited coating in the green state, ii) a reduction anneal that led to phase separation of the spinels to MnO and Cu metal, iii) A uniaxial press to densify the two phase coating, and iii) an anneal at 850°C in air for 100 hr to oxidize the Cu to CuO that reacts with the MnO to give a single phase spinel, which sinters into a dense coating. A pressure of 10,000 psi was found to be optimal close the post-EPD deposition pores and to close the pores in the post-reduced samples by deforming the Cu phase. Pressures in excess of 10,000 psi led to mechanical instabilities in the coatings. All three compositions were found to have electrical conductivities in excess of 100 S/cm in the 750-800°C range. Oxidation of the coated samples showed that the $CuMn_{1.8}O_4$ sample had the smallest weight gain. The Cr diffusion profile showed that this composition had the highest Cr diffusivity in the coating.

Acknowledgements

The DOE SECA Program is acknowledged for partial support of travel costs through Grant DE-FE0023325 to the TMS 2015 conference, where this paper was presented.

Bibliography

1. W.Z. Zhu, and S.C. Deevi, "Development of interconnect material for solid oxide fuel cells," *Materials Science and Engineering A,* 348 (2003), 227-243.

2. M.R. Bateni, P. Wei, X. Deng, and A. Petric, "Spinel coatings for UNS 430 stainless steel interconnects," *Surface Coatings and Technology*, 201 (2007), 4677-4684.

3. J.W. Fergus, "Lanthanum chromite-based materials for solid oxide fuel cell interconnects," *Solid State Ionics*, 171 (2004), 1-15.

4. Nguyen Quang Minh and Takehiko Takahashi, *Science and Technology of Ceramic Fuel Cells* (Amsterdam, Elsevier, 1995).

5. N.M. Sammes, R. Ratnaraj, and M.G. Fee, "The effect of sintering on the mechanical properties of SOFC ceramic interconnect materials," *Journal of Materials Science*, 29 (1994), 4319-4324.

6. S.W. Paulik, S. Baskaran, and T.R. Armstrong, "Mechanical properties of calcium- and strontium substituted lanthanum chromite," *Journal of Materials Science*, 33 (2000), 2397-2404.

7. W.Z. Zhu, and S.C. Deevi, "Opportunity of metallic interconnects for solid oxide fuel cells: a status on contact resistance," *Materials Research Bulletin*, 38 (2003), 957–972.

8. S. Elangovan et al., "Evaluation of Ferritic Stainless Steel for Use as Metal Interconnects for Solid Oxide Fuel Cells," *Journal of Materials Engineering and Performance*, 13 (2004), p265.

9. T. Horita et al., "Evaluation of Fe-Cr alloys as interconnects for reduced operation temperature SOFCs," *Journal of the Electrochemical Society*, 150 (2003), A243-A248.

10. C. Collins et al., "Chromium volatility of coated and uncoated steel interconnects for SOFCs," *Surface Coatings and Technology*, (201) 2006 4467–4470.

11. S.C. Paulson, and V.I. Birss, "Chromium Poisoning of LSM-YSZ SOFC Cathodes: I. Detailed Study of the Distribution of Chromium Species at a Porous, Single-Phase Cathode," *Journal of the Electrochemical Society*, 151 (2004), A1961-A1968.

12. K. Hilpert et al., "Chromium Vapor Species over Solid Oxide Fuel Cell Interconnect Materials and Their Potential for Degradation Processes", *Journal of the Electrochemical Society*, 143 (1996), 3642-3647.

13. H. Ling, and A.Petric, "Electrical and Thermal Properties of Spinels", *Solid Oxide Fuel Cells (SOFC IX)*, ed. S.C. Singhal and J. Mizusaki (Pennington, NJ, The Electrochemical Society, 2005), p1866.

14. A. Petric, and H. Ling, "Electrical Conductivity and Thermal Expansion of Spinels at Elevated Temperatures," *Journal of the American Ceramic Society*, 90 (2007), 1515–1520.

15. R.W. Powers, "The electrophoretic forming of beta alumina-ceramic," *Journal of the Electrochemical Society*, 122 (1975), 490-500.

16. H. Negishi, H. Yanagishita, and H. Yokokawa, "Electrophoretic Deposition of Solid Oxide Fuel Cell Material Powders," *Electrophoretic Deposition, Fundamentals and Applications*, ed. A.R. Boccaccini, O. van der Biest, J.B. Talbot (Pennington, NJ, The Electrochemical Society, 2002) 214-221.

17. I. Zhitomirsky, L. Gar-Or, "Electrophoretic deposition of hydroxyapatite," *Journal of Materials Science: Materials in Medicine*, 8 (1997) 213-219.

18. J. Li, and I. Zhitomirsky, "Cathodic electrophoretic deposition of manganese dioxide films," *Colloids and Surfaces A: Physicochemical and Engineering Aspects*, 234 (2009), 248–253.

19. P. Sarkar, and P.S. Nicholson, "Electrophoretic Deposition (EPD): Mechanisms, kinetics, and application to ceramics," *Journal of the American Ceramic Society*, 79 (1996) 1987–2002.

20. Z. Yang et al., "Conductive protection layers on oxidation resistant alloys for SOFC interconnect applications," *Surface Coatings and Technology*, 201 (2006), 4476–4483.

21. L. Besra, and M. Liu, "A Review on the fundamentals and applications of electrophoretic deposition (EPD)," *Progress in Materials Science*, 52 (2007), 1-61.

22. B.E. Martin, and A. Petric, "Electrical properties of copper–manganese spinel solutions and their cation valence and cation distribution," *Journal of Physics and Chemistry of Solids*, 68 (2007) 2262–2270.

23. W. Huang et al., "Evaluation of electrophoretically deposited CuMn1.8O4 spinel coatings on crofer 22 APU for solid oxide fuel cell interconnects," *Journal of the Electrochemical Society*, 155 (2008), 405-411.

HIGH-TEMPERATURE SYSTEMS FOR
ENERGY CONVERSION AND STORAGE

Innovation in Energy Conversion and Storage I

Session Chairs:

Soumendra Basu
Jung Choi

Energy Technology 2015: Carbon Dioxide Management and Other Technologies
Edited by: Animesh Jha, Cong Wang, Neale R. Neelameggham, Donna P. Guillen, Li Li, Cynthia K. Belt, Randolph Kirchain, Jeffrey S. Spangenberger, Frank Johnson, Andrew J. Gomes, Amit Pandey, and Peter Hosemann
TMS (The Minerals, Metals & Materials Society), 2015

DEGRADATION OF $Sm_2Zr_2O_7$ THERMAL BARRIER COATING CAUSED BY CALCIUM-MAGNESIUM-ALUMINUM-SILICON OXIDE (CMAS) DEPOSITION

Honglong Wang[1], Zhizhi Sheng[1], Emily Tarwater[1], Xingxing Zhang[1], Sudip Dasgupta[1] and Jeffrey W. Fergus[1]

[1]Auburn University; 275 Wilmore Laboratories, Auburn University, AL, 36849, United States

Keywords: Thermal barrier coating, $Sm_2Zr_2O_7$, CMAS

Abstract

Rare earth zirconates are promising materials for use as thermal barrier coatings in gas turbine engines. Among the lanthanide zirconate materials, $Sm_2Zr_2O_7$ with the pyrochlore structure has lower thermal conductivity and better corrosion resistance against calcium-magnesium-aluminum-silicon oxide (CMAS). In this work, after reaction with CMAS, the pyrochlore structure transforms to the cubic fluorite structure and $Ca_2Sm_8(SiO_4)_6O_2$ forms in elongated grain.

Introduction

Higher working temperatures in gas turbine engines increase the energy efficiency. To protect the alloys from working temperatures as high as 1200°C, thermal barrier coatings are needed. Additionally, delamination of thermal barrier coatings caused by reaction with CMAS could degrade the protective function. Thus, the desired properties of thermal barrier coatings are low thermal conductivity and good corrosion resistant against CMAS. Yttria-stabilized zirconia (YSZ) produced in the metastable t' phase is currently the material most widely used for thermal barrier coatings [1]. However, the t' phase can transform to tetragonal phase when the temperature is above 1200°C. After long term usage, the monoclinic phase can form during and result in delamination of the coatings [2, 3]. Lanthanide zirconate materials have been shown to have lower thermal conductivity, better corrosion resistant and better phase stability at elevated temperatures as compared to the currently used YSZ coatings.

The thermal barrier system includes several layers: bond coat, thermally grown oxide layer (TGO) and thermal barrier coating [4]. For YSZ thermal barrier coatings, CMAS deposition accelerates transformation of the t' phase [5]. While for the lanthanide zirconate TBCs, the mechanism is different. Lanthanide zirconate TBCs can elevate the working temperature and CMAS will melt at that temperature. Molten CMAS can react with pyrochlore materials to form new phases after corrosion, which can decrease the effectiveness of the coating by increasing the thermal conductivity or lead to delamination of the TBC. In addition, changes in the thermal expansion coefficient can lead to residual stress and even cracking in the TBC [6]. Furthermore, reaction of molten CMAS that penetrates through pores in the coating can affect TGO and bond coat properties to promote creep, which may result in break away from the superalloy substrate [7]. Thus, the corrosion mechanism of CMAS on lanthanide zirconate TBCs needs attention. This paper focuses on $Sm_2Zr_2O_7$ and the corrosion results after deposition of CMAS glass.

Experimental

Preparation of pyrochlore ceramics

Pyrochlore $Sm_2Zr_2O_7$ ceramics were prepared using the co-precipitation method. Firstly, $ZrOCl_2 \cdot 8H_2O$ was dissolved in deionized water. Sm_2O_3 in stoichiometric amounts was dissolved in nitric acid separately and then added to the $ZrOCl_2 \cdot 8H_2O$ aqueous solution. The resulting mixture was stirred for 30 minutes to obtain a homogeneous solution and an aqueous ammonia solution was added. The pH was kept above 10 in this step. The precipitate was washed with deionized water and dried at 150°C for 10 hours. After desiccation, the powders were heated to 800°C and pressed under 100 MPa into discs, which were sintered at 1400°C and 1500°C.

Corrosion test

CaO-MgO-Al_2O_3-SiO_2 glass was used to simulate deposit ingested during application. The composition used in this study is shown in Table 1. The glass in this composition has a melting temperature of around 1240°C [8]. To synthesize the glass, powders for the constituent oxides in stoichiometric ratio were mixed and sintered at 1300°C for 5 hours to obtain a homogeneous glass. The glasses were crushed and ball milled for 24 hours to obtain glass particles. The glass powders were placed on the top of pyrochlore discs and then sintered at different temperatures from 1300°C to 1500°C.

Table 1. Composition of CMAS in molar percentage.

CaO	33
MgO	9
$AlO_{1.5}$	13
SiO_2	45

Characterization

Phase analysis was performed by XRD using a Bruker D8 diffractometer with Cu X-ray source. The working current and voltage were 40 mA and 40 kV. Morphology and composition analyses were carried out using a JEOL JSM-7000F scanning electron microscope (SEM) equipped with energy dispersive x-ray spectroscopy (EDS).

Result and discussion

Composition and phase analysis of pyrochlore ceramics

The XRD patterns of $Sm_2Zr_2O_7$ ceramic sintered at different temperatures and times are shown in Figure 1. All the peaks can be indexed to $Sm_2Zr_2O_7$ pyrochlore ceramic according to JCPDS card (#00-024-1012). The (311), (331) and (511) peaks are not present in spectra for the cubic fluorite structure (JCPDS # 01-078-1291), so the presence or absence of these peaks can be used to distinguish the pyrochlore and cubic fluorite structures.

Figure 2 shows that pyrochlore ceramics have relatively uniform grain sizes of around 1 μm. The elemental distribution of Sm and Zr is also relatively uniform as shown in Figure 3. Au is detected due to the gold sputter coating to obtain a conductive surface for the SEM analysis.

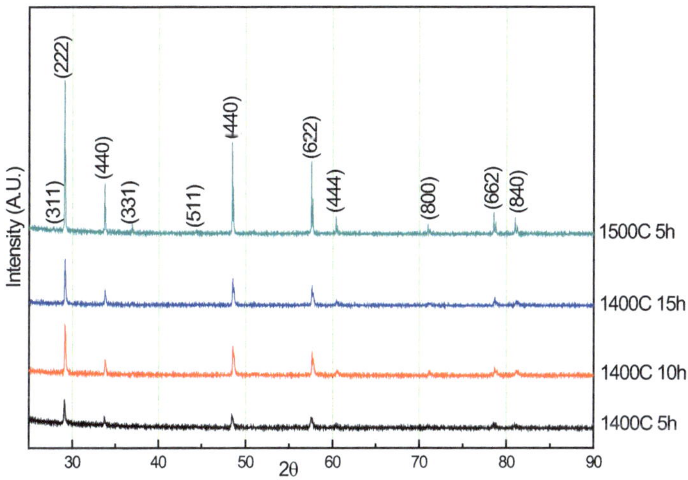

Figure 1. XRD spectra of $Sm_2Zr_2O_7$ ceramic prepared by the co-precipitation method.

Figure 2. Surface morphology of $Sm_2Zr_2O_7$, sintered at 1500°C for 5h.

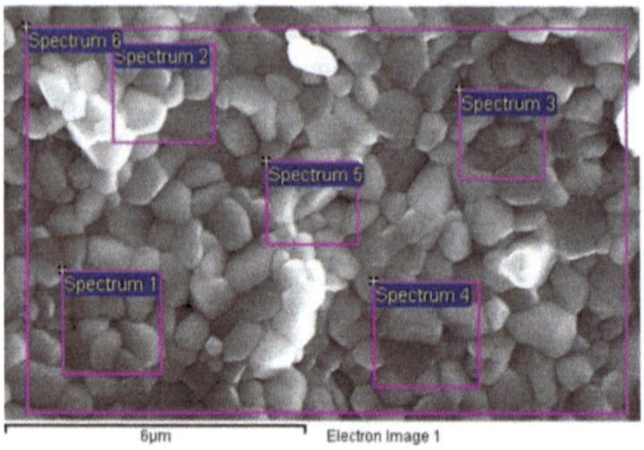

Atomic Percentage

	O	Zr	Sm	Au
Spectrum 1	53	23	17	7
Spectrum 2	60	20	14	6
Spectrum 3	54	22	17	7
Spectrum 4	53	24	17	7
Spectrum 5	57	21	16	6
Spectrum 6	55	22	17	7

Figure 3. Elemental percentage of $Sm_2Zr_2O_7$ ceramic, sintered at 1500°C for 5 hours.

Effects of CMAS glass exposure on pyrochlore ceramics

The XRD spectra of $Sm_2Zr_2O_7$ pellets after exposure to CMAS glass are shown in Figure 4 and Figure 5. The $Sm_2Zr_2O_7$ ceramics without CMAS exposure are also shown in the figures for comparison. The peaks are mostly the same before and after corrosion test, but the (311), (331) and (511) peaks for the pyrochlore structure are not present in the CMAS contacted surface, which suggests that the pyrochlore structure has transformed to fluorite structure. Additionally, all the peaks in the CMAS contacted surface shift to higher angles than those in the non-corroded surface. The small peaks at 31.6°, 31.9°, 32.7°, 46.4° can be attributed to $Ca_2Sm_8(SiO_4)_6O_2$ according to JCPDS #00-029-0365. The non-contacted surface is still the pyrochlore structure but small peaks indexed to $Ca_2Sm_8(SiO_4)_6O_2$ are also present. Although this surface did not contact with CMAS glass, CMAS glass can melt above 1300°C and be transported to the bottom surface along the side surface.

The surface morphology of $Sm_2Zr_2O_7$ after corrosion test at 1300°C for 5 hours is shown in Figure 6. After corrosion, in some regions, a new phase with elongated grain formed on the CMAS contacted surface, while in other regions, the morphology is similar to that of the non-corroded sample as shown in Figure 2. Further elemental composition analyses were carried out in Figure 7. In the elongated grains, only small amounts of Zr were detected and this phase is rich in Si, Sm and Ca. Additionally, the atomic ratio of Ca:Sm:Si is close to that in $Ca_2Sm_8(SiO_4)_6O_2$, which indicates that the elongated grain is $Ca_2Sm_8(SiO_4)_6O_2$ phase. While in Figure 7(a), only small amounts of Sm were detected. The detected Mg, Al, Si and Ca are likely from the CMAS glass background melted on the surface. The backgrounds in spectra 5 and 6 contain only small amounts of Zr and Sm and contain large amounts of Si, Ca, Mg and Al. This suggests that Sm was removed from pyrochlore structure and reacted to form $Ca_2Sm_8(SiO_4)_6O_2$. The Sm_2O_3-ZrO_2 phase diagram is shown Figure 8 and indicates that the pyrochlore structure would transform to fluorite structure due to a decrease in Sm concentration. The peak angles for the cubic fluorite phase in the XRD spectra are higher than those of the pyrochlore structure and the higher angles indicate a decrease in the lattice parameter which is consistent with an increase in the zirconium content, since Zr^{4+} is smaller than Sm^{3+} as shown in Table 2

Table 2. Ionic radius of different elements with coordination number = VIII [9].

	Sm^{3+}	Zr^{4+}	Ca^{2+}
Ionic Radius (pm)	107.9	84	112

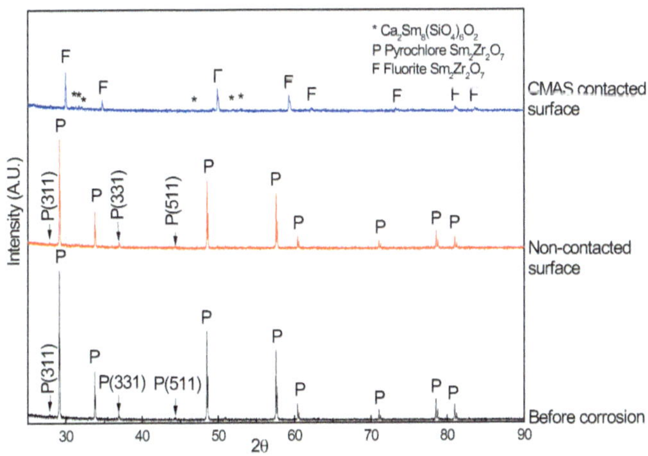

Figure 4. XRD spectra of $Sm_2Zr_2O_7$ after corrosion test at 1300C for 5h.

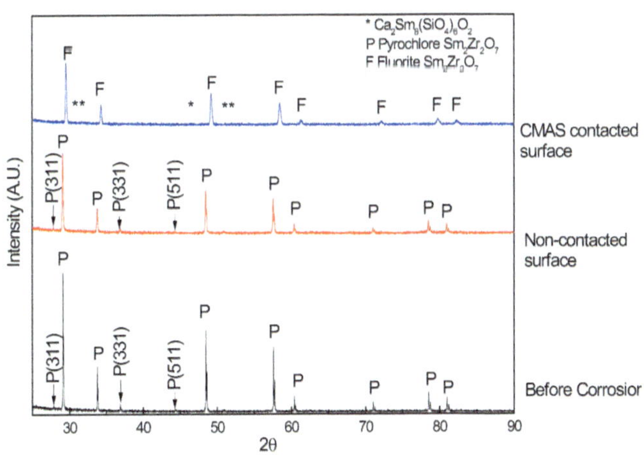

Figure 5. XRD spectra of $Sm_2Zr_2O_7$ after corrosion test at 1400°C for 5 hours.

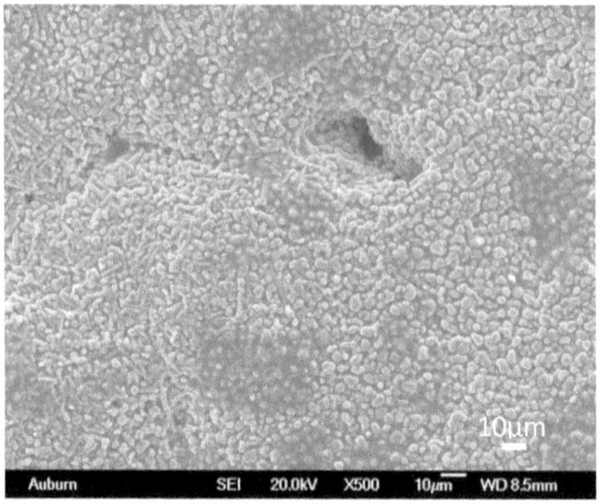

Figure 6. Surface morphology of $Sm_2Zr_2O_7$ ceramics after corrosion test at 1300°C for 5 hours.

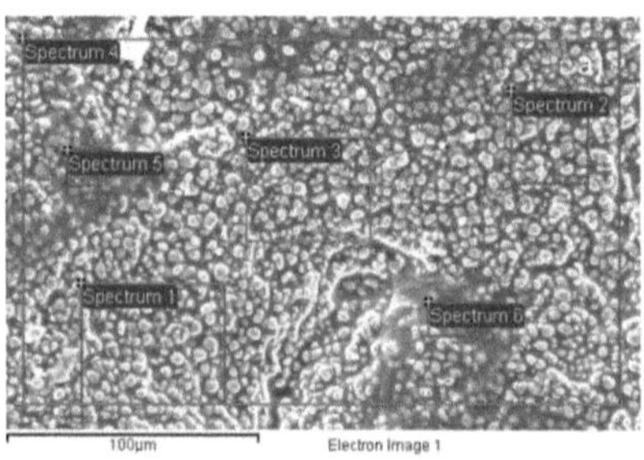

Atomic Percentage for a)

	O	Mg	Al	Si	Ca	Zr	Sm
Spectrum 1	62	2	5	8	5	13	3
Spectrum 2	62	2	5	8	6	12	3
Spectrum 3	63	3	5	8	5	11	3
Spectrum 4	61	3	5	8	7	11	2
Spectrum 5	58	4	6	10	11	7	2
Spectrum 6	44	3	7	13	19	7	2

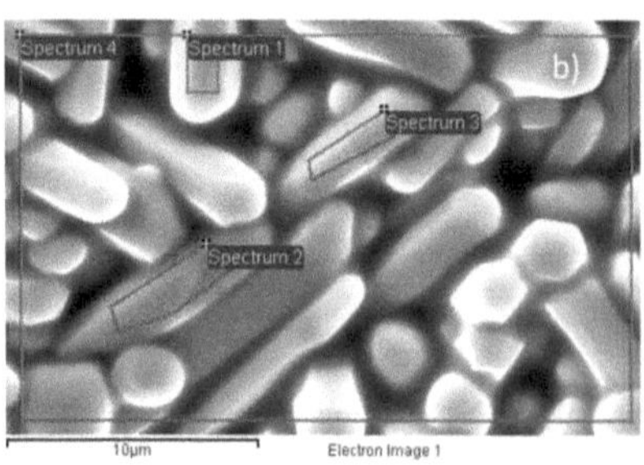

Atomic Percentage for b)

	O	Mg	Al	Si	Ca	Zr	Sm
Spectrum 1	58		1	15	6	1	15
Spectrum 2	66		1	13	5	1	11
Spectrum 3	62			14	5	1	14
Spectrum 4	60	2	3	12	6	4	10

Figure 7. Elemental composition of $Sm_2Zr_2O_7$ after corrosion test at 1300°C for 5 hours.

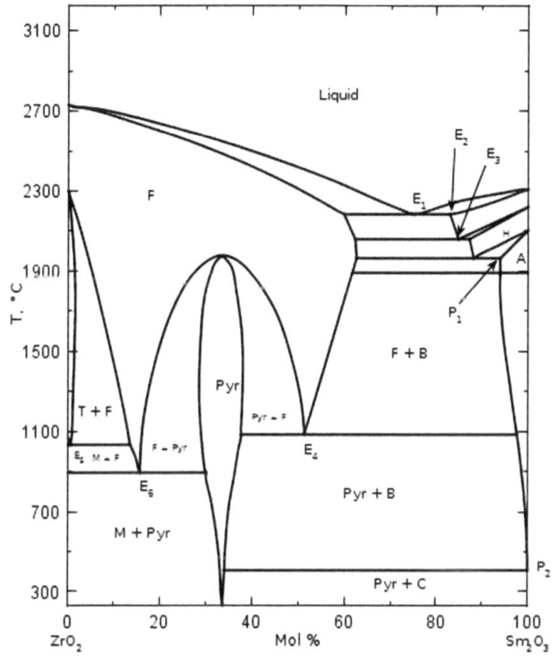

Figure 8. $SmO_{1.5}$-ZrO_2 phase diagram. T = tetragonal, F = cubic fluorite, M – monoclinic, Pyr = pyrochlore, C, B, H = Sm_2O_3 phases. [10]

Conclusion

$Sm_2Zr_2O_7$ pyrochlore ceramics were synthesized using the co-precipitation method and exposed to CMAS corrosion tests. After the corrosion tests, elongated grains of $Ca_2Sm_8(SiO_4)_6O_2$ formed and the pyrochlore structure transformed to the cubic fluorite structure due to the depletion in the Sm content.

Acknowledgement

This material is based upon work supported by the Department of Energy under Award Number DE-FE0011245. This report was prepared as an account of work sponsored by an agency of the United States Government.

Disclaimer: Neither the United States Government nor any agency thereof, nor any of their employees, makes any warranty, express or implied, or assumes any legal liability of responsibility for the accuracy, completeness, or usefulness of any information, apparatus, product, or process disclosed, or represents that its use would not infringe privately owned rights. Reference herein to any specific commercial product, process, or service by trade name, trademark, manufacturer, or otherwise does not necessarily constitute or imply its endorsement, recommendation, or favoring by the United States Government or any agency thereof. The views and opinions of the author expressed herein do not necessarily state or reflect those of the United States Government or any agency thereof.

References

[1] G. Witz, V. Shklover, W. Steurer, S. Bachegowda, and H. P. Bossmann, "High-Temperature Interaction of Yttria Stabilized Zirconia Coatings with CaO–MgO–Al_2O_3–SiO_2 (CMAS) Deposits," *Surface and Coatings Technology*.

[2] W. Pan, S. R. Phillpot, C. Wan, A. Chernatynskiy, and Z. Qu, "Low thermal conductivity oxides," *MRS Bulletin,* vol. 37, pp. 917-922, 2012.

[3] V. Lughi and D. R. Clarke, "High Temperature Aging of YSZ Coatings and Subsequent Transformation at Low Temperature," *Surface and Coatings Technology,* vol. 200, pp. 1287-1291, 11/21/ 2005.

[4] D. R. Clarke, M. Oechsner, and N. P. Padture, "Thermal-Barrier Coatings for more Efficient Gas-turbine Engines," *MRS Bulletin,* vol. 37, pp. 891-898, 2012.

[5] C. Mercer, S. Faulhaber, A. G. Evans, and R. Darolia, "A Delamination Mechanism for Thermal Barrier Coatings Subject to Calcium–Magnesium–Alumino-Silicate (CMAS) Infiltration," *Acta Materialia,* vol. 53, pp. 1029-1039, 2// 2005.

[6] A. G. Evans and J. W. Hutchinson, "The Mechanics of Coating Delamination in Thermal Gradients," *Surface and Coatings Technology,* vol. 201, pp. 7905-7916, 6/25/ 2007.

[7] K. M. Grant, S. Krämer, G. G. E. Seward, and C. G. Levi, "Calcium–Magnesium Alumino-Silicate Interaction with Yttrium Monosilicate Environmental Barrier Coatings," *Journal of the American Ceramic Society,* vol. 93, pp. 3504-3511, 2010.

[8] S. Krämer, J. Yang, C. G. Levi, and C. A. Johnson, "Thermochemical Interaction of Thermal Barrier Coatings with Molten CaO–MgO–Al_2O_3–SiO_2 (CMAS) Deposits," *Journal of the American Ceramic Society,* vol. 89, pp. 3167-3175, 2006.

[9] R.D. Shannon, "Revised Effective Ionic Radii and Systematic Studies of Interatomic Distances in Halides and Chalcogenides," *Acta Crystall.*, vol. 32, pp. 751-767, 1976.

[10] Figure 14384 in *Phase Equilibrium Diagrams for Ceramic Systems*, Version 4.0, (American Ceramic Society, Westerville, OH, 2013).

Energy Technology 2015
Carbon Dioxide Management and Other Technologies

Structural Materials, Heat Transport Fluids, and Novel System Designs for High Power and Process Heat Generation

Energy Technology 2015
Carbon Dioxide Management and Other Technologies

STRUCTURAL MATERIALS, HEAT TRANSPORT FLUIDS, AND NOVEL SYSTEM DESIGNS FOR HIGH POWER AND PROCESS HEAT GENERATION

Heat Transport Fluids I

Session Chairs:

Peter Hosemann
Bruce Pint

Energy Technology 2015: Carbon Dioxide Management and Other Technologies
Edited by: Animesh Jha, Cong Wang, Neale R. Neelameggham, Donna P. Guillen, Li Li, Cynthia K. Belt, Randolph Kirchain, Jeffrey S. Spangenberger, Frank Johnson, Andrew J. Gomes, Amit Pandey, and Peter Hosemann
TMS (The Minerals, Metals & Materials Society), 2015

MINIMUM SYSTEM ENTROPY PRODUCTION FOR THE FIGURE OF MERIT OF HIGH TEMPERATURE HEAT TRANSFER FLUID PROPERTIES

Peiwen Li*, Ye Zhang

Department of Aerospace and Mechanical Engineering
The University of Arizona
Tucson, AZ 85721, USA
*Email: peiwen@email.arizona.edu

Abstract

A high temperature heat transfer fluid in a solar thermal power plant is used to receive and take the heat from the solar concentrator and then give to another fluid (such as water or air) in a thermal power cycle. With regard to their thermal and transport properties, high temperature heat transfer fluids such as synthetic oil, various molten salts, and liquid metals are different. Therefore, a general criterion is needed for evaluation of the merit of fluids regarding their transport properties. The present work considers the system entropy production as the criterion to define figure of merit (FOM). For the required amount of heat transfer rate in a heat transfer system, the fluid with properties that make the system entropy production the minimum is considered to be a winner. The system entropy production includes two parts, due to heat transfer (addition and removal) and pressure loss in the heat exchangers and fluid circulation pipes. Analysis and relevant equations are presented in this work. Several heat transfer fluids known well-known in solar thermal industry including synthetic oils, molten nitrate salts, and halide salts are compared using this standard.

Key Words: Heat Transfer Fluid, Figure of Merit, Heat Transfer System, Entropy Production Rate, Solar Thermal Energy

Nomenclature

A total heat transfer area in a heat exchanger (m^2)
Cp heat capacity [J/(kg K)]
d diameter of fluid tubes (m)
D_h hydraulic diameter of tubes (m)
f Darcy friction coefficient
h heat transfer coefficient [W/(m^2 K)]
k thermal conductivity [W/(m K)]
L length of fluid tube (m)
$\dot{m}$ mass flow rate (kg/s)
n number of flow tubes
Pr Prandtl number
$\dot{Q}$ Heat transfer rate or thermal power (W)
Re Reynolds number
s specific entropy [J/(K kg)]
$\dot{S}_{gen}$ entropy generation rate [J/(K s)]
T Temperature (K or °C)
U velocity (m/s)
v specific volume (m^3/kg)
$\dot{V}$ volume flow rate (m^3/s)

Greek symbols

ΔP pressure loss (Pa)
μ viscosity (Pa s)
ρ density (kg/m^3)

Subscripts

c circulation loop
gen generation
h, high temperature
HE heat exchanger
HE-h high temperature heat exchanger

HE-l	low temperature heat exchanger	*m*	average temperature
i, in	inlet	*o, out*	outlet
l	low temperature	*1, 2*	locations

1. Introduction

It is well known that concentrated solar thermal energy in a high temperature is desired for a solar thermal power plant to have high energy conversion efficiency [1]. To achieve the high temperature, a heat transfer fluid (HTF) being able to work at high temperature is typically needed to receive the heat from a solar thermal concentrator and then deliver the heat to another fluid which is the working substance in a thermal power cycle.

Up to now, there have been three generations of HTFs developed particularly for concentrated solar thermal power systems (CSP). The first generation HTF may be represented by the non-toxic petroleum-based oil (such as Xceltherm® 600-C_{20} paraffin oil), which has an upper limit temperature of 315 °C [2]. The second-generation of HTFs are known as synthetic organic oils (such as Therminol®VP-1 which is a mixture of Diphenyl oxide and Biphenyl in weight percentage of 73.5% versus 26.5%), which has an upper limit temperature of 400 °C [3]. To further improve the working temperature in a CSP system, the third generation of HTF is based on nitrate molten salts [4], which has upper limit temperatures of 580 °C (such as mixture of sodium nitrate-sodium nitrite-potassium nitrate, $NaNO_3$–$NaNO_2$–KNO_3, 7-49-44 mol% also known as Hitec® salt) [5, 6]. A great effort is currently undergoing in the solar energy community to develop the fourth generation HTFs (either eutectic high temperature molten salts [7, 8] or eutectic metal mixtures [9, 10]) to approach a working temperature of at least 800 °C.

The requirement about the properties to the HTF is multifold. First and foremost, it should be able to work as a fluid in a wide range of temperatures at low vapor pressures and low corrosion rates to the pipes and containers in the system. Second, the fluid must have low corrosion to the pipes and containers of the system. Thirds, the fluid must have favorable transport properties including a high thermal conductivity, a large heat capacity, and a low viscosity. Last, but not least, the cost of the fluid should be acceptable by industrial customers.

Researchers have done some work to evaluate and compare different types of heat transfer fluids at different temperatures [11, 12], however, there is no universal figure of merit yet been proposed. A summit criterion of choosing a fluid may be converged to the economic value of the produced electricity over the lifetime of a power plant due to the use of the fluid and the associated materials of pipes and containers. As the first step, it is important to evaluate transport properties of different fluids at given temperatures. Therefore, exergy loss should be examined, which is attributed to the entropy production in the process [13, 14]. The exergy loss can be essentially translated into the loss of output electrical power from the CSP power plant. The entropy production in the heat transfer system that uses a HTF should include that from heat transfer processes as well as from the pressure loss due to friction in the circulation of the HTF. Therefore, the system entropy production from using a HTF can be defined as a figure of merit (FOM).

It is worth noting that since the system entropy production is related to the reduction of electrical power from the CSP plant, the FOM presented in this work is thus a basic parameter needed for economic and investment analysis to the CSP system. Details of analysis to the entropy production are presented in the following sections.

2. Theoretical Analysis of System Entropy Generation

The properties of a fluid including, thermal conductivity, viscosity, density, and heat capacity, jointly determine how well the heat transfer can be performed in a thermal system with the minimum loss of exergy. Therefore, the objective of the following analysis is to find the relationship of the system entropy production to all the relevant fluid properties.

The application of a HTF in a concentrated solar thermal power plant is associated to two heat transfer processes and also the circulation of the HTF in the system. As show in Fig. 1, the system has heat addition to the HTF from the solar field and heat removal through a heat exchanger to deliver the heat to another fluid in the thermal power cycle of the power plant.

One may define the desired fluid temperatures as T_i at inlet and T_o at outlet of the solar collection heat exchanger. This defines the average temperature, $(T_i + T_o)/2$, of the HTF that receives heat from solar collector, and also that of the HTF when delivering the heat to a cold fluid. A good HTF should render the heat transfer to be so accomplished that the high temperature in the solar collector T_h should not be too high, and also the T_l in the hear delivery heat exchanger should not be too low. For the desired amount of heat transfer rate $\dot{Q}$, one also does not want to use too much of pumping power to achieve the goal of the heat transfer. Therefore, to estimate the performance of a HTF in accomplishing the heat transfer function, the entropy production in the entire process has to be considered.

It is assumed that the pipes and heat exchangers in the system illustrated in Fig. 1 have been decided based on the properties of a given HTF and the demanded heat transfer rate, which is the total thermal power for a CSP plant. The entropy production rates can be obtained for every HTFs of interest for comparison. The one that makes the minimum entropy production in the system will be elected as the best candidate. It also needs to be noted that the current FOM of HTFs is about the transport properties. The fluids to be evaluated should first have good material compatibility with devices, and fluid vapor pressures are low and not of concern at studied temperatures. Another issue is the restriction of the density of a HTF for application in a solar power tower, where too high density of the fluid will cause unacceptable high pressure in the pipe at ground level when the HTF is pumped to the tower of a significant height.

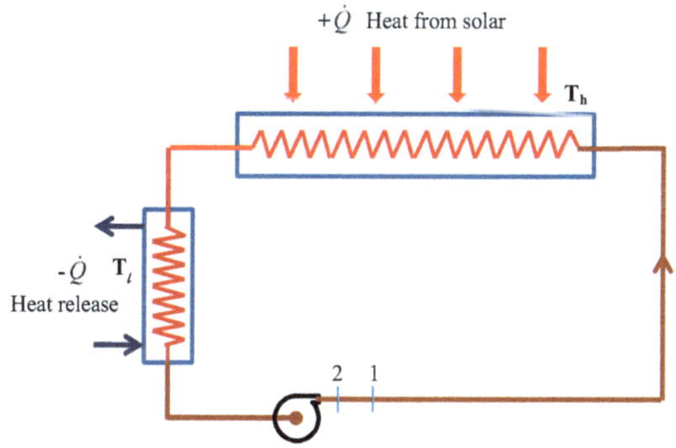

Figure 1. Flow loop that a HTF circulated inside

From the knowledge of second law of thermodynamics, the steady state entropy production rate in the system (a control volume from location 1 to location 2 including the pipes and heat exchangers) should be expressed as:

$$s_2\dot{m} = s_1\dot{m} + \frac{\dot{Q}_{in}}{T_h} - \frac{\dot{Q}_{out}}{T_l} - \dot{m}\frac{v}{T_m}\Delta P + \dot{m}s_{gen} \tag{1}$$

where the system inlet is at location 1 and outlet is at location 2; T_h is the temperature of the heat source (concentrated solar energy) that is added to the system, and T_l is for a low temperature heat source that receives the heat from the HTF system.

Since the HTF circulates in the system, we can choose point 1 and 2 to be the same point, meaning the fluid circulated from a starting point and back to an ending point which is the same as the starting point. Therefore, there must have $s_2\dot{m} = s_1\dot{m}$, which results in the following expression for the system entropy production rate:

$$\dot{m}s_{gen} = \dot{S}_{gen} = \frac{\dot{Q}_{out}}{T_l} - \frac{\dot{Q}_{in}}{T_h} + \dot{m}\frac{v}{T_m}\Delta P = \frac{\dot{Q}_{out}}{T_l} - \frac{\dot{Q}_{in}}{T_h} + \frac{\dot{V}}{T_m}\Delta P \tag{2}$$

where $\dot{S}_{gen}$ is the system entropy production at a steady state, $\dot{V}$ is the volume flow rate of the HTF; $\dot{Q}_{in}$ and $\dot{Q}_{out}$ are the absolute values of the heat rates adding to and delivered out of the system, respectively. It is clear to see from Eq. (2) that the entropy production is due to the two heat transfer processes, and the pressure loss in the system that is overcome by the pump. Ideally, the heat added into the system and that removed from the system has the same absolute value. For convenience, we have $\dot{Q}_{out} = \dot{Q}_{in} = \dot{Q}$ in the analyses hereafter.

The fluid temperature at the inlet of the solar collection field is defined as T_i and at the exit of the solar collection field as T_o. When considering the heat removal in the heat exchanger to deliver the heat to the fluid in the thermal power cycle, it is easy to understand that the temperature of the HTF at the inlet and exit of this heat exchanger is T_o and T_i, respectively, which is just the opposite to that in the solar collection field. The average temperatures of the HTF in the solar collection field and the heat exchanger are therefore the same, which is $T_m = 0.5(T_o + T_i)$. Therefore, the average temperature of the HTF in the circulation should also be $T_m = 0.5(T_o + T_i)$. The temperature differences that drive the heat exchange in the solar collection field and in the heat exchanger are $T_h - 0.5(T_o + T_i)$ and $0.5(T_o + T_i) - T_l$, respectively.

Considering the heat transfer processes in the solar collection field and in the heat exchanger, as seen in Fig. 1, there are:

$$\dot{Q} = A_h h_h \left[T_h - 0.5(T_o + T_i)\right] \tag{3}$$

$$\dot{Q} = A_l h_l \left[0.5(T_o + T_i) - T_l\right] \tag{4}$$

where A_h and A_l are heat transfer surface areas; h_h and h_l are heat transfer coefficients. From these two equations, we can find that,

$$T_h = \frac{\dot{Q}}{A_h h_h} + 0.5(T_o + T_i) \tag{5}$$

$$T_l = 0.5(T_o + T_i) - \frac{\dot{Q}}{A_l h_l} \tag{6}$$

Substitute Eqs. (5) and (6) into Eq. (2), the system entropy production rate is obtained, which is

$$\dot{S}_{gen}=\frac{\dot{Q}}{0.5(T_o+T_i)-\dot{Q}/(A_l h_l)}-\frac{\dot{Q}}{0.5(T_o+T_i)+\dot{Q}/(A_h h_h)}+\frac{\dot{V}}{0.5(T_o+T_i)}\Delta P \tag{7}$$

For the purpose of evaluating the transport properties of a HTF, this analysis needs to prescribe the heat rate $\dot{Q}$, the desired temperatures of T_i and T_o. With the given transport properties of the HTF, the flow rates can thus be decided as

$$\dot{Q}=\dot{m}Cp(T_o-T_i) \tag{8}$$

$$\dot{m}v=\frac{\dot{m}}{\rho}=\dot{V} \tag{9}$$

$$\dot{V}=\frac{\dot{Q}}{\rho Cp(T_o-T_i)} \tag{10}$$

Substitute Eq. (10) into Eq. (7), the entropy production rate is further expressed as

$$\dot{S}_{gen}=\frac{\dot{Q}}{0.5(T_o+T_i)-\dot{Q}/(A_l h_l)}-\frac{\dot{Q}}{0.5(T_o+T_i)+\dot{Q}/(A_h h_h)}+\frac{\dot{Q}}{0.5\rho Cp(T_o+T_i)(T_o-T_i)}\Delta P \tag{11}$$

where the first two terms are due to heat transfer processes, which is names as $\dot{S}_{gen-ht}$, and the last term is due to pressure loss when the HTF circulates in the system (including in heat exchangers and in circulation pipes), which may be defined as $\dot{S}_{gen-fl}$.

From Eq. (11), it is also understandable that a large value of h_l and h_h will result in less entropy production, while the pressure loss ΔP will contribute to higher entropy production. Furthermore, it is also easy to understand that if the heat transfer is improved (either due to heat transfer surface enhancement or better properties of HTF), the entire heat transfer task can be achieved with the temperature T_l closer to T_h, which therefore is beneficial to the thermal energy efficiency in the thermal power plant. The pressure loss in the entire system includes the losses from two heat exchangers and from fluid circulation pipes, which may be defined as ΔP_{HE-h}, ΔP_{HE-l}, and ΔP_c, respectively. With the volume flow rate of the HTF available from Eq. (10), the pressure losses in the system can be conveniently obtained.

In the next section, the heat transfer coefficients, h_h and h_l, will be expressed in terms of the fluid properties, the prescribed temperatures, as well as the heat rate, which jointly determines the flow rate and velocities of the fluid in given devices.

2.1 Entropy production pate in heat transfer processes

2.1.1 Heat transfer due to turbulent flow

For the heat transfer of turbulent flow in pipes with Re close to 1.0×10^4 or above, $0.7\leq \Pr\leq 160$, and length-diameter ratio $L/d\geq 10$, we take the famous Dittus-Boelter equation [15] and obtain heat transfer coefficients as follows:

$$Nu_h = \frac{h_h D_h}{k} = 0.023 \mathrm{Re}_h^{0.8} \mathrm{Pr}^{0.4} \qquad \text{(fluid is heated)} \tag{12}$$

$$Nu_l = \frac{h_l D_h}{k} = 0.023 \mathrm{Re}_l^{0.8} \mathrm{Pr}^{0.3} \qquad \text{(fluid is cooled)} \tag{13}$$

Assume heat transfer design analysis already found the number (n) of tubes (in diameter of d) for the heat exchanger, the flow velocity in the tubes of a heat exchanger is obtained as:

$$u = \frac{\dot{V}}{n(0.25\pi d^2)} = \frac{\dot{Q}}{\rho Cp(T_o - T_i)n(0.25\pi d^2)} = \frac{\dot{Q}}{\rho Cp(T_o - T_i)Nd^2} \tag{14}$$

where we define $N = 0.25n\pi$. Using Eq. (14), the Reynolds number for the flow in the heat transfer tubes is obtained as

$$\mathrm{Re} = \frac{\dot{Q}}{\rho Cp(T_o - T_i)Nd^2}\frac{\rho d}{\mu} = \frac{\dot{Q}}{Cp(T_o - T_i)N \cdot d \cdot \mu} \tag{15}$$

Bring the expression for Reynolds numbers into Eqs. (12) and (13), the heat transfer coefficients are obtained as follows:

$$h_h = \frac{k}{d_h} 0.023 \left[\frac{\dot{Q}}{\rho Cp(T_o - T_i)N_h d_h^2} \frac{\rho d_h}{\mu} \right]^{0.8} \left(\frac{\mu Cp}{k} \right)^{0.4} = 0.023 \left[\frac{\dot{Q}}{N_h(T_o - T_i)} \right]^{0.8} \frac{1}{d_h^{1.8}} \frac{k^{0.6}}{Cp^{0.4}\mu^{0.4}} \tag{16}$$

$$h_l = \frac{k}{d_l} 0.023 \left[\frac{\dot{Q}}{\rho Cp(T_o - T_i)N_l d_l^2} \frac{\rho d_l}{\mu} \right]^{0.8} \left(\frac{\mu Cp}{k} \right)^{0.3} = 0.023 \left[\frac{\dot{Q}}{N_l(T_o - T_i)} \right]^{0.8} \frac{1}{d_l^{1.8}} \frac{k^{0.7}}{Cp^{0.5}\mu^{0.5}} \tag{17}$$

Substituting Eqs. (16) and (17) into Eq. (11), the entropy generation due to heat transfer in the system is

$$\dot{S}_{gen-ht}^{Turbulent} = \frac{\dot{Q}}{0.5(T_o + T_i) - \dfrac{\dot{Q}}{4N_l^{0.2}L_l}\dfrac{1}{0.023}\left[\dfrac{(T_o - T_i)}{\dot{Q}}\right]^{0.8}\dfrac{d_l^{0.8}Cp^{0.5}\mu^{0.5}}{k^{0.7}}} - \frac{\dot{Q}}{0.5(T_o + T_i) + \dfrac{\dot{Q}}{4N_h^{0.2}L_h}\dfrac{1}{0.023}\left[\dfrac{(T_o - T_i)}{\dot{Q}}\right]^{0.8}\dfrac{d_h^{0.8}Cp^{0.4}\mu^{0.4}}{k^{0.6}}} \tag{18}$$

2.1.2 Heat transfer due to laminar flow

For laminar flow in heat exchangers, the Nusselt number for a constant wall heat flux case is a constant (which gives Nu_h=Nu_l=4.36). Therefore, the heat transfer coefficient is obtained to be $h_h = h_l = 4.36 \cdot k/d$. This allows us to calculate the entropy production due to laminar flow heat transfer, which gives

$$\dot{S}_{gen-ht}^{\text{Laminar}} = \frac{\dot{Q}}{0.5(T_o + T_i) - \frac{\dot{Q}}{4.36k4N_h L_h}} - \frac{\dot{Q}}{0.5(T_o + T_i) + \frac{\dot{Q}}{4.36k4N_l L_l}} \tag{19}$$

2.2 Entropy production pate due to pressure losses

2.2.1 Flow with $2300 \le \text{Re} \le 2\times10^4$

The following equation [15] is chosen to calculate the pressure loss due to turbulent flow in the heat transfer tubes in heat exchangers.

$$\Delta P_{HE} = \frac{f \cdot \rho u^2 L}{2d} \tag{20}$$

where f is Darcy friction factor. Taking Blasius equation ($f = 4*0.079\,\text{Re}^{-0.25}$) for Darcy friction factor, the pressure drop is obtained as

$$\Delta P_{HE} = 0.079\frac{2\cdot\rho u^2 L}{d}\left(\frac{ud\rho}{\mu}\right)^{-0.25} = 0.158\frac{\rho^{0.75}u^{1.75}L\mu^{0.25}}{d^{1.25}} \tag{21}$$

where L is the length of the tube or pipe subject to study. Substituting velocity u (by Eq. (14)) into Eq. (21), there is

$$\Delta P_{HE} = 0.158\frac{\rho^{0.75}L\mu^{0.25}}{d^{1.25}}\left[\frac{\dot{Q}}{\rho Cp(T_o - T_i)Nd^2}\right]^{1.75} = 0.158\frac{L\mu^{0.25}}{\rho Cp^{1.75}d^{4.75}}\left[\frac{\dot{Q}}{(T_o - T_i)N}\right]^{1.75} \tag{22}$$

For the pressure loss in the fluid circulation pipes (in a total length of L_c and pipe diameter of d_c), the same expression of pressure loss is

$$\Delta P_c = 0.158\frac{\rho^{0.75}L_c\mu^{0.25}}{d_c^{1.25}}\left[\frac{\dot{Q}}{\rho Cp(T_o - T_i)N_c d_c^2}\right]^{1.75} = 0.158\frac{L_c\mu^{0.25}}{\rho Cp^{1.75}d_c^{4.75}}\left[\frac{\dot{Q}}{(T_o - T_i)N_c}\right]^{1.75} \tag{23}$$

where N_c is definition as $N_c = 0.25n_c\pi$, and n_c is the number of pipes that are used to circulate the fluid.

The total system pressure drop includes three parts, from fluid circulation pipes, as well as from the heat exchangers (one has heat added to the HTF, and the other has heat removed from the HTF). Therefore, the total pressure drop in heat exchangers and in circulation pipe is

$$\Delta P = \Delta P_{HE-h} + \Delta P_{HE-l} + \Delta P_c \tag{24}$$

Substituting pressure losses into the third term in Eq. (11), the entropy production rate due to pressure loss is

$$\dot{S}_{gen-fl}^{T-1} = \frac{\dot{Q}}{0.5(T_o + T_i)(T_o - T_i)} \frac{0.158\mu^{0.25}}{\rho^2 Cp^{2.75}} \left[\frac{\dot{Q}}{(T_o - T_i)}\right]^{1.75} \left(\frac{L_h}{d_h^{4.75} N_h^{1.75}} + \frac{L_l}{d_l^{4.75} N_l^{1.75}} + \frac{L_c}{d_c^{4.75} N_c^{1.75}}\right) \quad (25)$$

where $\dot{S}_{gen-fl}^{T-1}$ designates the first case of turbulent flow, where there is $2300 \le \text{Re} \le 2 \times 10^4$.

2.2.2 Flow with $\text{Re} \ge 2 \times 10^4$

Under the condition of $\text{Re} \ge 2 \times 10^4$, the equation for pressure loss of turbulent flow in the heat transfer tubes of heat exchangers is

$$\Delta P_{HE} = 0.184 \frac{\rho u^2 L}{2d} \left(\frac{ud\rho}{\mu}\right)^{-0.2} = 0.092 \frac{\rho^{0.8} u^{1.8} L \mu^{0.2}}{d^{1.2}} \quad (26)$$

Bring the expression of velocity given in Eq. (14) into the above expression, there is

$$\Delta P_{HE} = 0.092 \frac{\rho^{0.8} L \mu^{0.2}}{d^{1.2}} \left[\frac{\dot{Q}}{\rho Cp(T_o - T_i) N d^2}\right]^{1.8} = 0.092 \frac{L \mu^{0.2}}{\rho Cp^{1.8} d^{4.8}} \left[\frac{\dot{Q}}{(T_o - T_i) N}\right]^{1.8} \quad (27)$$

Similarly, the pressure drop in a circulation pipe with $\text{Re} \ge 2 \times 10^4$ is

$$\Delta P_c = 0.092 \frac{\rho^{0.8} L_c \mu^{0.2}}{d_c^{1.2}} \left[\frac{\dot{Q}}{\rho Cp(T_o - T_i) N_c d_c^2}\right]^{1.8} = 0.092 \frac{L_c \mu^{0.2}}{\rho Cp^{1.8} d_c^{4.8}} \left[\frac{\dot{Q}}{(T_o - T_i) N_c}\right]^{1.8} \quad (28)$$

In case that the Reynolds numbers of the flow in heat exchangers and in HTF circulation pipes are all larger than 2×10^4, the entropy production rate due to all pressure losses is

$$\dot{S}_{gen-fl}^{T-2} = \frac{\dot{Q}}{0.5(T_o + T_i)(T_o - T_i)} \frac{0.092\mu^{0.2}}{\rho^2 Cp^{2.8}} \left[\frac{\dot{Q}}{(T_o - T_i)}\right]^{1.8} \left(\frac{L_h}{d_h^{4.8} N_h^{1.8}} + \frac{L_l}{d_l^{4.8} N_l^{1.8}} + \frac{L_c}{d_c^{4.8} N_c^{1.8}}\right) \quad (29)$$

where $\dot{S}_{gen-fl}^{T-2}$ designates the second case of turbulent flow when there is $\text{Re} \ge 2 \times 10^4$.

2.2.3 With laminar flow in heat exchangers and circulation pipes

Although the flow of HTF in a solar thermal power plant can rarely be laminar flow, for completion, an analysis is still provided here. The Darcy friction factor for laminar flow is $f = 64/\text{Re}$. Bring the expression of velocity into the Reynolds number, the pressure losses in a tube of heat exchanger and in a circulation tube are given as follows, respectively.

$$\Delta P_{HE} = \frac{\rho u^2 L}{2d} \cdot 64 \cdot \left(\frac{ud\rho}{\mu}\right)^{-1} = 32 \frac{L\mu}{d^2} \frac{\dot{Q}}{\rho Cp(T_0 - T_i) N d^2} \quad (30)$$

$$\Delta P_c = \frac{\rho u_c^2 L_c}{2d_c} \cdot 64 \cdot \left(\frac{u_c d_c \rho}{\mu}\right)^{-1} = 32 \frac{L_c \mu}{d_c^2} \frac{\dot{Q}}{\rho Cp(T_0 - T_i) N_c d_c^2} \tag{31}$$

The entropy production rate due to all pressure losses in heat exchangers and circulation tubes, if all in laminar flow, is expressed as

$$\dot{S}^L_{gen-fl} = \frac{\dot{Q}}{0.5(T_o + T_i)(T_o - T_i)} \frac{32\mu}{\rho^2 Cp^2} \left[\frac{\dot{Q}}{(T_o - T_i)}\right] \left(\frac{L_h}{d_h^4 N_h} + \frac{L_l}{d_l^4 N_l} + \frac{L_c}{d_c^4 N_c}\right) \tag{32}$$

where the definitions of N_h, N_l, N_c, L_h, L_l, and L_c have been defined in previous discussions.

2.2.4 Different Reynolds numbers in different devices

It needs to be noted that the Reynolds number of the flow in heat exchangers and in HTF circulation tube can be different depending on the flow rate and designs of the devices. One has to check the Reynolds numbers in each device and choose the proper terms from Eqs. (25), Eq. (29), and Eq.(32) accordingly to calculate the total entropy generation rate due to pressure losses.

Equations for heat transfer coefficient and pressure loss in better precision may also be applied in the calculations presented above. Nevertheless, the basic analysis to the system entropy generation will not change.

3. Discussion and Evaluation of Several Heat Transfer Fluids

In all the above expressions of entropy production, the given parameters are the system heat transfer rate $\dot{Q}$, the desired working temperatures of fluid, T_i, T_o, and dimensions of the devices including d_h, d_l, d_c, L_h, L_l, L_c, n_h, n_l, and n_c. In fact, the dimensions (L_h, L_l, n_h, n_l) of the devices must be obtained through design of the heat exchangers, which is based on constrains of given diameters of the heat transfer tubes, tolerable maximum temperature of the tube wall against the desired temperatures of fluid T_i and T_o.

For heat transfer of turbulent flow, it is interesting to examine Eq. (18) that the clusters of properties $d_l^{0.8} Cp^{0.5} \mu^{0.5} / k^{0.7}$ and $d_h^{0.8} Cp^{0.4} \mu^{0.4} / k^{0.6}$ actually influence the entropy production in the heat transfer processes. For heat transfer of laminar flow, the cluster of properties contribute to entropy production in the heat transfer processes is $1/k$. Large values of theses cluster of properties will result in lager entropy production in the system.

For the entropy production due to pressure loss in the heat exchangers and in the circulation pipe, the property clusters are $\mu^{0.25}/(\rho^2 Cp^{2.75} d^{4.75})$ and $\mu^{0.2}/(\rho^2 Cp^{2.8} d^{4.8})$ for turbulent flow with Reynolds number less than 2×10^4, and larger than 2×10^4, respectively. For laminar flow the cluster of properties to entropy production due to pressure loss is $\mu/(\rho^2 Cp^2 d^4)$. Large values of theses cluster of properties will result in large entropy production in the system. The clusters of HTF properties contributing to lager entropy generation due to heat transfer and pressure drop are summarized in Table 1.

It is important to note that a large viscosity always contribute to a large entropy production either from heat transfer or from pressure loss. A large thermal conductivity contributes to a less entropy production in heat transfer process. Whereas a large heat capacity contributes to a large entropy production due to heat transfer, it also contributes to less entropy production due to pressure loss in the flow. Although generally the pressure loss causes a relatively small proportion of the total entropy production, one has to actually calculate the total entropy production from both heat transfer and pressure loss to evaluate the merit of a HTF.

Table 1 Clusters of properties contributing entropy production in the system

Heat transfer		Pressure drop	
Fluid is heated	Fluid is cooled		
$\mathrm{Re} \geq 1\times10^4$: $d_h^{0.8}Cp^{0.4}\mu^{0.4}/k^{0.6}$	$\mathrm{Re} \geq 1\times10^4$: $d_l^{0.8}Cp^{0.5}\mu^{0.5}/k^{0.7}$	$\mathrm{Re} \leq 2\times10^4$: $\mu^{0.25}/(\rho^2Cp^{2.75}d^{4.75})$	$\mathrm{Re} \geq 2\times10^4$: $\mu^{0.2}/(\rho^2Cp^{2.8}d^{4.8})$
$\mathrm{Re} \leq 2300$: $1/k$	$\mathrm{Re} \leq 2300$: $1/k$	$\mathrm{Re} \leq 2300$: $\mu/(\rho^2Cp^2d^4)$	

As has been mentioned before, the system entropy production rate is viewed as the FOM for a HTF. To show a comparison of the FOM of different HTFs, analysis was conducted to a system that transfers thermal energy of 50 MW_t from a solar concentrator to another fluid. The following cases of given conditions of temperatures and HTFs are selected for the study. Representing the history of the development of HTFs and there application, from low to higher temperature duty, these examples include:

(1) The average working temperature is 200 °C with T_i and T_o being 150 °C and 250 °C, respectively. Two HTFs, Xceltherm-600 [2] and Therminol VP-1 [3] are compared.

(2) The average working temperature is 340 °C with T_i and T_o being 290 °C and 390 °C, respectively. Comparison is made between HTFs of Therminol VP-1 [3] and nitrate molten salt (KNO_3-$NaNO_2$-$NaNO_3$).

(3) The average working temperature is 450 °C with T_i and T_o being 400 °C and 500 °C, respectively. Comparison is made between HTFs of nitrate molten salt (KNO_3-$NaNO_2$-$NaNO_3$) [5] and halide molten salt (NaCl-KCl-$ZnCl_2$) [16].

(4) The average working temperature is 700°C, and T_i and T_o are 650 °C and 750 °C, respectively. Entropy production rate using halide molten salt (NaCl-KCl-$ZnCl_2$) [16] is studied, which is to be compared to itself when the working temperatures are the same as that in case (3).

The calculated results of entropy production and the Carnot efficiency (η) based on the temperatures (T_l) of the heat delivered to the power cycle for all the above cases are listed in Tables 2-5, together with the given conditions of temperature, the property of the fluids, and the dimensions of the heat exchangers. For a clear comparison, Fig. 2 gives the entropy production rate when transferring a heat of 50 MWt at different levels of temperatures as well as using different types of HTFs.

It is clear that the heat transfer system at low temperature level has high entropy generation rate. Comparison of the entropy generation rate between every two different HTFs at the same operating temperature of $(T_i+T_o)/2$ shows which HTF has better performance. Obviously, Therminol VP-1 is better than the most earliest HTF, Xceltherm-600, at working temperature of 200 °C level. While working at temperature level of 340 °C, Therminol VP-1 and Nitrate molten salts are comparable. Going to a higher temperature of 450 °C, Nitrate molten salt and Halide molten salt are compared. Because of the lower thermal conductivity, the latter one has lower performance. When the operating temperature is at 700 °C level, the entropy production with the use of the molten halide salt reaches the lowest level of all the cases at different temperatures. This clearly shows the benefit of operating the heat transfer system at higher temperatures. Another important parameter to examine is the Carnot efficiency (η)

based on the temperatures (T_l) for each case. With the use of the Halide molten salt working at temperature of 700 °C, the efficiency (η) reaches 62.5%; whereas at a temperature of 450 °C, using Nitrate molten salt, this efficiency is only 51.8%.

The analysis can also be applied to studies to look at the gain of the reduced entropy production if some of the properties of a HTF can be improved through material engineering. For example, in Table 5, targeted higher properties of thermal conductivity and heat capacity were investigated to show the increase of the Carnot efficiency. The entropy production was reduced due to the improved properties, and at the same time the efficiency was also improved.

Table 2 Properties of heat transfer fluids at average temperature of 200 °C and the results of entropy production rate.

	Xceltherm-600 at 200°C	Therminol VP-1 200°C
ρ kg/m^3	743.2	913
μ m Pa s	0.623	0.395
k W/m-K	0.1216	0.114
Cp kJ/kg-K	2.61	2.048
T_h K	573	552
T_l K	373	399
$\dot{S}_{gen}$ J/(K-s)	46793	34872
$\eta = 1 - 300/T_l$	0.196	0.247
T_i=423 K; T_o=523 K; d_l=d_h=0.02 m; n_l=n_h=820; L_h=7.7 m; L_l=10.0 m; d_c=0.5 m; n_c=4; L_c=200 m		

Table 3 Properties of heat transfer fluids at average temperature of 340 °C and the results of entropy production rate

	Therminol VP-1 340°C	KNO_3-$NaNO_2$-$NaNO_3$ (7-49-44 %mole) 340 °C
ρ kg/m^3	773	1833.921
μ m Pa s	0.185	2.56394
k W/m-K	0.089	0.51
Cp kJ/kg-K	2.425	1.89
T_h K	723	713
T_l K	523	526
$\dot{S}_{gen}$ J/(K-s)	26448	24956
$\eta = 1 - 300/T_l$	0.426	0.429
T_i=563 K; T_o=663 K; d_l=d_h=0.02 m; n_l=n_h=820; L_h=5.1 m; L_l=7.3 m; d_c=0.5 m; n_c=4; L_c=200 m		

Table 4 Properties of heat transfer fluids at average temperature of 450 °C and the results of entropy production rate

	KNO_3-$NaNO_2$-$NaNO_3$@450°C	NaCl-KCl-$ZnCl_2$ @450 °C [5]
ρ kg/m^3	1751.454	2150
μ m Pa s	1.375	7.0
k W/m-K	0.51	0.32

Cp kJ/kg-K	1.7596	0.81
T_h K	793	853
T_l K	623	511
$\dot{S}_{gen}$ J/(K-s)	17206	39287
$\eta = 1 - 300/T_l$	0.518	0.413
T_i=673 K; T_o=773 K; d_l=d_h=0.02 m; n_l=n_h=820; L_h=5.5 m; L_l=4.5 m; d_c=0.5 m; n_c=4; L_c=200 m		

Table 5 Properties of heat transfer fluid at average temperature of 700 °C and the results of entropy production rate.

	$NaCl\text{-}KCl\text{-}ZnCl_2$ 700 °C	
ρ kg/m^3	1950	
μ m Pa s	4.0	
k W/m-K	0.31	Targeted higher value: 0.51
Cp kJ/kg-K	0.90	Targeted higher value: 1.50
T_h K	1083	
T_l K	800	
$\dot{S}_{gen}$ J/(K-s)	16355	14745
$\eta = 1 - 300/T_l$	0.625	0.632
T_i=923 K; T_o=1023 K; d_l=d_h=0.02 m; n_l=n_h=820; L_h=5.5 m; L_l=4.5 m; d_c=0.5 m; n_c=4; L_c=200 m		

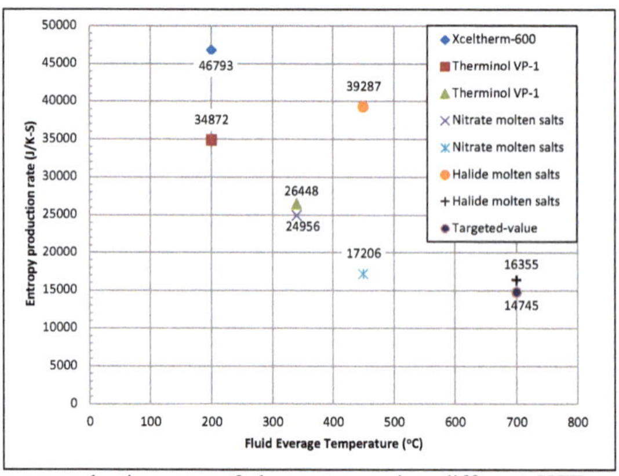

Figure 2. The entropy production rate of the system using different HTFs at different operating temperatures for the duty of transferring 50 MWt heat.

4. Concluding Remarks

This paper tries to provide a standard of evaluating the FOM of heat transfer fluids that serves the function of transferring heat from a concentrated solar collector and deliver to a thermal power plant.

Entropy production of the system is used to appreciate the FOM of different HTFs at different temperature levels.

To be noted is that through analysis, the role and contribution of each of the transport properties to the minimized system entropy generation can be found, and a figure of merit (FOM) including all the transport properties was able to be defined. The FOM will allow comparison and selection of different heat transfer fluids.

It is worth noting that the system entropy production as a FOM helps evaluate the transport properties of HTFs. Since the entropy production is related to the exergy destruction (defined as $T_0\dot{S}_{gen}$) or the reduction of useful work, this analysis can be eventually associated to cost analysis to the system, which will be further pursued by the authors.

The basic equations of entropy production can also be applied to evaluate different measures for heat transfer enhancement which is at the expenses of increased pressure loss.

Acknowledgement

The authors are pleased to acknowledge the financial support for this research project from Department of Energy (DOE) Grant No. DE-EE0005942. We also thank the University of Arizona for providing the experimental facilities.

References

[1] Jon T. Van Lew, Peiwen Li, Cho Lik Chan, Wafaa Karaki, Jake Stephens, Analysis of Heat Storage and Delivery of a Thermocline Tank Having Solid Filler Material general on thermal storage, Journal of Solar Energy Engineering, MAY 2011, Vol. 133 / 021003-1.

[2] XCELTHERM® 600 Engineering Properties, Radco Industries, (http://www.radcoind.com/).

[3] Therminol VP-1, http://www.solutioa.com

[4] Manohar S. Sohal, Matthias A. Ebner, Piyush Sabharwall, Phil Sharpe, Engineering Database of Liquid Salt Thermophysical and Thermochemical Properties, March 2010, Report # INL/EXT-10-18297, by Idaho National Laboratory Idaho Falls, Idaho 83415.

[5] R. W. Bradshaw and N. P. Siegel (2008). "Molten nitrate salt development for thermal energy storage in parabolic trough solar power systems," Proceedings of ES2008, Energy Sustainability 2008, August, 10-14, 2008, Jacksonville, FL, pp 1-7.

[6] A. Baraka, A. I. Abdel-Rohman, and A. A. El Hosary, (1976). "Corrosion of Mild Steel in Molten Sodium Nitrate-Potassium Nitrate Eutectic," British Corrosion Journal, 11(1), 43-46.

[7] Yu-ting Wu, Nan Ren, Tao Wang, Chong-fang Ma, Experimental study on optimized composition of mixed carbonate salt for sensible heat storage in solar thermal power plant, Solar Energy 85 (2011) 1957–1966.

[8] K. Vignarooban, P. Pugazhendhi, C. Tucker, D. Gervasio and A.M. Kannan, Corrosion Resistance of Hastelloys in Molten Metal-Chloride Heat-Transfer Fluids for Concentrating Solar Power Applications, Solar Energy 103 (2014) 62-69.

[9] J. Pacio, Cs. Singer, Th. Wetzel, R. Uhlig, Thermodynamic evaluation of liquid metals as heat transfer fluids in concentrated solar power plants, Applied Thermal Engineering, Volume 60, Issues 1–2, 2 October 2013, Pages 295–302.

[10] J. Pacio, Th. Wetzel, Assessment of liquid metal technology status and research paths for their use as efficient heat transfer fluids in solar central receiver systems, Solar Energy, Vol. 93, July 2013, Pages 11–22.

[11] Nicholas Boerema, Graham Morrison, Robert Taylor, Gary Rosengarten, Liquid sodium versus Hitec as a heat transfer fluid in solar thermal central receiver systems, Solar Energy, Vol. 86, Issue 9, 2012, Pages 2293–2305.

[12] Manfred Becker, Comparison of heat transfer fluids for use in solar thermal power stations, Electric Power Systems Research, Volume 3, Issues 3–4, 1980, Pages 139–150.

[13] Adrian Bejan, Entropy Generation Minimization –The method of thermodynamic optimization of finite-size systems and finite-time processes, 1995, CRC Press.

[14] Adrian Bejan, Advanced Engineering Thermodynamics, 3rd Edition, 2006, Wiley.

[15] Incropera, F. P., and DeWitt, D. P., 2002, Introduction to Heat Transfer, 4th Ed., Wiley, New York.

[16] Kai Wang, Edgar Molina, Ghazal Dehghani, Ben Xu, Peiwen Li, Qing Hao, Pierre Lucas, Mohamad H. Kassaee, Sheldon M. Jeter, Amyn S. Teja, Experimental Investigation to the Properties of Eutectic Salts by NaCl-KCl-ZnCl2 for Application as High Temperature Heat Transfer Fluids, ES-FuelCell2014-6578, Proceedings of the ASME 2014, 8th International Conference on Energy Sustainability, June 30-July 2, 2014, Boston, MA, USA.

AUTHOR INDEX
Energy Technology 2015

SUBJECT INDEX
Energy Technology 2015

X

Z

MIX
Papier aus verantwortungsvollen Quellen
Paper from responsible sources
FSC® C105338

If you have any concerns about our products,
you can contact us on
ProductSafety@springernature.com

In case Publisher is established outside the EU,
the EU authorized representative is:
Springer Nature Customer Service Center GmbH
Europaplatz 3, 69115 Heidelberg, Germany

Printed by Libri Plureos GmbH
in Hamburg, Germany